STRATHCLYDE UNIVERSITY LIBRARY
30125 00302992 2

KV-038-646

This book is to be returned on or before the last date stamped below.

05 JAN 2005

17 FEB 1987
31 MAR 1987
12 MAY 1987
2 FEB 1988

23/4/86
16 July 86
24 JUN 1987

30 JAN 1985
15 MAR 1985
12 AUG 1987

LIBREX

The scientific management of hazardous wastes

C.B. COPE
W.H. FULLER
S.L. WILLETTS

CAMBRIDGE UNIVERSITY PRESS
Cambridge
London New York New Rochelle
Melbourne Sydney

05860144

Published by the Press Syndicate of the University of Cambridge
The Pitt Building, Trumpington Street, Cambridge CB2 1RP
32 East 57th Street, New York, NY 10022, USA
296 Beaconsfield Parade, Middle Park, Melbourne 3206, Australia

© Cambridge University Press 1983

First published 1983

Printed in Great Britain at the University Press, Cambridge

Library of Congress catalogue card number: 82-14650

British Library cataloguing in publication data

Cope, C.B.
The scientific management of hazardous wastes.
1. Factory and trade waste
I. Title II. Fuller, W.H. III Willetts, S
628.5′4 TD897

ISBN 0 521 25100 1

SE

D
628.54
COP

Our interest in conclusions has been so great that the method of reaching them has been neglected: it mattered little how much prejudice or blind acceptance of authority was connected with them, so long as they were understood and remembered.

F.M. McMurry, 1909

Contents

Disclaimer

Any views or opinions expressed within this book have been arrived at individually (rather than collectively) by the contributors as a result of their research studies and experiences in the waste management field. They do not necessarily represent the official views of any of the organisations with whom the contributors are associated.

Mention of any trade name, commercial product or service does not constitute endorsement or recommendation for use.

Acknowledgements

The authors and publishers of this book wish to acknowledge their gratitude to many individuals and organisations who have assisted in its preparation by providing information, illustrations and facilities.

Birmingham Radiation Centre
The University of Birmingham
The University of Aston in Birmingham
The University of Arizona
Custodian Pollution Consultants Ltd
Steetley Construction Materials Ltd
The Institute of Wastes Management
American Society of Civil Engineers
United States Environmental Protection Agency
Leigh Analytical Services Ltd
Harwell Laboratories
International Harvester Company
Stablex Ltd
American Colloid Co.
Border (UK) Ltd
Effluent Services Ltd
D.J. Broady Ltd
T.R. International (Chemicals) Ltd
Confederation of British Industry
Beckman-RICC Ltd
The Incinerator Company Ltd
Polymeric Treatments Ltd
American Society of Testing and Materials
American Fisheries Society
Marcel Dekker Publishers Inc.
Niagara Falls Water Treatment Plant
Institute of Water Pollution Control
Water Research Centre
Brookhaven National Laboratory
University of Illinois
Dr R.C. Keen
Department of the Environment
The Controller of Her Majesty's Stationery Office
County Surveyors' Society
Society of County Treasurers
Wuidart Engineering Ltd
Volclay Ltd
National Association of Waste Disposal Contractors

Introduction

The problems of the safe management and disposal of hazardous industrial waste have become widely recognised in the last few years. Public outcry and concern as a result of environmental damages caused by the uncontrolled dumping of hazardous wastes persuaded government authorities to devote resources to the scientific study of the environmental aspects involved. Much of this research has now been described in the scientific literature, though, naturally, the studies are continuing.

In this book, three independent scientists examine the results of these studies and the legislative framework which form the basis of hazardous waste disposal. The limits of our scientific knowledge are carefully defined and the ways in which this knowledge is extrapolated and applied to hazardous waste management are examined. Significant areas of uncertainty are identified and the authors have not been afraid to draw attention to the fallibility of certain interpretations.

For example, the current UK legislation is described and attention is drawn to loopholes in the laws and dual standards applied by the controlling authorities. In the chapters which examine British research work on disposal sites and test cell studies, anomalies are described in the methods used to collect and analyse samples. These anomalies may have led British authorities to underestimate the pollution hazards and environmental dangers which may result from current British disposal practices. The characteristic properties of landfill leachates are described and the near impossibility of treating leachates to conform to acceptable discharge standards is explained.

Clays and soils are often relied upon to attenuate the pollutants leaching from hazardous wastes. The book reviews research studies in this field and questions the reliability of the interpretations being made about British findings. American studies of geochemistry are examined in detail and the factors which affect pollutant attenuation are identified. Many of these factors are outside man's control in landfill situations. Environmental health aspects of hazardous waste are discussed, together with the methods used for measuring toxicity. The difficulties of studying toxic effects of pure chemicals are compounded when researchers have to pass judgement on heterogeneous mixtures of chemical wastes and their decomposition products.

Alternative methods to the landfill of hazardous wastes are examined and compared. Chemical treatment methods and incineration are reviewed and available capacities identified. The methods used for risk assessment are examined and their applicability to waste management is questioned. Costs and benefits are compared and the economic, risk and legal aspects of alternatives to landfill of hazardous waste are examined. The reasons why public acceptance and perception of risks will play an increasing role in future developments are discussed.

1

The nature of hazardous wastes and their recycling potential

Identifying the problem

There is no consensus of opinion which enables us to define hazardous waste and therefore we must begin by defining industrial wastes. Industrial wastes are the byproducts, spent residues and discarded components of industrial manufacturing processes which have no realisable value. They are produced in gaseous, liquid, solid and semi-solid forms. Excluded from consideration in this textbook are explosive residuals from military sources, spoil heaps from mining and quarrying and radioactive wastes.

A tentative definition being considered by the World Health Organisation defines hazardous wastes as those wastes which present either:

(a) short-term hazards, such as acute toxicity by ingestion, inhalation or skin absorption, corrosivity or other skin or eye contact hazards or the risk of fire and explosion or

(b) long-term environmental hazards including chronic toxicity upon repeated exposure, carcinogenicity (which may in some cases result from acute exposure but with a long latent period), resistance to detoxification processes such as biodegradation, the potential to pollute underground or surface waters or aesthetically objectionable properties such as offensive smells.

Difficulties of definition continue to exist, especially when it is recognised that a waste can be hazardous in one circumstance though not in another. Hazards arise not only from dangerous substances themselves but also from catalytic, synergistic or antagonistic reactions of two or more substances which are in themselves harmless. Catalytic and synergistic effects are not always predictable and, as we shall see, can produce circumstances in which pollutants can be mobilised to threaten valuable water supplies.

Table 1 [1] shows a list of categories of hazardous waste but it must be emphasised that this list is by no means comprehensive. Indeed, with industrial producers adding new chemical products at a rate which exceeds one per day, it would be impossible to produce an inclusive definitive list of hazardous wastes.

Table 1. *Categories of hazardous waste*

Type of waste	Groups and *Sub-groups*	Group code	Sub-group code
Inorganic acids	Hydrochloric acid	A10	
	Sulphuric acid	A20	
	Nitric acid	A30	
	Chromic acid	A40	
	Phosphoric acid	A50	
	Hydrofluoric acid	A60	
	Others	A90	
Organic acids and related compounds	All	B10	
	Aliphatic acids eg formic, acetic and oxalic acids		*B11*
	Aromatic acids eg benzoic, phthalic acids		*B12*
	Acid anhydrides eg acetic, phthalic anhydrides		*B13*
	Acid chlorides eg acetyl, benzoyl chlorides		*B14*
	Sulphonic acids		*B15*
	Others		*B19*
Alkalis	Alkali metal oxides and hydroxides, calcium oxide, proprietary alkaline cleaners	C10	
	Sodium and/or potassium hydroxides or oxides		*C11*
	Calcium oxide		*C12*
	Proprietary alkaline cleaners		*C13*
	Ammonia	C20	
	Others	C90	
	Calcium hydroxide		*C91*
	Sodium and/or potassium carbonates		*C92*

Table 1. (*cont.*)

Type of waste	Groups and *Sub-groups*	Group code	Sub-group code
Toxic metal compounds	Cadmium	D10	
	Mercury	D20	
	Lead	D30	
	Arsenic	D40	
	Others	D90	
	Copper		*D91*
	Zinc		*D92*
	Barium (water soluble forms)		*D93*
	Thallium		*D94*
	Nickel		*D95*
	Vanadium		*D96*
	Silver		*D97*
	Others		*D99*
Non-toxic metal compounds	Iron	E10	
	Others	E90	
	Ammonium salts		*E91*
	Titanium		*E92*
	Others		*E99*
Metals (Elemental)	Alkali, alkaline earth and other hazardous metals	F10	
	Sodium and potassium		*F11*
	Cadmium		*F12*
	Mercury		*F13*
	Aluminium		*F14*
	Magnesium		*F15*
	Other metals	F90	
Metal oxides	Hazardous oxides	G10	
	Cadmium oxide		*G11*
	Beryllium oxide		*G12*
	Others		*G19*
	Other oxides	G90	

Table 1. (*cont.*)

Type of waste	Groups and *Sub-groups*	Group code	Sub-group code
Inorganic compounds	Cyanides	H10	
	Sodium and potassium cyanides		*H11*
	Soluble complex cyanides		*H12*
	Ferro and ferri cyanides		*H13*
	Other cyanides		*H19*
	Others which liberate toxic gases on acidification	H20	
	Sulphides, selenides, tellurides and arsenides		*H21*
	Oxidizing compounds	H30	
	Hypochlorites and chlorites		*H31*
	Chlorates, perchlorates, bromates, iodates, periodates, persulphates and permanganates		*H32*
	Peroxides		*H33*
	Toxic compounds	H40	
	Chromates		*H41*
	Fluorides, silicofluorides, borofluorides		*H42*
	Arsenates and arsenites		*H43*
	Others	H90	
	Carbides and acetylides		*H91*
	Borates		*H92*
	Nitrites		*H93*
	Nitrates		*H94*
Other inorganic materials	Asbestos	J10	
	Slag including boiler and flue cleanings	J20	
	Mineral processing wastes	J30	
	Silt and dredgings	J40	
	Water (contaminated)	J50	
	Metal scrap	J60	
	Ferrous metal scrap		*J61*
	Non-ferrous metal scrap		*J62*
	Others	J90	

Table 1. (*cont.*)

Type of waste	Groups and *Sub-groups*	Group code	Sub-group code
Organic compounds	Hydrocarbons (not included in M)	K10	
	Aliphatic hydrocarbons		*K11*
	Aromatic hydrocarbons		*K12*
	Phenols, analogues and derivatives	K20	
	Chlorinated phenols and analogues		*K21*
	Peroxides	K30	
	Halogenated cleaning compounds	K40	
	Trichloroethylene		*K41*
	Perchloroethylene		*K42*
	Trichlorethane		*K43*
	Trichlorotrifluoroethane		*K44*
	Others		*K49*
	Halogenated compounds excluding cleaning compounds	K50	
	PCBs and analogues		*K51*
	Other halogenated hydrocarbons		*K52*
	Other halogenated organics eg chlorinated dioxins		*K53*
	Organo metallics	K60	
	Tetra ethyl lead		*K61*
	Tetra methyl lead		*K62*
	Others		*K69*
	Nitrogen, sulphur or phosphorus-containing compounds	K70	
	Amines and amides		*K71*
	Nitro compounds		*K72*
	Nitriles		*K73*
	Isocyanates		*K74*
	Other organo nitrogen compounds		*K75*
	Organophosphorus compounds		*K76*
	Organosulphur compounds		*K77*
	Oxygen containing compounds	K80	
	Esters		*K81*

Table 1. (*cont.*)

Type of waste	Groups and *Sub-groups*	Group code	Sub-group code
Organic compounds (*contd.*)	*Ethers*		*K82*
	Aldehydes and Ketones		*K83*
	Alcohols		*K84*
	Others	K90	
	Chelating compounds		*K91*
	Phthalates		*K92*
Polymeric materials and precursors	Precursors, monomers and products of incomplete polymerization	L10	
	Epoxy resins (not finished products)		*L11*
	Polyester resins (not finished products)		*L12*
	Phenol-formaldehyde resins (not finished products)		*L13*
	Finished products and manufacturing scrap	L20	
	Polyurethane		*L22*
	Other resins and polymeric materials		*L29*
	Scrap rubber (including tyres)	L30	
	Latex, latex and rubber solutions and suspensions	L40	
	Synthetic adhesive wastes	L50	
	Ion-exchange resin wastes	L60	
Fuel, oils and greases	Mineral oils	M10	
	Kerosene and derv	M20	
	Fuel oil	M30	
	Vegetable and other oils	M40	
	Oil/water mixtures	M50	
	Fats, waxes and greases	M60	
Fine chemicals and biocides	Pharmaceutical and cosmetic products	N10	
	Pharmaceutical products in retail containers		*N11*
	Pharmaceutical products in bulk and production containers		*N13*
	Biocides	N20	
	Pesticides		*N21*
	Herbicides		*N22*

Table 1. (*cont.*)

Type of waste	Groups and *Sub-groups*	Group code	Sub-group code
Fine chemicals and biocides (*contd.*)	*Fungicides*		*N23*
	Other biocides		*N29*
Miscellaneous chemical waste	Mixed organic compounds	P10	
	Mixed inorganic compounds	P20	
	Unidentified chemical waste	P30	
	Organics identified by trade names only*		*P31*
	Inorganics identified by trade names only*		*P32*
Filter materials, treatment sludge and contaminated rubbish	Used filter materials eg kieselguhr, carbon, filter cloths	Q10	
	Contaminated rubbish (including bags and sacks)	Q20	
	Empty used containers	Q30	
	Industrial effluent treatment sludge	Q40	
Interceptor wastes, tars, paint, dyes and pigments	Tank cleaning sludge (note K60 for lead content)	R10	
	Interceptor pit wastes (note M10–M30 for oil content)	R20	
	Printing industry wastes (ink manufacture and use)	R30	
	Dyestuffs waste	R40	
	Distillation residues	R50	
	Acid tars	R60	
	Tar, pitch, bitumen and asphalts	R70	
	Paint waste (manufacture and use)	R80	
Miscellaneous wastes	Tannery and fellmongers waste	S10	
	Tannery waste		*S11*
	Fellmongers waste		*S12*
	Cellulose wastes (natural and synthetic)	S20	
	Waste treated timber	S30	
	Soap and detergents	S50	
	Soap		*S51*
	Detergents		*S52*
	Other industrial wastes	S90	
Animal and food wastes	Animal processing wastes	T10	
	Carcasses and flesh		*T11*

*Where trade names are used the source of the material should be specified.

Table 1. (*cont.*)

Type of waste	Groups and *Sub-groups*	Group code	Sub-group code
Animal and food wastes (*contd.*)	*Blood, fat grease etc*		*T12*
	Excrement		*T13*
	Food processing wastes (including starch)	T20	
	Glue wastes	T30	

Source: Reference [1].

The term 'toxic waste' is often used in waste management circles and again the definition of the term causes some difficulty. However, it seems appropriate to comment that the term 'toxic waste' does not imply 'causing death'. Toxic effects should include growth retardation, decreased fullness of health and intellectual capability, detrimental changes in reproductive cycle with mortality of offspring, increased morbidity, pathological change, appearance of tumours, chronic disease symptoms and decreased longevity and can be applied to animal (and plant) life as well as humans.

We shall encounter, during the course of this book, other important definitions such as *notifiable waste* and *special waste* which have profound legal meanings. The term 'special waste' in Britain originates from the Control of Pollution Act (Special Waste) Regulations 1980. This definition is discussed in detail in Dr Willetts' contribution in Chapter 4 but it is important to note here that a special waste is described [2] by reference to:

(a) it containing any substance listed in Part 1, Schedule 1 to S1 1980 No. 1709 *and* by reason of the presence of such substance;

(i) is dangerous to life by way of carcinogenicity, corrosivity to tissues or toxicity,

(ii) has a flash point less than 21°C,

(b) is a prescribed medicinal product.

These regulations are quite new in Britain and the Department of the Environment is reviewing them at the present time. The definition is very restrictive since it ignores water pollution, damage to animals and vegetation and site sterilisation. Critics of these regulations include The National Water Council, The Scottish River Purification Boards Association, Friends of The Earth, the Association of County Councils in evidence to the House of Lords Inquiry [3]. This author has drawn attention to the differences between these regulations and American ones [4].

The United States Federal Register [5] begins by defining solid waste as

> any garbage, refuse, sludge from a waste water treatment plant, water supply treatment plant or air pollution control facility and other discarded material including solid, liquid, semi-solid or contained gaseous material resulting from industrial, commercial or mining and agricultural operations, and from community activities but does not include solid or dissolved materials in domestic sewage or solid or dissolved materials in irrigation return flows, or industrial discharges which are point sources subject to permits under Section 402 of the Federal Water Pollution Control Act: or certain special nuclear or byproduct material as defined by the Atomic Energy Act of 1954.

The United States Environmental Protection Agency (USEPA) employs extensive and rigorous criteria to determine whether a waste is hazardous [5]. A waste is hazardous if:

(1) it is not excluded in listings which identify domestic sewage, point source discharges to surface waters, overburden waste from mines and sewage sludge from publicly owned treatment works;

(2) it exhibits any of the characteristics of hazardous waste in terms of ignitability, corrosivity, reactivity or extractive procedure toxicity.

Ignitability of liquids is determined by a Pensky–Martens Closed Cup Tester or a Setaflash Closed Cup Tester in accordance with prescribed methods. A flash point of less than 60°C (140°F) qualifies liquid wastes in this category. Solid wastes which cause fire under friction at standard temperature and pressure or by absorption of moisture or spontaneous chemical change or solids which, when ignited, burn so vigorously and persistently that they cause a hazard are also included in this category. Ignitable compressed gases and specific oxidizers are also included in this category.

Corrosive wastes are liquid wastes with a pH less than or equal to 2 or greater than or equal to 12.5 or which corrode steel under specified test conditions at a rate greater than 6.35 mm per year.

Reactive wastes are those which are normally unstable and readily undergo violent change without detonating, or react violently with water, or form potentially explosive mixtures with water, or when mixed with water generate toxic gases, vapours or fumes in a quantity sufficient to present a danger to human health or the environment, or is a cyanide or sulphide which, when exposed to pH conditions between 2 and 12.5, can generate toxic gases, or are capable of detonation or explosive decomposition or reaction at normal temperatures and pressures.

The extractive procedure toxicity (EP toxicity) is a test in which 100 grams of a representative sample are obtained by a prescribed sampling procedure and separated into solid and liquid components by a prescribed method. Solid residues are segregated according to particle size. Residues which pass through a 9.5 mm standard sieve are weighed and placed in an extractor with 16 times its weight of de-ionised water. Residues which have larger particle size are first ground and then subjected to the same procedure. The extractor mixes the contents so that all sample surfaces are wetted. The pH is measured and if greater than 5.0 it is reduced to 5.0 ± 0.2 by adding 0.5 N acetic acid. The pH is monitored at prescribed intervals and if it rises above 5.2 more acetic acid is added to maintain this condition. The limit of acid addition is set at 4 ml of acid per gram of solid. The mixture is agitated for 24 hours at 20–40°C. At the end of this time the components are separated. Any liquid retained from the sample preparation procedure is added to the solution obtained from the extractive procedure. The aqueous solution is analysed for arsenic, barium, cadmium, chromium, lead, mercury, selenium, silver and six other organic halogen compounds by prescribed methods. The maximum concentration of these components is specified. In order to allow for attenuation and dilution, USEPA consider that if the concentration of any pollutant in this extract exceeds the Federal Drinking Water Standard by 100-fold then the waste is classified as hazardous. USEPA chose these conditions of test to simulate the acidic leaching medium which occurs in actively decomposing landfills. USEPA recognises that codisposal of industrial waste in a municipal landfill generates more aggressive leachate media than other landfills and describes codisposal as a 'mismanagement scenario'.

(3) it contains any of the toxic constituents listed in Table 2 unless after consideration of any of the following factors the 'Administrator' considers that the waste is not capable of posing a substantial present or potential hazard to human health or the environment when improperly treated, stored, transported or disposed of or otherwise managed;

(i) the nature of the toxicity presented by the constituent,

(ii) the concentration of the constituent in the waste,

(iii) the potential of the constituent or any toxic degradation product of the constituent to migrate from the waste into the environment under improper management,

(iv) the persistence of the constituent or any toxic degradation product of the constituent,

Table 2. *Hazardous constituents*

- Acetaldehyde
- (Acetato)phenylmercury
- Acetonitrile
- 3-(alpha-Acetonylbenzyl)-4-hydroxycoumarin and salts
- 2-Acetylaminofluorene
- Acetyl chloride
- 1-Acetyl-2-thiourea
- Acrolein
- Acrylamide
- Acrylonitrile
- Aflatoxins
- Aldrin
- Allyl alcohol
- Aluminum phosphide
- 4-Aminobiphenyl
- 6-Amino-1,1a,2,8,8a,8b-hexahydro-8-(hydroxymethyl)-8a-methoxy-5-methylcarbamate azirino(2',3':3,4) pyrrolo(1,2-a)indole-4,7-dione (ester) (Mitomycin C)
- 5-(Aminomethyl)-3-isoxazolol
- 4-Aminopyridine
- Amitrole
- Antimony and compounds, N.O.S.[1]
- Aramite
- Arsenic and compounds, N.O.S.
- Arsenic acid
- Arsenic pentoxide
- Arsenic trioxide
- Auramine
- Azaserine
- Barium and compounds, N.O.S.
- Barium cyanide
- Benz[c]acridine
- Benz[a]anthracene
- Benzene
- Benzenearsonic acid
- Benzenethiol
- Benzidine
- Benzo[a]anthracene
- Benzo[b]fluoranthene
- Benzo[j]fluoranthene
- Benzo[a]pyrene
- Benzotrichloride
- Benzyl chloride
- Beryllium and compounds, N.O.S.
- Bis(2-chloroethoxy)methane
- Bis(2-chloroethyl) ether
- N,N-Bis(2-chloroethyl)-2-naphthylamine
- Bis(2-chloroisopropyl) ether
- Bis(chloromethyl) ether
- Bis(2-ethylhexyl) phthalate
- Bromoacetone
- Bromomethane
- 4-Bromophenyl phenyl ether
- Brucine
- 2-Butanone peroxide
- Butyl benzyl phthalate
- 2-sec-Butyl-4,6-dinitrophenol (DNBP)
- Cadmium and compounds, N.O.S.
- Calcium chromate
- Calcium cyanide
- Carbon disulfide
- Chlorambucil
- Chlordane (alpha and gamma isomers)
- Chlorinated benzenes, N.O.S.
- Chlorinated ethane, N.O.S.
- Chlorinated naphthalene, N.O.S.
- Chlorinated phenol, N.O.S.
- Chloroacetaldehyde
- Chloroalkyl ethers
- p-Chloroaniline
- Chlorobenzene
- Chlorobenzilate
- 1-(p-Chlorobenzoyl)-5-methoxy-2-methylindole-3-acetic acid
- p-Chloro-m-cresol
- 1-Chloro-2,3-epoxybutane
- 2-Chloroethyl vinyl ether
- Chloroform
- Chloromethane
- Chloromethyl methyl ether
- 2-Chloronaphthalene
- 2-Chlorophenol
- 1-(o-Chlorophenyl)thiourea
- 3-Chloropropionitrile
- alpha-Chlorotoluene
- Chlorotoluene, N.O.S.
- Chromium and compounds, N.O.S.
- Chrysene
- Citrus red No. 2
- Copper cyanide
- Creosote
- Crotonaldehyde
- Cyanides (soluble salts and complexes), N.O.S.
- Cyanogen
- Cyanogen bromide
- Cyanogen chloride
- Cycasin
- 2-Cyclohexyl-4,6-dinitrophenol
- Cyclophosphamide
- Daunomycin
- DDD
- DDE
- DDT
- Diallate
- Dibenz[a,h]acridine
- Dibenz[a,j]acridine
- Dibenz[a,h]anthracene(Dibenzo[a,h] anthracene)
- 7H-Dibenzo[c,g]carbazole
- Dibenzo[a,e]pyrene
- Dibenzo[a,h]pyrene
- Dibenzo[a,i]pyrene
- 1,2-Dibromo-3-chloropropane
- 1,2-Dibromoethane
- Dibromomethane
- Di-n-butyl phthalate
- Dichlorobenzene, N.O.S.
- 3,3'-Dichlorobenzidine
- 1,1-Dichloroethane
- 1,2-Dichloroethane
- trans-1,2-Dichloroethane
- Dichloroethylene, N.O.S.
- 1,1-Dichloroethylene
- Dichloromethane
- 2,4-Dichlorophenol
- 2,6-Dichlorophenol
- 2,4-Dichlorophenoxyacetic acid (2,4-D)
- Dichloropropane
- Dichlorophenylarsine
- 1,2-Dichloropropane
- Dichloropropanol, N.O.S.
- Dichloropropene, N.O.S.
- 1,3-Dichloropropene
- Dieldrin
- Diepoxybutane
- Diethylarsine
- 0,0-Diethyl-S-(2-ethylthio)ethyl ester of phosphorothioic acid
- 1,2-Diethylhydrazine
- 0,0-Diethyl-S-methylester phosphorodithioic acid
- 0,0-Diethylphosphoric acid, 0-p-nitrophenyl ester
- Diethyl phthalate
- 0,0-Diethyl-0-(2-pyrazinyl)phosphorothioate
- Diethylstilbestrol
- Dihydrosafrole
- 3,4-Dihydroxy-alpha-(methylamino)-methyl benzyl alcohol
- Di-isopropylfluorophosphate (DFP)
- Dimethoate
- 3,3'-Dimethoxybenzidine
- p-Dimethylaminoazobenzene
- 7,12-Dimethylbenz[a]anthracene
- 3,3'-Dimethylbenzidine
- Dimethylcarbamoyl chloride
- 1,1-Dimethylhydrazine
- 1,2-Dimethylhydrazine
- 3,3-Dimethyl-1-(methylthio)-2-butanone-0-((methylamino) carbonyl)oxime
- Dimethylnitrosoamine
- alpha,alpha-Dimethylphenethylamine
- 2,4-Dimethylphenol
- Dimethyl phthalate
- Dimethyl sulfate
- Dinitrobenzene, N.O.S.
- 4,6-Dinitro-o-cresol and salts
- 2,4-Dinitrophenol
- 2,4-Dinitrotoluene
- 2,6-Dinitrotoluene Di-n-octyl phthalate
- 1,4-Dioxane
- 1,2-Diphenylhydrazine
- Di-n-propylnitrosamine
- Disulfoton
- 2,4-Dithiobiuret
- Endosulfan
- Endrin and metabolites
- Epichlorohydrin
- Ethyl cyanide
- Ethylene diamine
- Ethylenebisdithiocarbamate (EBDC)
- Ethyleneimine
- Ethylene oxide
- Ethylenethiourea
- Ethyl methanesulfonate
- Fluoranthene
- Fluorine
- 2-Fluoroacetamide
- Fluoroacetic acid, sodium salt
- Formaldehyde
- Glycidylaldehyde
- Halomethane, N.O.S.
- Heptachlor
- Heptachlor epoxide (alpha, beta, and gamma isomers)
- Hexachlorobenzene
- Hexachlorobutadiene
- Hexachlorocyclohexane (all isomers)
- Hexachlorocyclopentadiene
- Hexachloroethane
- 1,2,3,4,10,10-Hexachloro-1,4,4a,5,8,8a-hexahydro-1,4:5,8-endo,endo-dimethanonaphthalene
- Hexachlorophene
- Hexachloropropene
- Hexaethyl tetraphosphate
- Hydrazine
- Hydrocyanic acid
- Hydrogen sulfide
- Indeno(1,2,3-c,d)pyrene
- Iodomethane
- Isocyanic acid, methyl ester
- Isosafrole
- Kepone
- Lasiocarpine
- Lead and compounds, N.O.S.
- Lead acetate
- Lead phosphate
- Lead subacetate
- Maleic anhydride
- Malononitrile
- Melphalan
- Mercury and compounds, N.O.S.
- Methapyrilene
- Methomyl
- 2-Methylaziridine
- 3-Methylcholanthrene
- 4,4'-Methylene-bis-(2-chloroaniline)
- Methyl ethyl ketone (MEK)
- Methyl hydrazine
- 2-Methyllactonitrile
- Methyl methacrylate
- Methyl methanesulfonate
- 2-Methyl-2-(methylthio)propionaldehyde-o-(methylcarbonyl) oxime
- N-Methyl-N'-nitro-N-nitrosoguanidine
- Methyl parathion
- Methylthiouracil
- Mustard gas
- Naphthalene
- 1,4-Naphthoquinone
- 1-Naphthylamine
- 2-Naphthylamine
- 1-Naphthyl-2-thiourea
- Nickel and compounds, N.O.S.
- Nickel carbonyl
- Nickel cyanide
- Nicotine and salts
- Nitric oxide
- p-Nitroaniline
- Nitrobenzene
- Nitrogen dioxide
- Nitrogen mustard and hydrochloride salt
- Nitrogen mustard N-oxide and hydrochloride salt
- Nitrogen peroxide
- Nitrogen tetroxide
- Nitroglycerine
- 4-Nitrophenol
- 4-Nitroquinoline-1-oxide
- Nitrosamine, N.O.S.
- N-Nitrosodi-N-butylamine
- N-Nitrosodiethanolamine
- N-Nitrosodiethylamine
- N-Nitrosodimethylamine
- N-Nitrosodiphenylamine
- N-Nitrosodi-N-propylamine
- N-Nitroso-N-ethylurea
- N-Nitrosomethylethylamine
- N-Nitroso-N-methylurea

[1] The abbreviation N.O.S. signifies those members of the general class "not otherwise specified" by name in this listing.

Table 2. (*cont.*)

N-Nitroso-N-methylurethane
N-Nitrosomethylvinylamine
N-Nitrosomorpholine
N-Nitrosonornicotine
N-Nitrosopiperidine
N-Nitrosopyrrolidine
N-Nitrososarcosine
5-Nitro-o-toluidine
Octamethylpyrophosphoramide
Oleyl alcohol condensed with 2 moles ethylene oxide
Osmium tetroxide
7-Oxabicyclo[2.2.1]heptane-2,3-dicarboxylic acid
Parathion
Pentachlorobenzene
Pentachloroethane
Pentachloronitrobenzene (PCNB)
Pentacholorophenol
Phenacetin
Phenol
Phenyl dichloroarsine
Phenylmercury acetate
N-Phenylthiourea
Phosgene
Phosphine
Phosphorothioic acid, O,O-dimethyl ester, O-ester with N,N-dimethyl benzene sulfonamide
Phthalic acid esters, N.O.S.
Phthalic anhydride
Polychlorinated biphenyl, N.O.S.
Potassium cyanide
Potassium silver cyanide
Pronamide
1,2-Propanediol
1,3-Propane sultone
Propionitrile
Propylthiouracil
2-Propyn-1-ol
Pryidine
Reserpine
Saccharin
Safrole
Selenious acid
Selenium and compounds, N.O.S.
Selenium sulfide
Selenourea
Silver and compounds, N.O.S.
Silver cyanide
Sodium cyanide
Streptozotocin
Strontium sulfide
Strychnine and salts
1,2,4,5-Tetrachlorobenzene
2,3,7,8-Tetrachlorodibenzo-p-dioxin (TCDD)
Tetrachloroethane, N.O.S.
1,1,1,2-Tetrachloroethane
1,1,2,2-Tetrachloroethane
Tetrachloroethene (Tetrachloroethylene)
Tetrachloromethane
2,3,4,6-Tetrachlorophenol
Tetraethyldithiopyrophosphate
Tetraethyl lead
Tetraethylpyrophosphate
Thallium and compounds, N.O.S.
Thallic oxide
Thallium (I) acetate
Thallium (I) carbonate
Thallium (I) chloride
Thallium (I) nitrate
Thallium selenite
Thallium (I) sulfate
Thioacetamide
Thiosemicarbazide
Thiourea
Thiuram
Toluene
Toluene diamine
o-Toluidine hydrochloride
Tolylene diisocyanate
Toxaphene
Tribromomethane
1,2,4-Trichlorobenzene
1,1,1-Trichloroethane
1,1,2-Trichloroethane
Trichloroethene (Trichloroethylene)
Trichloromethanethiol
2,4,5-Trichlorophenol
2,4,6-Trichlorophenol
2,4,5-Trichlorophenoxyacetic acid (2,4,5-T)
2,4,5-Trichlorophenoxypropionic acid (2,4,5-TP) (Silvex)
Trichloropropane, N.O.S.
1,2,3-Trichloropropane
0,0,0-Triethyl phosphorothioate
Trinitrobenzene
Tris(1-azridinyl)phosphine sulfide
Tris(2,3-dibromopropyl) phosphate
Trypan blue
Uracil mustard
Urethane
Vanadic acid, ammonium salt
Vanadium pentoxide (dust)
Vinyl chloride
Vinylidene chloride
Zinc cyanide
Zinc phosphide

Source: Reference [5].

(v) the potential for the constituent or any toxic degradation product to degrade into non-harmful constituents and the rate of degradation,
(vi) the degree to which the constituent or any degradation product of the constituent bioaccumulates in ecosystems,
(vii) the plausible types of improper management to which the waste could be subjected,
(viii) the quantities of the waste generated at individual generation sites or on a regional or national basis,
(ix) the nature and severity of the human health and environmental damage that has occurred as a result of the improper management of wastes containing the constituent,
(x) action taken by other governmental agencies based on the health or environmental hazard posed by the waste or its constituents,
(xi) such other factors as may be appropriate.

The substances listed in Table 2 have been shown in scientific studies to have toxic, carcinogenic, mutagenic or teratogenic effects on human or other life forms.

(4) It is listed in prescribed lists. These lists are shown in Table 3 which identifies hazardous wastes from specific and non-specific sources, Table 4 which identifies discarded commercial chemical products, off-specification

Table 3. *Hazardous waste from specific and non-specific sources*

Industry and EPA hazardous waste No.	Hazardous waste	Hazard code
Wood Preservation: K001	Bottom sediment sludge from the treatment of wastewaters from wood preserving processes that use creosote and/or pentachlorophenol	(T)
Inorganic Pigments:		
K002	Wastewater treatment sludge from the production of chrome yellow and orange pigments	(T)
K003	Wastewater treatment sludge from the production of molybdate orange pigments	(T)
K004	Wastewater treatment sludge from the production of zinc yellow pigments	(T)
K005	Wastewater treatment sludge from the production of chrome green pigments	(T)
K006	Wastewater treatment sludge from the production of chrome oxide green pigments (anhydrous and hydrated)	(T)
K007	Wastewater treatment sludge from the production of iron blue pigments	(T)
K008	Oven residue from the production of chrome oxide green pigments	(T)
Organic Chemicals:		
K009	Distillation bottoms from the production of acetaldehyde from ethylene	(T)
K010	Distillation side cuts from the production of acetaldehyde from ethylene	(T)
K011	Bottom stream from the wastewater stripper in the production of acrylonitrile	(R, T)
K012	Still bottoms from the final purification of acrylonitrile in the production of acrylonitrile	(T)
K013	Bottom stream from the acetonitrile column in the production of acrylonitrile	(R, T)
K014	Bottoms from the acetronitrile purification column in the production of acrylonitrile	(T)
K015	Still bottoms from the distillation of benzyl chloride	(T)
K016	Heavy ends or distillation residues from the production of carbon tetrachloride	(T)
K017	Heavy ends (still bottoms) from the purification column in the production of epichlorohydrin	(T)
K018	Heavy ends from fractionation in ethyl chloride production	(T)
K019	Heavy ends from the distillation of ethylene dichloride in ethylene dichloride production	(T)
K020	Heavy ends from the distillation of vinyl chloride in vinyl chloride monomer production	(T)
K021	Aqueous spent antimony catalyst waste from fluoromethanes production	(T)
K022	Distillation bottom tars from the production of phenol/acetone from cumene	(T)
K023	Distillation light ends from the production of phthalic anhydride from naphthalene	(T)
K024	Distillation bottoms from the production of phthalic anhydride from naphthalene	(T)
K025	Distillation bottoms from the production of nitrobenzene by the nitration of benzene	(T)
K026	Stripping still tails from the production of methyl ethyl pyridines	(T)
K027	Centrifuge residue from toluene diisocyanate production	(R, T)
K028	Spent catalyst from the hydrochlorinator reactor in the production of 1,1,1-trichloroethane	(T)
K029	Waste from the product stream stripper in the production of 1,1,1-trichloroethane	(T)
K030	Column bottoms or heavy ends from the combined production of trichloroethylene and perchloroethylene	(T)
Pesticides:		
K031	By-products salts generated in the production of MSMA and cacodylic acid	(T)
K032	Wastewater treatment sludge from the production of chlordane	(T)
K033	Wastewater and scrub water from the chlorination of cyclopentadiene in the production of chlordane	(T)
K034	Filter solids from the filtration of hexachlorocyclopentadiene in the production of chlordane	(T)
K035	Wastewater treatment sludges generated in the production of creosote	(T)
K036	Still bottoms from toluene reclamation distillation in the production of disulfoton	(T)
K037	Wastewater treatment sludges from the production of disulfoton	(T)
K038	Wastewater from the washing and stripping of phorate production	(T)
K039	Filter cake from the filtration of diethylphosphorodithoric acid in the production of phorate	(T)
K040	Wastewater treatment sludge from the production of phorate	(T)
K041	Wastewater treatment sludge from the production of toxaphene	(T)
K042	Heavy ends or distillation residues from the distillation of tetrachlorobenzene in the production of 2,4,5-T	(T)
K043	2,6-Dichlorophenol waste from the production of 2,4-D	(T)
Explosives:		
K044	Wastewater treatment sludges from the manufacturing and processing of explosives	(R)
K045	Spent carbon from the treatment of wastewater containing explosives	(R)
K046	Wastewater treatment sludges from the manufacturing, formulation and loading of lead-based initiating compounds	(T)
K047	Pink/red water from TNT operations	(R)
Petroleum Refining:		
K048	Dissolved air flotation (DAF) float from the petroleum refining industry	(T)
K049	Slop oil emulsion solids from the petroleum refining industry	(T)
K050	Heat exchanger bundle cleaning sludge from the petroleum refining industry	(T)
K051	API separator sludge from the petroleum refining industry	(T)
K052	Tank bottoms (leaded) from the petroleum refining industry	(T)
Leather Tanning Finishing:		
K053	Chrome (blue) trimmings generated by the following subcategories of the leather tanning and finishing industry: hair pulp/chrome tan/retan/wet finish; hair save/chrome tan/retan/wet finish; retan/wet finish; no beamhouse; through-the-blue; and shearling.	(T)
K054	Chrome (blue) shavings generated by the following subcategories of the leather tanning and finishing industry: hair pulp/chrome tan/retan/wet finish; hair save/chrome tan/retan/wet finish; retan/wet finish; no beamhouse; through-the-blue; and shearling.	(T)
K055	Buffing dust generated by the following subcategories of the leather tanning and finishing industry: hair pulp/chrome tan/retan/wet finish; hair save/chrome tan/retan/wet finish; retan/wet finish; no beamhouse; and through-the-blue.	(T)
K056	Sewer screenings generated by the following subcategories of the leather tanning and finishing industry: hair pulp/chrome tan/retan/wet finish; hair save/chrome tan/retan/wet finish; retan/wet finish; no beamhouse; through-the-blue; and shearling.	(T)
K057	Wastewater treatment sludges generated by the following subcategories of the leather tanning and finishing industry: hair pulp/chrome tan/retan/wet finish; hair save/chrome tan/retan/wet finish; retan/wet finish; no beamhouse; through-the-blue and shearling.	(T)
K058	Wastewater treatment sludges generated by the following subcategories of the leather tanning and finishing industry: hair pulp/chrome tan/retan/wet finish; hair save/chrome tan/retan/wet finish; and through-the-blue.	(R, T)
K059	Wastewater treatment sludges generated by the following subcategory of the leather tanning and finishing industry: hair save/non-chrome tan/retan/wet finish.	(R)
Iron and Steel:		
K060	Ammonia still lime sludge from coking operations	(T)
K061	Emission control dust/sludge from the electric furnace production of steel	(T)
K062	Spent pickle liquor from steel finishing operations	(C, T)
K063	Sludge from lime treatment of spent pickle liquor from steel finishing operations	(T)
Primary Copper: K064	Acid plant blowdown slurry/sludge resulting from the thickening of blowdown slurry from primary copper production	(T)
Primary Lead: K065	Surface impoundment solids contained in and dredged from surface impoundments at primary lead smelting facilities	(T)
Primary Zinc:		
K066	Sludge from treatment of process wastewater and/or acid plant blowdown from primary zinc production	(T)
K067	Electrolytic anode slimes/sludges from primary zinc production	(T)
K068	Cadmium plant leach residue (iron oxide) from primary zinc production	(T)
Secondary Lead: K069	Emission control dust/sludge from secondary lead smelting	(T)
Generic		
F001	The spent halogenated solvents used in degreasing, tetrachloroethylene, trichloroethylene, methylene chloride, 1,1,1-trichloroethane, carbon tetrachloride, and the chlorinated fluorocarbons; and sludges from the recovery of these solvents in degreasing operations.	(T)
F002	The spent halogenated solvents, tetrachloroethylene, methylene chloride, trichloroethylene, 1,1,1-trichloroethane, chlorobenzene, 1,1,2-trichloro-1,2,2-trifluoroethane, o-dichlorobenzene, trichlorofluoromethane and the still bottoms from the recovery of these solvents.	(T)
F003	The spent non-halogenated solvents, xylene, acetone, ethyl acetate, ethyl benzene, ethyl ether, n-butyl alcohol, cyclohexanone, and the still bottoms from the recovery of these solvents.	(I)
F004	The spent non-halogenated solvents, cresols and cresylic acid, nitrobenzene, and the still bottoms from the recovery of these solvents	(T)
F005	The spent non-halogenated solvents, methanol, toluene, methyl ethyl ketone, methyl isobutyl ketone, carbon disulfide, isobutanol, pyridine and the still bottoms from the recovery of these solvents.	(I, T)
F006	Wastewater treatment sludges from electroplating operations	(T)
F007	Spent plating bath solutions from electroplating operations	(R, T)
F008	Plating bath sludges from the bottom of plating baths from electroplating operations	(R, T)
F009	Spent stripping and cleaning bath solutions from electroplating operations	(R, T)
F010	Quenching bath sludge from oil baths from metal heat treating operations	(R, T)
F011	Spent solutions from salt bath pot cleaning from metal heat treating operations	(R, T)
F012	Quenching wastewater treatment sludges from metal heat treating operations	(T)
F013	Flotation tailings from selective flotation from mineral metals recovery operations	(T)
F014	Cyanidation wastewater treatment tailing pond sediment from mineral metals recovery operations	(T)
F015	Spent cyanide bath solutions from mineral metals recovery operations	(R, T)
F016	Dewatered air pollution control scrubber sludges from coke ovens and blast furnaces	(T)

Source: Reference [5].

Table 4. *Discarded commercial chemical products, off-specification species, containers and spill residues thereof*

Hazardous waste No.	Substance [1]
	(Acetato)phenylmercury see P092
	Acetone cyanohydrin see P069
P001	3-(alpha-Acetonylbenzyl)-4-hydroxycoumarin and salts
P002	1-Acetyl-2-thiourea
P003	Acrolein
	Agarin see P007
	Agrosan GN 5 see P092
	Aldicarb see P069
	Aldifen see P048
P004	Aldrin
	Algimycin see P092
P005	Allyl alcohol
P006	Aluminum phosphide (R)
	ALVIT see P037
	Aminoethylene see P054
P007	5-(Aminomethyl)-3-isoxazolol
P008	4-Aminopyridine
	Ammonium metavanadate see P119
P009	Ammonium picrate (R)
	ANTIMUCIN WDR see P092
	ANTURAT see P073
	AQUATHOL see P088
	ARETIT see P020
P010	Arsenic acid
P011	Arsenic pentoxide
P012	Arsenic trioxide
	Athrombin see P001
	AVITROL see P008
	Azindene see P054
	AZOFOS see P061
	Azophos see P061
	BANTU see P072
P013	Barium cyanide
	BASENITE see P020
	BCME see P016
P014	Benzenethiol
	Benzoepin see P050
P015	Beryllium dust
P016	Bis(chloromethyl) ether
	BLADAN-M see P071
P017	Bromoacetone
P018	Brucine
P019	2-Butanone peroxide
	BUFEN see P092
	Butaphene see P020
P020	2-sec-Butyl-4,5-dinitrophenol
P021	Calcium cyanide
	CALDON see P020
P022	Carbon disulfide
	CERESAN see P092
	CERESAN UNIVERSAL see P092
	CHEMOX GENERAL see P020
	CHEMOX P.E. see P020
	CHEM-TOL see P090
P023	Chloroacetaldehyde
P024	p-Chloroaniline
P025	1-(p-Chlorobenzoyl)-5-methoxy-2-methylindole-3-acetic acid
P026	1-(o-Chlorophenyl)thiourea
P027	3-Chloropropionitrile
P028	alpha-Chlorotoluene
P029	Copper cyanide
	CRETOX see P108
	Coumadin see P001
	Coumafen see P001
P030	Cyanides
P031	Cyanogen
P032	Cyanogen bromide
P033	Cyanogen chloride
	Cyclodan see P050
P034	2-Cyclohexyl-4,6-dinitrophenol
	D-CON see P001
	DETHMOR see P001
	DETHNEL see P001
	DFP see P043
P035	2,4-Dichlorophenoxyacetic acid (2,4-D)
P036	Dichlorophenylarsine
	Dicyanogen see P031
P037	Dieldrin
	DIELDREX see P037
P038	Diethylarsine
P039	0,0-Diethyl-S-(2-(ethylthio)ethyl)ester of phosphorothioic acid
P040	0,0-Diethyl-0-(2-pyrazinyl)phosphorothioate
P041	0,0-Diethyl phosphoric acid, 0-p-nitrophenyl ester
P042	3,4-Dihydroxy-alpha-(methylamino)-methyl benzyl alcohol
P043	Di-isopropylfluorophosphate
	DIMETATE see P044
	1,4:5,8-Dimethanonaphthalene, 1,2,3,4,10,10-hexachloro-1,4,4a,5,8,8a-hexahydro endo, endo see P060
P044	Dimethoate
P045	3,3-Dimethyl-1-(methylthio)-2-butanone-O-[(methylamino)carbonyl] oxime
P046	alpha,alpha-Dimethylphenethylamine
	Dinitrocyclohexylphenol see P034
P047	4,6-Dinitro-o-cresol and salts
P048	2,4-Dinitrophenol
	DINOSEB see P020
	DINOSEBE see P020
	Disulfoton see P039
P049	2,4-Dithiobiuret
	DNBP see P020
	DOLCO MOUSE CEREAL see P108
	DOW GENERAL see P020
	DOW GENERAL WEED KILLER see P020
	DOW SELECTIVE WEED KILLER see P020
	DOWICIDE G see P090
	DYANACIDE see P092
	EASTERN STATES DUOCIDE see P001
	ELGETOL see P020
P050	Endosulfan
P051	Endrin
	Epinephrine see P042
P052	Ethylcyanide
P053	Ethylenediamine
P054	Ethyleneimine
	FASCO FASCRAT POWDER see P001
	FEMMA see P091
P055	Ferric cyanide
P056	Fluorine
P057	2-Fluoroacetamide
P058	Fluoroacetic acid, sodium salt
	FOLODOL-80 see P071
	FOLODOL M see P071
	FOSFERNO M 50 see P071
	FRATOL see P058
	Fulminate of mercury see P065
	FUNGITOX OR see P092
	FUSSOF see P057
	GALLOTOX see P092
	GEARPHOS see P071
	GERUTOX see P020
P059	Heptachlor
P060	1,2,3,4,10,10-Hexachloro-1,4,4a,5,8,8a-hexahydro-1,4:5,8-endo, endo-dimethanonaphthalene
	1,4,5,6,7,7-Hexachloro-cyclic-5-norbornene-2,3-dimethanol sulfite see P050
P061	Hexachloropropene
P062	Hexaethyl tetraphosphate
	HOSTAQUICK see P092
	HOSTAQUIK see P092
	Hydrazomethane see P068
P063	Hydrocyanic acid
	ILLOXOL see P037
	INDOCI see P025
	Indomethacin see P025
	INSECTOPHENE see P050
	Isodrin see P060
P064	Isocyanic acid, methyl ester
	KILOSEB see P020
	KOP-THIODAN see P050
	KWIK-KIL see P108
	KWIKSAN see P092
	KUMADER see P001
	KYPFARIN see P001
	LEYTOSAN see P092
	LIQUIPHENE see P092
	MALIK see P050
	MAREVAN see P001
	MAR-FRIN see P001
	MARTIN'D MAR-FRIN see P001
	MAVERAN see P001
	MEGATOX see P005
P065	Mercury fulminate
	MERSOLITE see P092
	METACID 50 see P071
	METAFOS see P071
	METAPHOR see P071
	METAPHOS see P071
	METASOL 30 see P092
P066	Methomyl
P067	2-Methylaziridine
	METHYL-E 605 see P071
P068	Methyl hydrazine
	Methyl isocyanate see P064
P069	2-Methyllactonitrile
P070	2-Methyl-2-(methylthio)propionaldehyde-o-(methylcarbonyl) oxime
	METHYL NIRON see P042
P071	Methyl parathion
	METRON see P071
	MOLE DEATH see P108
	MOUSE-NOTS see P108
	MOUSE-RID see P108
	MOUSE-TOX see P108
	MUSCIMOL see P007
P072	1-Naphthyl-2-thiourea
P073	Nickel carbonyl
P074	Nickel cyanide
P075	Nicotine and salts
P076	Nitric oxide
P077	p-Nitroaniline
P078	Nitrogen dioxide
P079	Nitrogen peroxide
P080	Nitrogen tetroxide
P081	Nitroglycerine (R)
P082	N-Nitrosodimethylamine
P083	N-Nitrosodiphenylamine
P084	N-Nitrosomethylvinylamine
	NYLMERATE see P092
	OCTALOX see P037
P085	Octamethylpyrophosphoramide
	OCTAN see P092
P086	Oleyl alcohol condensed with 2 moles ethylene oxide
	OMPA see P085
	OMPACIDE see P085
	OMPAX see P085
P087	Osmium tetroxide
P088	7-Oxabicyclo[2.2.1]heptane-2,3-dicarboxylic acid
	PANIVARFIN see P001
	PANORAM D-31 see P037
	PANTHERINE see P007
	PANWARFIN see P001
P089	Parathion
	PCP see P090
	PENNCAP-M see P071
	PENOXYL CARBON N see P048
P090	Pentachlorophenol
	Pentachlorophenate see P090
	PENTA-KILL see P090
	PENTASOL see P090
	PENWAR see P090
	PERMICIDE see P090
	PERMAGUARD see P090
	PERMATOX see P090
	PERMITE see P090
	PERTOX see P090
	PESTOX III see P085
	PHENMAD see P092
	PHENOTAN see P020
P091	Phenyl dichloroarsine
	Phenyl mercaptan see P014
P092	Phenylmercury acetate
P093	N-Phenylthiourea
	PHILIPS 1861 see P008
	PHIX see P092
P094	Phorate
P095	Phosgene
P096	Phosphine
P097	Phosphorothioic acid, 0,0-dimethyl ester, 0-ester with N,N-dimethyl benzene sulfonamide
	Phosphorothioic acid 0,0-dimethyl-0-(p-nitrophenyl) ester see P071
	PIED PIPER MOUSE SEED see P108
P098	Potassium cyanide
P099	Potassium silver cyanide
	PREMERGE see P020
P100	1,2-Propanediol
	Propargyl alcohol see P102
P101	Propionitrile
P102	2-Propyn-1-ol
	PROTHROMADIN See P001
	QUICKSAM see P092
	QUINTOX see P037
	RAT AND MICE BAIT see P001
	RAT-A-WAY see P001
	RAT-B-GON see P001
	RAT-O-CIDE #2 see P001
	RAT-GUARD see P001
	RAT-KILL see P001
	RAT-MIX see P001
	RATS-NO-MORE see P001
	RAT-OLA see P001
	RATOREX see P001
	RATTUNAL see P001
	RAT-TROL see P001
	RO-DETH see P001
	RO-DEX see P108
	ROSEX see P001
	ROUGH & READY MOUSE MIX see P001
	SANASEED see P108
	SANTOBRITE see P090
	SANTOPHEN see P090
	SANTOPHEN 20 see P090
	SCHRADAN see P085
P103	Selenourea
P104	Silver Cyanide
	SMITE see P105
	SPARIC see P020
	SPOR-KIL see P092
	SPRAY-TROL BRAND RODEN-TROL see P001
	SPURGE see P020
P105	Sodium azide
	Sodium coumadin see P001
P106	Sodium cyanide
	Sodium fluoroacetate see P056
	SODIUM WARFARIN see P001
	SOLFARIN see P001
	SOLFOBLACK BB see P048
	SOLFOBLACK SB see P048
P107	Strontium sulfide
P108	Strychnine and salts
	SUBTEX see P020
	SYSTAM see P085
	TAG FUNGICIDE see P092
	TEKWAISA see P071
	TEMIC see P070
	TEMIK see P070
	TERM-I-TROL see P090
P109	Tetraethyldithiopyrophosphate
P110	Tetraethyl lead
P111	Tetraethylpyrophosphate
P112	Tetranitromethane
	Tetraphosphoric acid, hexaethyl ester see P062
	TETROSULFUR BLACK PB see P048
	TETROSULPHUR PBR see P048
P113	Thallic oxide
	Thallium peroxide see P113
P114	Thallium selenite
P115	Thallium (I) sulfate
	THIFOR see P092
	THIMUL see P092
	THIODAN see P050
	THIOFOR see P050
	THIOMUL see P050
	THIONEX see P050
	THIOPHENIT see P071
P116	Thiosemicarbazide
	Thiosulfan tionel see P050
P117	Thiuram
	THOMPSON'S WOOD FIX see P090
	TIOVEL see P050
P118	Trichloromethanethiol
	TWIN LIGHT RAT AWAY see P001
	USAF RH-8 see P069
	USAF EK-4890 see P002
P119	Vanadic acid, ammonium salt
P120	Vanadium pentoxide
	VOFATOX see P071
	WANADU see P120
	WARCOUMIN see P001
	WARFARIN SODIUM see P001
	WARFICIDE see P001
	WOFOTOX see P072
	YANOCK see P057
	YASOKNOCK see P058
	ZIARNIK see P092
P121	Zinc cyanide
P122	Zinc phosphide (R,T)
	ZOOCOUMARIN see P001

[1] The Agency included those trade names of which it was aware; an omission of a trade name does not imply that the omitted material is not hazardous. The material is hazardous if it is listed under its generic name.

Source: Reference [5].

species, containers and spill residues thereof and Table 5 which identifies commercial chemical products and intermediates categorised as toxic wastes and subject to small quantity exclusions.

The USEPA approach can be seen to be much more definitive than the present UK 'Special Waste' Regulations. In particular it identifies sampling and analytical methods and highlights specific sources of wastes, brand names etc. Furthermore, by including the EP toxicity characteristics, the American Regulations seek to protect against ground and surface water pollution, a feature sadly missing from the UK Special Waste Regulations.

Irrespective of the current debate on governmental definitions there exists a real practical problem in controlling wastes as they arrive at their disposal point. The waste producer, the disposal site operator and the regulatory authorities need to determine whether a particular waste consignment contains detectable hazardous materials or not. The limited time available for such analysis dictates that testing procedures need to be broadly based and quick.

A medical research team [6] has developed field test kits for the detection of arsenic, barium, copper, cyanide, mercury and explosives in waste extracts and a British research worker has introduced a patented test kit called the 'Haz Test Kit', for this purpose [7]. It permits the identification of hazards (rather than chemicals) and can be used by untrained staff to identify poisonous, explosive, flammable material etc. Fig. 1 shows the easily-portable kit.

Hazardous waste production

Notwithstanding all the difficulties summarised above relating to the definition of hazardous waste it is desirable to quantify its production rate. In Britain estimates include the following: [3]

	Million tonnes/yr
Department of The Environment	3.7
Harwell Laboratory	4.0
Institute of Waste Management	5.0
House of Lords Report	5.54

and this compares with about 18 million tonnes per year of household and commercial wastes [8].

In the USA hazardous waste production has been estimated at 10 million tonnes per year [9] compared with 190 million tonnes per year of residential, commercial and institutional wastes [10].

Table 5. *Commercial chemical products and intermediates designated as toxic wastes unless subject to small quantity exclusion*

Hazardous Waste No.	Substance[1]
	AAF see U005
U001	Acetaldehyde
U002	Acetone (I)
U003	Acetonitrile (I,T)
U004	Acetophenone
U005	2-Acetylaminoflourene
U006	Acetyl chloride (C,T)
U007	Acrylamide
	Acetylene tetrachloride see U209
	Acetylene trichloride see U228
U008	Acrylic acid (I)
U009	Acrylonitrile
	AEROTHENE TT see U226
	3-Amino-5-(p-acetamidophenyl)-1H-1,2,4-triazole, hydrate see U011
U010	6-Amino-1,1a,2,8,8a,8b-hexahydro-8-(hydroxymethyl)8-methoxy-5-methylcarbamate azinno(2',3':3,4) pyrrolo(1,2-a) indole-4, 7-dione (ester)
U011	Amitrole
U012	Aniline (I)
U013	Asbestos
U014	Auramine
U015	Azaserine
U016	Benz[c]acridine
U017	Benzal chloride
U018	Benz[a]anthracene
U019	Benzene
U020	Benzenesulfonyl chloride (C,R)
U021	Benzidine
	1,2-Benzisothiazolin-3-one, 1,1-dioxide see U202
	Benzo[a]anthracene see U018
U022	Benzo[a]pyrene
U023	Benzotrichloride (C,R,T)
U024	Bis(2-chloroethoxy)methane
U025	Bis(2-chloroethyl) ether
U026	N,N-Bis(2-chloroethyl)-2-naphthylamine
U027	Bis(2-chloroisopropyl) ether
U028	Bis(2-ethylhexyl) phthalate
U029	Bromomethane
U030	4-Bromophenyl phenyl ether
U031	n-Butyl alcohol (I)
U032	Calcium chromate
	Carbolic acid see U188
	Carbon tetrachloride see U211
U033	Carbonyl fluoride
U034	Chloral
U035	Chlorambucil
U036	Chlordane
U037	Chlorobenzene
U038	Chlorobenzilate
U039	p-Chloro-m-cresol
U040	Chlorodibromomethane
U041	1-Chloro-2,3-epoxypropane
	CHLOROETHENE NU see U226
U042	Chloroethyl vinyl ether
U043	Chloroethene
U044	Chloroform (I,T)
U045	Chloromethane (I,T)
U046	Chloromethyl methyl ether
U047	2-Chloronaphthalene
U048	2-Chlorophenol
U049	4-Chloro-o-toluidine hydrochloride
U050	Chrysene
	C.I. 23060 see U073
U051	Cresote
U052	Cresols
U053	Crotonaluehyde
U054	Cresylic acid
U055	Cumene
	Cyanomethane see U003
U056	Cyclohexane (I)
U057	Cyclohexanone (I)
U058	Cyclophosphamide
U059	Daunomycin
U060	DDD
U061	DDT
U062	Diallate
U063	Dibenz[a,h]anthracene
	Dibenzo[a,h]anthracene see U063
U064	Dibenzo[a,i]pyrene
U065	Dibromochloromethane
U066	1,2-Dibromo-3-chloropropane
U067	1,2-Dibromoethane
U068	Dibromomethane
U069	Di-n-butyl phthalate
U070	1,2-Dichlorobenzene
U071	1,3-Dichlorobenzene
U072	1,4-Dichlorobenzene
U073	3,3'-Dichlorobenzidine
U074	1,4-Dichloro-2-butene
	3,3'-Dichloro-4,4'-diaminobiphenyl see U073
U075	Dichlorodifluoromethane
U076	1,1-Dichloroethane
U077	1,2-Dichloroethane
U078	1,1-Dichloroethylene
U079	1,2-trans-dichloroethylene
U080	Dichloromethane
	Dichloromethylbenzene see U017
U081	2,4-Dichlorophenol
U082	2,6-Dichlorophenol
U083	1,2-Dichloropropane
U084	1,3-Dichloropropene
U085	Diepoxybutane (I,T)
U086	1,2-Diethylhydrazine
U087	0,0-Diethyl-S-methyl ester of phosphorodithioic acid
U088	Diethyl phthalate
U089	Diethylstilbestrol
U090	Dihydrosafrole
U091	3,3'-Dimethoxybenzidine
U092	Dimethylamine (I)
U093	p-Dimethylaminoazobenzene
U094	7,12-Dimethylbenz[a]anthracene
U095	3,3'-Dimethylbenzidine
U096	alpha,alpha-Dimethylbenzylhydroperoxide (R)
U097	Dimethylcarbamoyl chloride
U098	1,1-Dimethylhydrazine
U099	1,2-Dimethylhydrazine
U100	Dimethylnitrosoamine
U101	2,4-Dimethylphenol
U102	Dimethyl phthalate
U103	Dimethyl sulfate
U104	2,4-Dinitrophenol
U105	2,4-Dinitrotoluene
U106	2,6-Dinitrotoluene
U107	Di-n-octyl phthalate
U108	1,4-Dioxane
U109	1,2-Diphenylhydrazine
U110	Dipropylamine (I)
U111	Di-n-propylnitrosamine
	EBDC see U114
	1,4-Epoxybutane see U213
U112	Ethyl acetate (I)
U113	Ethyl acrylate (I)
U114	Ethylenebisdithiocarbamate
U115	Ethylene oxide (I,T)
U116	Ethylene thiourea
U117	Ethyl ether (I,T)
U118	Ethylmethacrylate
U119	Ethyl methanesulfonate
	Ethylnitrile see U003
	Firemaster T23P see U235
U120	Fluoranthene
U121	Fluorotrichloromethane
U122	Formaldehyde
U123	Formic acid (C,T)
U124	Furan (I)
U125	Furfural (I)
U126	Glycidylaldehyde
U127	Hexachlorobenzene
U128	Hexachlorobutadiene
U129	Hexachlorocyclohexane
U130	Hexachlorocyclopentadiene
U131	Hexachloroethane
U132	Hexachlorophene
U133	Hydrazine (R,T)
U134	Hydrofluoric acid (C,T)
U135	Hydrogen sulfide
	Hydroxybenzene see U188
U136	Hydroxydimethyl arsine oxide
	4,4'-(Imidocarbonyl)bis(N,N-dimethyl)aniline see U014
U137	Indeno(1,2,3-cd)pyrene
U138	Iodomethane
U139	Iron Dextran
U140	Isobutyl alcohol
U141	Isosafrole
U142	Kepone
U143	Lasiocarpine
U144	Lead acetate
U145	Lead phosphate
U146	Lead subacetate
U147	Maleic anhydride
U148	Maleic hydrazide
U149	Malononitrile
	MEK Peroxide see U160
U150	Melphalan
U151	Mercury
U152	Methacrylonitrile
U153	Methanethiol
U154	Methanol
U155	Methapyrilene
	Methyl alcohol see U154
U156	Methyl chlorocarbonate
	Methyl chloroform see U226
U157	3-Methylcholanthrene
	Methyl chloroformate see U156
U158	4,4'-Methylene-bis-(2-chloroaniline)
U159	Methyl ethyl ketone (MEK) (I,T)
U160	Methyl ethyl ketone peroxide (R)
	Methyl iodide see U138
U161	Methyl isobutyl ketore
U162	Methyl methacrylate (R,T)
U163	N-Methyl-N'-nitro-N-nitrosoguanidine
U164	Methylthiouracil
	Mitomycin C see U010
U165	Naphthalene
U166	1,4-Naphthoquinone
U167	1-Naphthylamine
U168	2-Naphthylamine
U169	Nitrobenzene (I,T)
	Nitrobenzol see U169
U170	4-Nitrophenol
U171	2-Nitropropane (I)
U172	N-Nitrosodi-n-butylamine
U173	N-Nitrosodiethanolamine
U174	N-Nitrosodiethylamine
U175	N-Nitrosodi-n-propylamine
U176	N-Nitroso-n-ethylurea
U177	N-Nitroso-n-methylurea
U178	N-Nitroso-n-methylurethane
U179	N-Nitrosopiperidine
U180	N-Nitrosopyrrolidine
U181	5-Nitro-o-toluidine
U182	Paraldehyde
	PCNB see U185
U183	Pentachlorobenzene
U184	Pentachloroethane
U185	Pentachloronitrobenzene
U186	1,3-Pentadiene (I)
	Perc see U210
	Perchlorethylene see U210
U187	Phenacetin
U188	Phenol
U189	Phosphorous sulfide (R)
U190	Phthalic anhydride
U191	2-Picoline
U192	Pronamide
U193	1,3-Propane sultone
U194	n-Propylamine (I)
U196	Pyridine
U197	Quinones
U200	Reserpine
U201	Resorcinol
U202	Saccharin
U203	Safrole
U204	Selenious acid
U205	Selenium sulfide (R,T)
	Silvex see U233
U206	Streptozotocin
	2,4,5-T see U232
U207	1,2,4,5-Tetrachlorobenzene
U208	1,1,1,2-Tetrachloroethane
U209	1,1,2,2-Tetrachloroethane
U210	Tetrachloroethene
	Tetrachloroethylene see U210
U211	Tetrachloromethane
U212	2,3,4,6-Tetrachlorophenol
U213	Tetrahydrofuran (I)
U214	Thallium (I) acetate
U215	Thallium (I) carbonate
U216	Thallium (I) chloride
U217	Thallium (I) nitrate
U218	Thioacetamide
U219	Thiourea
U220	Toluene
U221	Toluenediamine
U222	o-Toluidine hydrochloride
U223	Toluene diisocyanate
U224	Toxaphene
	2,4,5-TP see U233
U225	Tribromomethane
U226	1,1,1-Trichloroethane
U227	1,1,2-Trichloroethane
U228	Trichloroethene
	Trichloroethylene see U228
U229	Trichlorofluoromethane
U230	2,4,5-Trichlorophenol
U231	2,4,6-Trichlorophenol
U232	2,4,5-Trichlorophenoxyacetic acid
U233	2,4,5-Trichlorophenoxypropionic acid alpha, alpha, alpha- Trichlorotoluene see U023
	TRI-CLENE see U228
U234	Trinitrobenzene (R,T)
U235	Tris(2,3-dibromopropyl) phosphate
U236	Trypan blue
U237	Uracil mustard
U238	Urethane
	Vinyl chloride see U043
	Vinylidene chloride see U078
U239	Xylene

[1] The Agency included those trade names of which it was aware; an omission of a trade name does not imply that it is not hazardous. The material is hazardous if it is listed under its generic name.

Source: Reference [5].

Fig. 1. The HAZ-KIT for the identification of hazards in waste. (Source: Reference [7].)

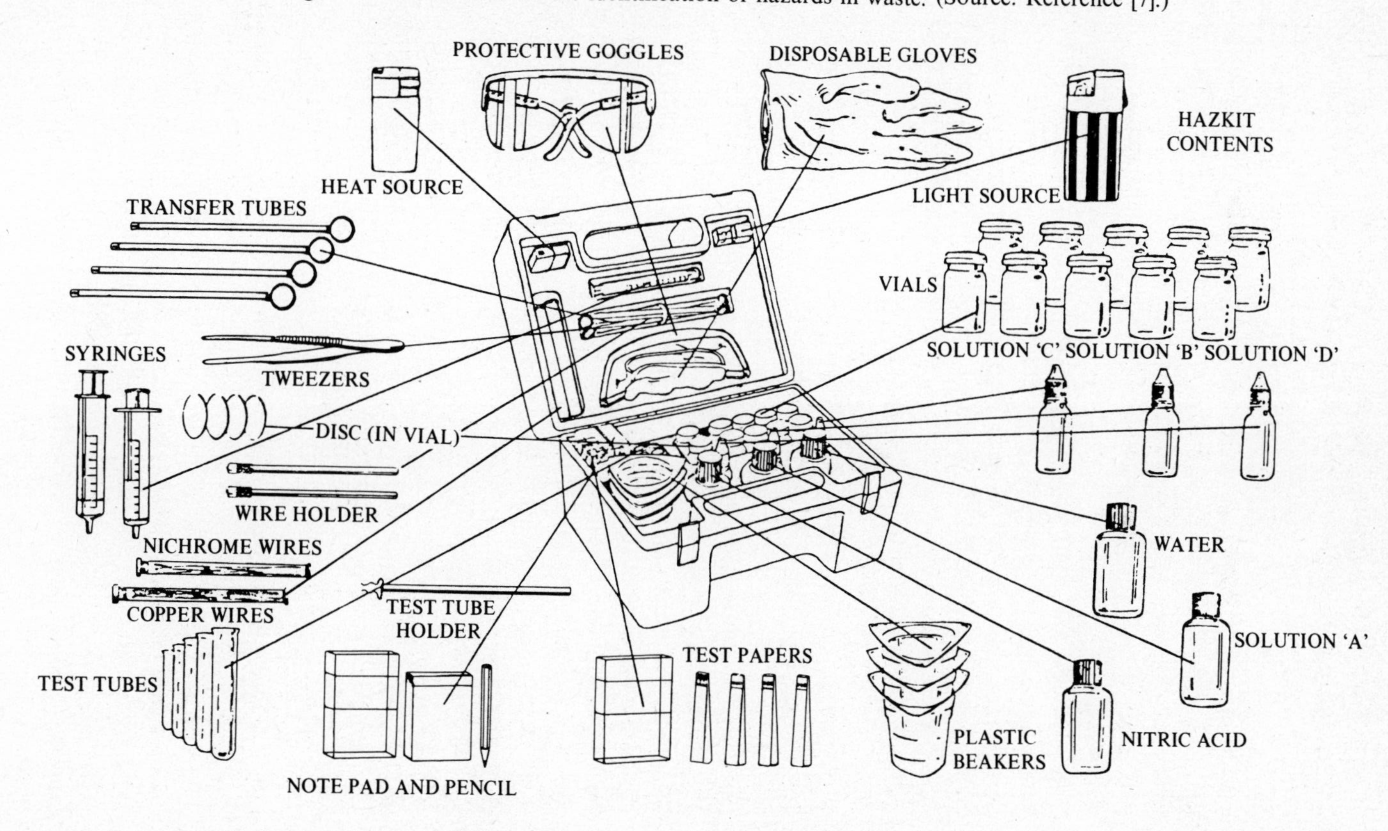

Why industry produces waste

In the western world all manufacturing industries are geared to produce and sell a series of products from a limited range of raw materials. Variations in the composition or source of the raw materials may occur but management must compensate for such changes and produce products of known composition at the required rate. It is the skill of the manager in making these compensations and marketing his products to the consumer which is the wealth creating activity upon which our capitalist society depends [11].

The free forces of competition demand that the manager seeks to find the lowest unit cost for each step in the production operation which he controls. To achieve this aim he discards impure raw materials, intermediate byproducts and off-specification products since he knows or suspects that effort devoted to the use, recovery or sale of such materials is not a wealth creating activity and does not serve to reduce the unit cost of his product. Thus waste is an essential feature of an industrial society. Our wealth producers are our waste producers. Indeed, our wealthiest nations produce the greatest quantity of industrial waste.

The economics of waste recycling

During recent decades the word 'recycling' has become fashionable in waste management circles and the implication is that this new concept can overcome all our problems in the waste management field. We have seen more references to this subject in local government journals than almost any other subject and conferences and exhibitions on this theme have been widely promoted. It is clear that in planning future waste management strategies we need to recognise the importance of recycling and waste recovery [11, 12].

Despite the fact that recycling options are frequently described by glamorous jargon, in reality only five basic schemes exist. It is desirable to recognise the characteristics of these schemes.

Scheme 1

This scheme, which is the simplest recycling scheme of all, is shown in Fig. 2. A factory receives a range of raw materials and produces a range of products. These products must conform to particular specifications and be produced at a rate which is dictated by the market place. An intrinsic part of the manufacturing process produces a waste product which can be collected and transferred back to the process to supplement the raw material feed. The term 'precycling' has been coined in some circles to describe this activity [3].

This type of scheme has the following noteworthy features:

(i) There is no selling or marketing effort involved.

(ii) There is no significant transport element required.

(iii) The factory management, who control both the production rate and re-use aspects, are the best-qualified people to align the rate of production with the rate of re-use.

A typical example of this type of operation is a plastic extrusion factory where waste products can be readily returned to the extruder.

In general, recycling operations of this kind are already widely used and the growth of new applications which would reduce the quantity of industrial waste produced is not likely to be a major factor affecting waste management policy in the future.

Scheme 2

Details of Scheme 2 are shown in Fig. 3. In this scheme the characteristics of the factory remain as outlined for Scheme 1 but the waste requires some form of storage or processing before it can be recycled to supplement the raw materials used. In most manufacturing processes the generation of waste is not 'time controlled'. That is to say that once management has decided to produce a particular product, then the waste is produced as a natural consequence. We cannot choose when we wish to produce waste and when we do not. The recycling of this waste then requires the construction of storage facilities to help balance production

Fig. 2. Waste recycling scheme 1. (Source: Reference [11].)

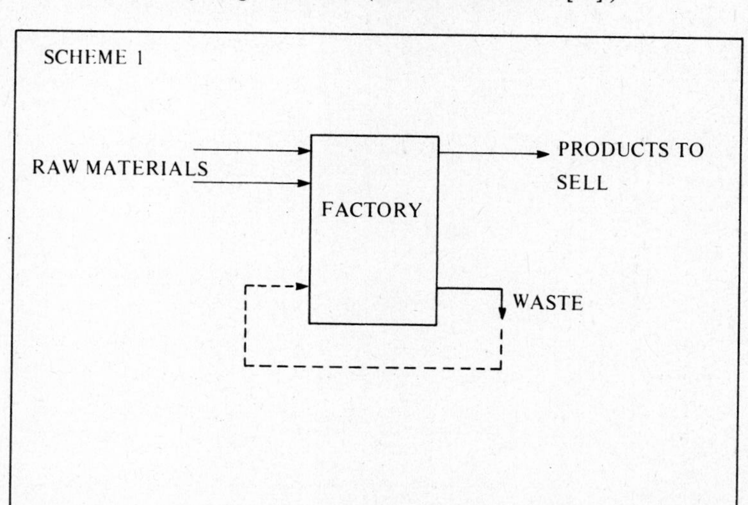

rate with the rate of re-use. In other cases, equipment is needed to condition the waste so that it becomes acceptable to the manufacturing process. The essential features of Scheme 2 are therefore:

(i) Capital investment is required.
(ii) No selling or marketing effort is required.
(iii) Transport is not required.

The capital investment is normally not separately accounted for. The justification for the capital may well be closely studied but once the plant is installed it becomes an intrinsic part of the manufacturing operation. No-one regularly measures or controls the 'profitability' of the particular component.

A typical example can be seen in paper mill operations where broke from the paper making machine requires re-pulping before it can be recycled. A broke hydrapulper is installed for this purpose.

Recycling operations of this kind are already widely used. Designs for new factories are now likely to include such facilities in new expansion programmes but the introduction of such schemes to existing manufacturing plants is not likely to produce noticeable changes which affect the waste disposal industry.

Scheme 3

Details of this scheme are shown in Fig. 4. Factory A has characteristics identical to those in previous examples but produces a waste

Fig. 3. Waste recycling scheme 2. (Source: Reference [11].)

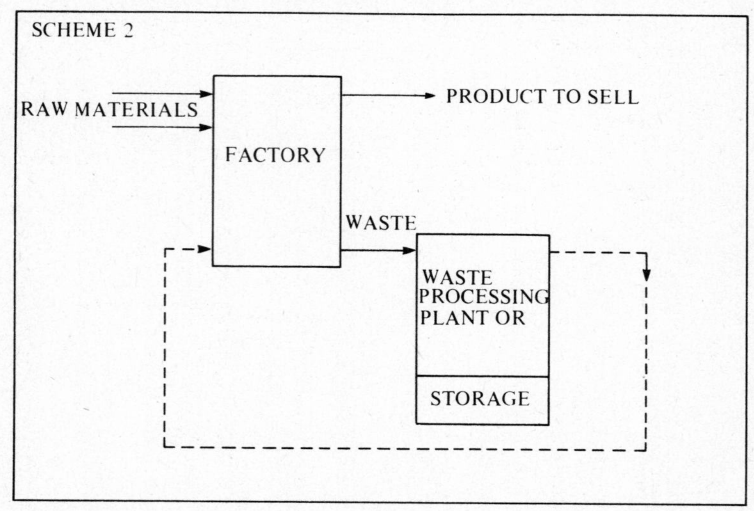

which can be used by Factory B as a supplement to its raw materials. This scheme has the following essential features:

(i) Selling and marketing effort is required. The magnitude of this is considerable because not only has first contact between A and B to be established but it has to be maintained to compensate for new conditions.

(ii) It is recognised that the quality of the waste from A varies as changes occur in the manufacturing process and these changes naturally affect its value as far as B is concerned. Management somewhere, usually at Factory B, needs to measure these quality changes and adjust its manufacturing process to take account of them. Certainly no change in the quality of Factory B's end product can be tolerated so extra management effort has to be devoted to monitor and control the changes in the waste.

(iii) As we have seen earlier, waste generation is not time controlled. Inevitably, Factory B will want waste when Factory A cannot produce it and/or Factory A will want to transfer more waste than Factory B can accept. Again management effort is required to try to dovetail production and re-use capabilities and, naturally, this frequently leads to the need for storage capability.

Thus we can see from (i) and (ii) that more management effort is required and the onus usually falls on Factory B. Managers in this factory are naturally judged by the profitability of their operations and are conscious of the fact that they can buy raw materials from a supplier as an alternative to re-using waste. The raw materials will be conveniently packaged,

Fig. 4. Waste recycling scheme 3. (Source: Reference [11].)

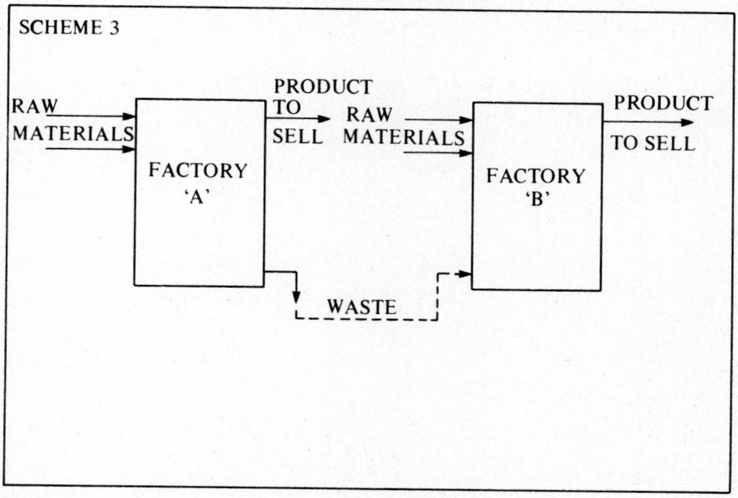

supplied when needed, have a known constant composition and require less management supervision. Under these circumstances, it seems likely that managers at Factory B might prefer to buy new raw materials rather than re-use waste.

(iv) Transport is required in some form to transfer the waste from Factory A to Factory B. This will usually be provided by a third party, a transport contractor, in the form of road or rail transport. The transport contractor is familiar with this type of operation since to him it is no different to the transport of raw materials and products. His contract terms will protect him against excessive loading and unloading times and he can be reasonably assured of his profit.

(v) Management at Factory A is also judged by profitability requirement and might be tempted to think that its waste is no longer a waste at all but another product which it can sell to Factory B to supplement its income. One can readily imagine the tactful negotiations between A and B to establish the market price of the waste.

A typical example of this type of recycling scheme is a chemical manufacturer who produces a waste alkali which has uses at another factory as a supplementary raw material. A few recycling schemes of this kind have been operating for several years but, in the absence of external subsidies, we are unlikely to see sufficient growth in this area which would have any impact on the quantities of industrial waste generated.

Fig. 5. Waste recycling scheme 4. (Source: Reference [11].)

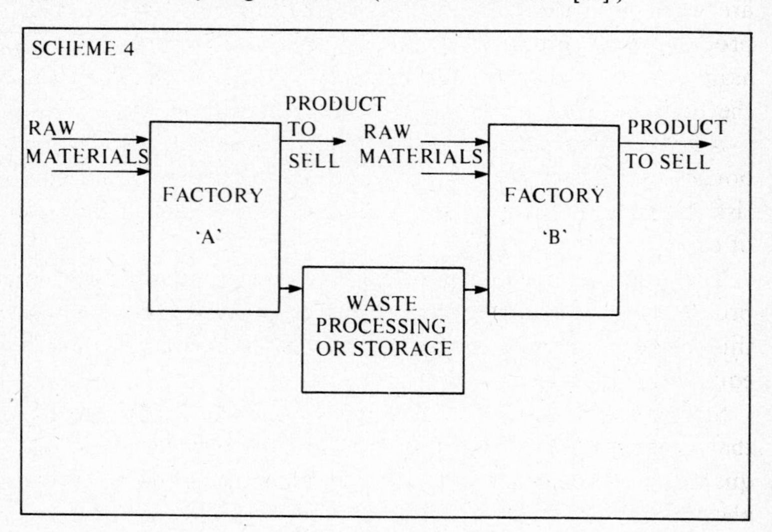

Scheme 4

Details of this scheme are shown in Fig. 5. This scheme closely resembles Scheme 3 except that intermediate storage or processing of the waste between Factory A and Factory B is necessary. The features of this scheme then include:

(i) Capital investment is necessary.

(ii) Selling and marketing effort is required.

(iii) Additional management control is required to compensate for quality changes in the recycled waste.

(iv) Additional management control is required to dovetail production and re-use rates.

(v) Transport is required. The processing or storage facility is usually installed at either Factory A or B (usually B) to avoid the need for two separate transport activities.

(vi) Questions concerning the value of the recycled waste remain to be settled by negotiation.

(vii) The question of finance for the capital investment needs careful consideration. Normally this decision falls upon managers at Factory B and they are likely to seek assurances from Factory A about the long-term availability of the waste, its control within acceptable chemical and physical limits, the agreed price and its escalation. If the capital sum is large, managers at Factory B may try to secure a long-term contract from Factory A but the manager at Factory A will point out that they have little control over its availability, rate of production and composition since these are indirectly determined by the fluctuating demands for their normal products (and products mix). Factory B managers might also seek assurances from the transport contractor concerning transport prices and they will find that they are equally difficult to obtain.

Against the certain knowledge that an alternative raw material can be bought by Factory B's management, one might expect that the business risk related to the investment will be unacceptably high in the vast majority of cases.

The example mentioned in Scheme 3 of a chemical manufacturer producing waste alkali which can be used at another factory would fall into this category if equipment such as storage tanks, pumps and pH control equipment were deemed necessary.

Since the businessman's risk seems unacceptably high it seems unlikely that new schemes of this kind will have any significant impact on the quantity of waste generated by our manufacturing industries in the years ahead.

Scheme 5

This scheme bears little resemblance to the schemes described above. It is shown in Fig. 6. It concerns the recycling of surplus chemicals and products produced by manufacturing industry. These materials may be small quantities of high quality products produced in excess of requirements or, alternatively, small quantities of sub-standard products which are unacceptable to the normal users. The features of this scheme include:

(i) Selling and marketing effort is required. The small quantities and varied nature of these wastes mean that the potential users are many and widely distributed throughout the land. The establishment of a regular sales team providing specialised help in the use of such products is consequently out of the question and sales effort has to be confined to the preparation and distribution of lists of wastes to a wide audience.

(ii) Potential purchasers who seek assurances on quality specification and packaging might expect to encounter difficulties.

(iii) There is little possibility of dovetailing production rate and re-usage rate.

(iv) Transport is an essential feature.

(v) The wastes are conveniently packaged in discrete containers such as drums or bags and can therefore be easily stored.

Nevertheless, the materials have to be warehoused somewhere and cost is incurred. It must be recognised that some materials may be warehoused for considerable periods of time but are never sold. This cost must be added to

Fig. 6. Waste recycling scheme 5. (Source: Reference [11].)

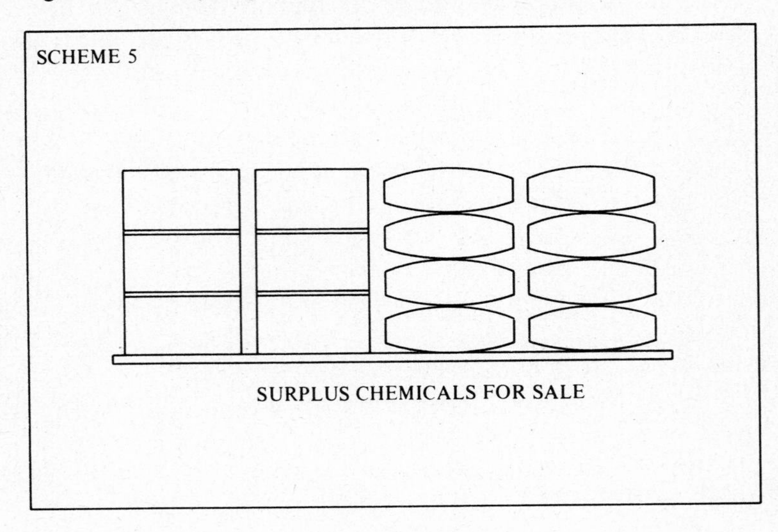

the warehousing cost of those materials that are sold. If these costs are properly included in the price calculations, it seems likely that many schemes would not be economically viable. However, there is normally a significant subsidy from other related operations and this is ignored in the pricing calculations.

A few private companies have been engaged in the recycling of surplus chemicals for many years now but the UK government's Waste Materials Exchange Scheme had to be discontinued since an on-going subsidy of taxpayers' money was shown to be essential to its continuance and this was deemed to be unacceptable.

The examination of the basic possibilities for industrial waste recycling has revealed many high-risk features which deter astute businessmen from entering this field. From this analysis it can be seen that the growth of recycling activities is a political development rather than an industrial one and, in the absence of external subsidies (e.g. government loans, tax incentives or voluntary contributions) the future of waste recycling is very limited. As long as we continue to measure industrial performance by the conventional assessment of profit, recycling is unlikely to have a major impact in reducing waste generation in the years ahead. Indeed, the concept of waste recycling, itself a contradiction in terms, is better politics than business.

References

[1] Department of the Environment, 'The Licensing of Waste Disposal Sites', *Waste Management Paper No. 4*, HMSO, London, 1976.

[2] Department of the Environment, 'Special Wastes: A Technical Memorandum Providing Guidance on their Definition', *Waste Management Paper No. 23*, HMSO, London, 1981.

[3] Select Committee on Science and Technology: House of Lords, *Hazardous Waste Disposal*, HMSO, London, Reports 273, Vols. I, II and III, 1981.

[4] Cope, C.B., Correspondence with Select Committee on Science and Technology: House of Lords. Available in House of Lords, Public Records Office, 1981.

[5] Federal Register, May 19, 1980, *Hazardous Waste and Consolidated Permit Regulations*, USEPA, Vol. 45, No. 98, Book 2.

[6] Snyder, R.E., M.E. Tonkin & A.M. McKissick, 'Development of Hazardous-Toxic Waste Analytical Screening Procedures', AD–A095506, Atlantic Research Corp, Alexandria Va, July 1980.

[7] Keen, R.C., *The Haz Test Kit*, Environmental Health Department, Bristol Polytechnic.

[8] Department of the Environment, 'Reclamation, Treatment and Disposal of Wastes', *Waste Management Paper No. 1*, HMSO, London, 1976.

[9] Powers, P.W., *How to Dispose of Toxic Substances and Industrial Wastes*, Noyes Data Corporation, New Jersey, 1976.

[10] Weiss, S., *Sanitary Landfill Technology*, Noyes Data Corporation, New Jersey, 1974.
[11] Cope, C.B., 'Progress in Waste Management', British Association for The Advancement of Science, 139th Annual Meeting, University of Aston in Birmingham, September, 1977.
[12] Cope, C.B., 'The Future Direction of Industrial Waste Disposal', Public Works Congress, Birmingham, November 1976.

2

Hazardous wastes legislation: planning and transport

Introduction

By all accounts, Oliver Cromwell did a pretty good job. Great Britain has a sound and enviable legislature in its parliament and, like it or not, believe it or not, rules and regulations by way of statute law are essential to society. Mind you, some would say that Guy Fawkes might have done a better job!

Parliament has powers to make and amend laws and very little, if anything, else. Our Civil Service is primarily responsible for the drafting of 'bills' which are then debated by MP's in the Commons and peers in the Lords and may eventually succeed to become 'acts' of parliament. An act then receives royal assent and may be put into force subsequently by a 'commencement order' or commencement orders. Such orders appear as 'statutory instruments'.

Several acts provide for 'regulations' or 'orders' to be made which may govern specific aspects within a particular piece of legislation and these, too, appear as statutory instruments.

Thus, central government in Great Britain has the prime function of making laws. Parliament itself is serviced by the Civil Service and this is structured on a ministry or department basis. For example, the Department of the Environment is the relevant agency responsible for 'the environment' and this includes functions related to planning, pollution and waste disposal. Each department or ministry is responsible to its Secretary of State who sits in parliament: thus, for waste disposal matters, it is the Secretary of State for the Environment who is accountable in parliament for the running of his Department of the Environment. Statutory instruments are signed by the Secretary of State and also, usually, by the Secretaries of State for Wales and for Scotland. Scottish law is sometimes peculiar and slightly at variance with England and Wales but this, really, is inconsequential as far as wastes are concerned.

Legislative administration

Having made enactments or laws, it is encumbent upon parliament to divest the implementation and administration thereof to 'local government' which includes county councils and district councils: the regional water authorities may also be considered here.

The central government departments or ministries undertake the task of ensuring that implementation and administration of acts of parliament are carried out by local government and this is usually prefaced by a 'circular' addressed to the chief executives of the councils involved. The circular is usually signed by an assistant secretary (i.e. civil servant) or assistant secretaries.

The circulars are advisory documents and very often seek to expand and explain the intentions of the Secretary(ies) of State as regards a piece of legislation. They give guidance on how to interpret, implement and administer the law.

Additionally, central government publishes 'white' and 'green' papers ('command papers') which set down government proposals for consultation or government policy and can commission reports from working parties. Many such documents will be referred to in the following discussion.

Legislative interpretation

The interpretation of the law, as written in acts of parliament, rests, ultimately, with the courts of justice. Decisions of the courts are known as 'case law' and it is a fact that the most interesting of legal cases revolve around the meaning and use of words. However well drafted a law may be, there will invariably be two ways of interpreting it. Consider the simple sign: 'Cars parked at owner's risk' – does the word 'owner' refer to the car or the car park? Does the sign: 'Please help yourself' in the supermarket really mean that the goods on display are free of charge?

Legislation is the raison d'être for the modern waste disposal industry: without the former, the latter would not exist. This is why it is vital to comprehend the law as it applies to waste disposal in any serious consideration of the subject.

It is salutary to reflect upon the Ten Commandments that are, in effect, a statement of the law and codes of conduct serving to govern the whole of societal organisation in the Christian world. Compared with our man-made legislation they are more precise, shorter and certainly more thought provoking. However, we do have to contend with parliamentary law and, hence, the need for this study.

Nevertheless, however good the intention, law is never black and white – if it were, half of us would be certain to be right every second time!

Historical perspective on legislation [1]

Dr Desmond Morris in *The Naked Ape* [2] reminds us that waste disposal is a problem of societal organisation: our primate kin drop their personal waste from trees and move on with little danger of contamination risks. Nomadic ways of life present, similarly, little problem: collections of people do present problems. Wherever man lives there is generated waste; both personal and indirect.

In remote areas burial of waste has always been easy and was, in fact, required practice in Israel as specified by Moses (*Deuteronomy* 23, vv. 12–13): 'Thou shalt have a place also without the camp, wither thou shalt go forth abroad. And thou shalt have a paddle upon the weapon; and it shall be, when thou wilt ease thyself abroad thou shalt dig therewith and shalt turn back and cover that which cometh from thee.' When, however, population increases in density so do piles of rubbish and the problems of collection and disposal become obvious. Formal legislation eventually became necessary in London because of human failing: the city was in such a mess. Thus, in 1297 the first order relating to waste disposal was made, by King Edward I, and it sought to ensure that, when the inhabitants of the city threw out their rubbish, it landed in the kennel (i.e. the gutter or centre of the road): in effect, each householder was obliged to keep the pathway in front of his tenement free from obstruction.

In 1309, the 'casting of filth' from houses into the streets and lanes of the city was prohibited by King Edward II. What householders did with their rubbish instead is a bit of a mystery but in 1349 came the 'Black Death' – the plague – and periodic recurrence of this horrible disease prompted the Order of 1354 (Edward III) which required that the filth accumulating in the kennels should be cleared out each week. 'Rakers' were employed in this behalf and they loaded the refuse onto 'tumbrils', or specially constructed horse-drawn carts: mostly, the rakers were conscripted convicts.

For some time, the streets of London were kept comparatively clean with the undoubted help of kites and ravens, that were designated 'protected', and scavenging pigs that roamed freely. The rakers delivered their collected rubbish to laystalls strategically provided on the advice of a Committee of

1387. Periodically, the laystalls were cleared out and the refuse loaded onto barges and taken down the Thames for dumping on the Essex marshes: this, of course, continues to this day.

The situation did not, however, improve and, in 1407, King Henry IV ordered the London householders to keep their rubbish indoors until the appointed day of the week when the rakers would call. This provided the basis of what should have been an orderly and organised waste collection service enforceable by law.

During these mediaeval attempts to clean up the London streets, those householders overlooking the Thames enjoyed a notable exemption from the laws: they had direct access to this well-used waste disposal facility. A Royal Order of 1357 addressed to the Mayor and Sherriffs of London informed them that Edward III had, when travelling along the Thames, 'beheld dung and laystalls and other filth accumulated . . . upon the bank of the said river . . . fumes and other abominable stenches arising therefrom; from the corruption of which, if tolerated, great peril . . . will it is feared arise'. The Order proclaimed that 'there shall henceforth no rubbish or filth be thrown or put into the rivers of Thames or Flete, or into the Fosses around the walls of the City, but all must be taken out of the City by carts'.

In 1372 the King had occasion once again to address the Mayor and Sherriffs complaining that the Thames was so choked with rubbish that 'great ships are not able, as of old they were wont, any longer to come up to the . . . City'.

An Ordinance was directed in 1383 against the residents of Wallbrook in the city who had actually managed to completely block the river with rubbish so as to make it impassable to boats!

First parliamentary legislation

It was not until 1388, however, that the first Sanitary Act was passed by parliament, sitting at Cambridge, that prohibited the throwing of 'dung, filth garbage, etc. into ditches, rivers or other waters and places within, about or nigh to any cities, boroughs or towns'. This, of course, applied to the whole of the kingdom and not just London.

The Great Fire of London in 1666 gave a temporary respite and transient purification of the city. It was, indeed, some time before complaints re-started: in 1741 Lord Tyrconnel described the streets of London as 'abounding with such heaps of filth as a savage would look on with amazement'. The decline continued and in 1832 two letters appeared in

An application for permission is addressed, usually in triplicate, to the local (district) authority together with proof that certain conditions, as specified in the Act, have been complied with. Section 26 requires that an application is accompanied by proof of publication of the details in a local newspaper and of advertising on the land. Article 8(1(b)) of the 1977 Order prescribes 'use of land, for the disposal of refuse or waste materials' as being subject to the provisions of Section 26 of the Act. The forms of notice are prescribed in the 1977 Order along these lines:

Schedule 3, Part I

The form of notice to be used under Section 26(2) requires publication in a local newspaper giving details of the type and location of proposed development and the address at which a copy of the application, and any plans, can be inspected by members of the public at all reasonable hours and free of charge for up to twenty days after the date of application. For waste disposal sites there may not be any suitable offices or premises at a site prior to development and, in such cases, it is quite in order to nominate the local planning authority offices as the address of deposition of the application and at which the public can inspect the details. Proof of advertisement should accompany the application.

Schedule 3, Part II

Schedule 3, Part II comprises a selection of three certificates (A, B and C), one of which is required to be completed under Section 26(2). Certificate A confirms that a notice has been posted on the land and remained there for seven days in a period of not more than one month prior to the making of the application; alternatively, Certificate B confirms that no notice was placed on the land owing to absence of legal right of access but that certain (to be specified) other actions were taken instead; alternatively, Certificate C confirms that a notice was posted but was removed or defaced and was not, therefore, on display for seven days and that certain (to be specified) other actions were taken instead. One such certificate should accompany the application.

Schedule 3, Part III

Schedule 3, Part III is the form of notice to be used for posting on the land under Section 26(3). Similar details as are required in the newspaper advertisement are to be displayed on the site and, again, the public invited to inspect the applications and plans at a nominated address in a period of up to twenty days from the date of posting of the notice.

Schedule 4, Part I

Schedule 4, Part I lays down the format of a choice of certificates relating to ownership. Anyone can make a planning application to develop land and this may not, necessarily, be the owner of the land. Section 27 of the Act requires, however, that certain steps are taken by a proposed developer to inform the owner(s) of the land about the proposed development. The first part comprises a choice of three certificates: Certificate A confirms that the applicant is the owner of the land for at least twenty days prior to the date of the application; alternatively, Certificate B confirms that due notice has been given to all persons who, twenty days before the date of the application, were owners of any part of the land; alternatively, Certificate C confirms either that an applicant has been unable to trace any of the relevant owners and describes the steps taken, including details of newspaper advertisements, to locate the same or that some of the owners have been traced and notices served but others have not and describes the efforts taken to trace them. For the purposes of the certificates to be completed in accordance with the Section 27 requirements, the term 'owner' is defined in the 1971 Act and the 1977 Order as 'a person having a freehold interest or a leasehold interest the unexpired term of which was not less than seven years'.

The second part requires confirmation either that none of the land constitutes or forms part of an agricultural holding or, alternatively, that notice has been given to any person who, during the previous twenty days, was a tenant of any agricultural holding any part of which was comprised in the land to which the application relates.

Schedule 4, Part II

Schedule 4, Part II comprises the form of notice for serving on other landowners inviting them to make representations about the proposed development if they so wish and the form of notice to be displayed in a local newspaper if the landowner(s) cannot be traced and with a similar invitation to make representation.

Planning consultations

When a valid application is received then the district planning authority will acknowledge such receipt and will indicate if it is a county matter and, if so, send it to the county authority within seven days. Despite waste disposal activities being prescribed county matters, the application will be addressed, for registration purposes, to the district authority. Normally, 28 days is allowed for the county to reply to the district but this can be extended by written agreement and the county can, and may, call for additional information.

The planning authority is obliged to take account of any representations it receives and will refer the application to various statutory consultees. Article 15(1(f(iii))) of the 1977 Order stipulates that the water authority will be consulted on applications involving 'the use of land for the deposit of any kind of refuse or waste'.

The choice of decisions available to an authority is as follows:

unconditional grant of permission;
conditional grant of permission;
referral to the Secretary of State (Section 35 of the 1971 Act and Article 19 of the 1977 Order);
failure to determine;
agreement for extension in time or deferral;
refusal of the application.

The time limit is eight weeks (Article 7 of the 1977 Order).

Other forms of application

It should be noted that an application may be made for 'outline' or for 'full' permission, but it is unlikely that the former would be applicable for waste disposal operations other than for treatment plant. Grant of permission for 'outline' development would usually require the settlement of 'reserved matters' within a certain time and these will be subject to a separate formal application.

Applications may be made under Section 32(1(b)) for the continuance of a use of land but without complying with a specified condition subject to which a previous planning permission was granted. This is a useful provision in cases where a permission, subject to conditions, exists but an applicant feels aggrieved at one, or some, of the conditions and is out of time for appeal to the Secretary of State (see later). A point of interest in this regard is that an application made under Section 32(1(b)) may not be subject to the advertising provisions of Section 26 of the Act on the basis that Article 8(1(b)) of the 1977 Order prescribes 'use of land for the disposal of refuse or waste materials' as being subject to those requirements: if a use is already permitted and it is merely deletion of a condition that is sought, then one could contend that the Section 26 provisions do not apply.

Where more than one condition is concerned it should be a matter of policy always to make separate application for each condition. Otherwise, if a local authority agrees to delete one, or more, condition(s) but refuses to delete another, or others, then it must refuse the whole application if such application relates to more than one condition.

Section 53 of the Act provides that a prospective developer may make application to a planning authority for a direction as to whether planning permission is required for a proposed use or change of use of land. In other words, if there is uncertainty, a direction can be sought from the local authority. Sections 24, 29(1), 31(1), 34(1) and (3) and 35–7 of the Act apply to such applications but Sections 26 and 27 do not: that is, the advertising and publication provisions do not apply but the usual consultations and procedures of the planning authority in determining a planning application do apply as does the logging of any Section 53 requests in a public register.

Section 94 of the Act provides for the application for an 'established use' certificate in appropriate cases. Basically it relates to use of land prior to 1 January 1964 and continued to the present day. Article 22 of the 1977 Order deals with the method of application for such certificates.

Appeal provisions

If an application for development is refused or granted subject to conditions unacceptable to the applicant, or remains undetermined at the end of the eight-week period, then an applicant can appeal to the Secretary of State for the Environment in accordance with Section 36 of the 1971 Act. Such an appeal must be made within six months of the date of notice or date of the expiry period. Appeals relating to applications for planning permission or for 'reserved matters' are made on prescribed forms obtainable from the Secretary of State and appeals relating to Section 53 determinations or to consents, agreements or approvals required by a planning condition are simply made in writing. The formal appeals are subject to the advertising and notification provisions under Section 27 and the required format can be found at Schedule 4, Parts I and III of the 1977 Order.

Appeals may be determined on an agreed basis of written representations or following a local, public enquiry. The former may seem attractive at first sight as public involvement is kept to a minimum and will usually involve vitriolic letters of condemnation and all manner of petitions but no personal appearance. Petitions are worthy of close study as there will inevitably be a Queen Victoria or Mickey Mouse amongst the signatories or an obvious duplication of handwriting purporting to be two different signatures from different addresses: it is easy to discredit such petitions. However, without a public enquiry there is no chance to establish key points and to cross-examine and such an opportunity may be essential.

General Development Order

Certain classes of development are exempted from the requirement for planning permission. One of these is Class XIX, 3: 'The deposit of refuse or waste materials by, or by licence of, a mineral undertaker in excavations made by such undertaker and already lawfully used for that purpose so long as the height of such deposit does not exceed the level of the land adjoining any such excavation.' (See also Section 22(3) of the Act.)

This may constitute a valuable provision for potential developers especially where a condition for future restoration is attached to a planning permission for mineral extraction.

Fees for applications

On 1 April 1981 the Secretary of State set down a statutory scale of fees to accompany certain classes of planning applications (SI 1981 No. 369 [32]). The scale of fees as set down in Schedule 1, Part II includes these categories:

	Category of development	*Fee payable*
(1)	*Operations*	
	(iv) The winning and working of minerals.	£20 for each 0.1 hectares subject to a maximum of £3000
	(vii) The carrying out of any operations . . . not coming within any of the above categories.	£20 for each 0.1 hectares of the site area subject to a maximum of £200
(2)	*Use of land*	
	(viii) The making of a material change in the use of a building or land.	£40
	(x) The continuation of a use of land . . . without compliance with a condition subject to which a previous planning permission has been granted.	£40

Some relevant appeal cases

It is always of interest and importance to study the relevant case studies before considering the worth of a proposed appeal. Also, it can be beneficial to quote case histories in evidence so as to highlight previous decisions on similar grounds if, of course, such decisions are beneficial to the case.

Some brief details of various appeal determinations relating to waste disposal are presented for interest.

(1) Catchpole Bros. v. Vale Royal District Council

(Ref: APP/5149/A/76/11552 – North West Regional Office)

Catchpole appealed against the refusal of the Council to grant permission for the tipping of soil, clay and building materials into Helsby Quarry, Cheshire.

The Council argued that infilling of the quarry would prejudice proposals for after-use. The Secretary of State thought this to be a bit tenuous but decided that 'tipping on the site . . . would detract considerably from the amenities . . . and would be unlikely within a reasonable period of time to achieve the objective of restoring the land'.

The appeal was dismissed.

(2) Sowter Bros. v. Ashfield District Council

(Ref: APP/5344/A/77/1048 – East Midlands Region)

Sowter appealed against the refusal of the Council to grant permission for the tipping of inert materials in a disused railway cutting at Kirby-in-Ashfield and the reclamation of the site for agricultural use.

The appellants claimed that the proposal was an 'engineering operation' to recover land for agricultural use and so, in any event, was permitted development under Class VI(1) of the 1977 General Development Order.

The Secretary of State commented that 'the view is now taken that the word "engineering" . . . would not cover the tipping of refuse on land or even the tipping of soil and similar excavated material for the purpose of raising the level of the land to render it more suitable for agricultural use'.

The appeal was dismissed.

(3) J. Rowland v. Renfrew County Council

(Ref: P/PPA/RN/371 – Scottish Development Department)

Rowland appealed against the refusal of the Council to permit tipping on land at Caudrens Farm, Paisley.

The Council put forward a variety of reasons for its refusal and the two most prominent were the risk to aircraft from birds and, to a lesser extent, from smoke. The site was within $1\frac{1}{4}$ miles of and crossed the flight path to Glasgow airport.

The Reporter commented that, although the plans envisaged the tipping of inert wastes only, it would not be possible practically to exclude edible material attractive to herring gulls from the input materials and the risk of bird strike was unacceptable.

The Secretary of State agreed with the Reporter and the appeal was dismissed.

(4) *Laporte Industries Ltd v. Peak Park Planning Board*
(Ref: APP/5175/A/74/593 and APP/5175/A/74/6513 – Becket House, London)

Laporte appealed against the refusal of the Board to allow either the disposal of fluorspar waste or the winning of fluorspar, lead and barytes at Longstone Edge. The two proposals were interrelated and considered together.

The Inspector considered that the disposal of wastes was the central issue since the winning of minerals could not proceed without adequate facilities for tailings disposal. Consideration was given to three factors:

(i) Were the proposals vital to the public interest?
(ii) Were there reasonable alternative disposal facilities?
(iii) Would the environmental impact of the disposal operations outweigh any national and local consequences of refusal of the proposals?

The Inspector concluded that fluorspar was vital to the national economy, that suitable alternative disposal facilities did not exist and that the proposed disposal site could be screened from normal sight-lines. A comment was also made about the high standard of previous restoration schemes undertaken by the Company.

The Secretary of State agreed with the Inspector and allowed the appeal subject to conditions of landscaping, screening, restoration, etc., all with time limits.

(5) *Ainsworth (Plant) Ltd v. West Yorkshire Metropolitan County Council*
(Ref: APP/5114/A/78/04845 – Yorkshire and Humberside Regional Office)

Ainsworth appealed against the refusal of the Council to allow the filling and regrading of a former opencast colliery at Howden Clough Road, Morley.

The site was part reclaimed and the Council preferred the continuation of the existing reclamation scheme which would have lasted 18 months. The company's new plan was to increase tipping and to extend the life to 8–10 years. The Inspector concluded that release to agriculture in 18 months was not significantly more advantageous than release in 8–10 years in this particular area and that tipping space was needed in the locality. He recommended that the appeal be allowed subject to various conditions, one of which related to the removal and storage of topsoil and subsoil and to the final restoration treatment.

The Secretary of State, however, disagreed with the need for a condition relating to soil preservation and final restoration on the basis that such a condition was not a relevant planning issue but was more properly a site licence issue. Thus, the appeal was allowed, subject to certain conditions but not including reference to restoration treatment.

(6) PD Pollution Control Ltd v. Greater Manchester Council
(Ref: APP/5081/A/76/10795 and APP/5081/A/77/10401 – North West Regional Office)

PD (Pollution Control) appealed against the refusal of the Council to allow two separate schemes, namely a revision of tipping contours and the tipping of wastes with subsequent landscaping. The area was already partly in use as a tip.

The Inspector thought the proposals to be 'admirably thought out', presenting an opportunity to transform a desolate landscape within an industrial area for future public benefit. He recommended that permission be granted subject to conditions on traffic movement restrictions and time limits.

The Secretary of State was of the opinion that the revision of contours in the first appeal constituted a 'substantially different' development taking place to that envisaged in the (original) planning applications. This first appeal was dismissed.

The second appeal was allowed subject to conditions. The Secretary commented that he had no power to make a condition that the site be available to the public on completion.

(7) Land Reclamation Co. Ltd v. Basildon District Council
(Ref: APP/5209/78/159–64 and APP/5209/78/903; 5361 – Tollgate House, Bristol)

The Company appealed against six enforcement notices served by the Council and relating to alleged breaches of conditions attached to the planning permission in force and to the carrying out of engineering (etc.) operations without permission and also appealed against the failure of the Council to determine two applications for the construction of a site access road within the statutory period. The notices related to liquid wastes in general.

Various grounds of appeal were subsequently withdrawn and one of the appeals itself (Ref: . . ./903) was withdrawn at a public enquiry.

This is a case of major importance and the report is lengthy and complex dealing with various points of law at depth. Briefly, the material points were:

(i) The Council alleged that tipping was an 'operation' whereas it should have alleged 'a material change of use'.

(ii) The Council referred to breaches of a planning condition 'after the end of 1963' whereas the Act provides only for the service of such an enforcement notice within 'the period of four years from the date of the breach' (Section 87(3(b))).

(iii) The Inspector found it difficult to reconcile the wording of the planning permission that related firstly to 'refuse or waste material' implying some distinction between the two and subsequently used the terms interchangeably. Having regard to Section 22(3(b)) of the Act, he decided that there was no material distinction but that liquid wastes could not be excluded by either term.

On legal merits, all six enforcement notices were quashed, some of them having been declared defective anyway.

The Secretary of State made reference to the national importance of the site in question and detailed, at some length, the hydrogeological, waste balance, liquid input, pollution potential and site operational aspects.

The formal decision was to quash all six enforcement notices and to declare that eight conditions on the 1972 permission had been discharged and to permit the application for the construction of the access road.

(8) Leigh Interests Ltd v. Dudley Metropolitan Borough Council
(Ref: APP/5106/A/78/05302 – West Midlands Regional Office)

Leigh appealed against the refusal of the Council to permit the infilling of a site at Coseley.

The Council's case for refusal primarily dwelt on the likely impact of operations on the surrounding housing. The site in question was surrounded on all sides by residential properties having back gardens abutting the site boundary.

The Inspector concluded that the site should be restored but suggested imposition of most of the conditions put forward by the Council at the public enquiry.

The Secretary of State allowed the appeal but deleted most of the prospective planning conditions on the grounds that 'minimising risks of water pollution and danger to public health and to safeguarding the amenities of the locality' were more properly dealt with under the site licensing provisions of the Control of Pollution Act 1974.

(9) Leigh Interests Ltd v. Dudley Metropolitan Borough Council
(Ref: APP/5106/A/78/4024 – Becket House, London)

Leigh appealed against six conditions attached to the grant of permission relating to removal of an ash bank and subsequent regrading using imported fill.

The Secretary of State agreed with Leigh that some of the conditions under appeal were more properly applicable to a site licence and amended them accordingly.

Of particular interest in this case was that the Secretary of State examined a condition on the permission that was not subject to appeal, namely: 'The materials to be tipped shall be non-toxic, non-biodegradable and non-oily and shall consist of builders' waste of a granular structure, or be material of a free-draining nature.' He deleted this condition in its entirety since he took the view that the nature of the infill material was a matter for a site licence.

(10) The Queen v. Derbyshire County Council
(High Court of Justice, Queen's Bench Division; Judgement dated 18 January 1979)

North East Derbyshire District Council applied to bring up and quash a disposal site licence issued on 10 August 1978 by Derbyshire County Council to Cambro Waste Products Ltd.

North East Derbyshire District Council submitted that no planning permission was in force and that the County Council should have refused a licence under Section 5(2) of the Control of Pollution Act.

The site comprised 21 acres forming part of a larger mineral extraction. In 1969 planning permission was granted for winning and working of fire clays, shales and coal over 130 acres. Condition 4 required overburden together with 'fill' to be used for infilling, grading and levelling of worked out parcels. Condition 8 required phasing of mineral extraction and subsequent infilling in six-acre parcels. Reasons given for the conditions included 'to secure the progressive and effective restoration . . . so as to avoid the creation of permanently derelict land'.

Mineral working went forward without the six-acre parcel restoration and Condition 8 was not observed. 21 acres remained unfilled. Mineral working ceased in 1978.

Cambro took option to purchase in 1978 and received assurance from the County Council that planning permission was valid for tipping. They applied for and were granted a site licence by the County Council.

Both Cambro and the County Council submitted that the 1969 planning permission was valid. Cambro raised the question of estoppel, i.e. the County Council had previously given written assurance that planning permission was valid and could not, therefore, go back on this assurance.

North East Derbyshire District Council contended that the 1969 conditions required infilling to be progressive and ancillary to mineral extraction. The use of land, therefore, was originally envisaged as mixed and use of land solely for tipping without mineral extraction amounted to a change of use. Completion of mineral extraction in 1978 amounted to the abandonment of planning permission.

The Judge determined that there had been an irremediable breach of Condition 8 and it was no longer possible to enforce the infilling and restoration programme as envisaged. It did not follow that this breach of Condition 8 had also made Condition 4 ineffective: indeed, Condition 4 contemplated something to be done *after* mineral extraction had ceased, namely infilling, albeit envisaging a final parcel of six acres.

Condition 4 was still enforceable by the planning authority, i.e. the County Council, in this case of mineral working. Cambro were seeking to comply with this condition voluntarily: 'If the refilling of the land is something which, by law, they could be required to do, I ask theoretically, how can it be said that it is not something which they are permitted to do?'

The North East Derbyshire District Council quoted a previous decision as authority in this case, namely the purchase of a vacant quarry for infilling which, it was decided, amounted to a change of use. However, the Judge said this was irrelevant as a vacant quarry did not have a restoration condition requiring importation of fill.

North East Derbyshire District Council quoted the site licence conditions that required clay sealing and bunding of the site to prevent pollution. They argued that this constituted an engineering or other operation and thus required specific planning permission. The Judge decided that waste disposal, including clay, was a use of land and not an engineering or other operation and that the condition merely stipulated that the waste clay be deposited in a certain manner.

North East Derbyshire District Council quoted the site licence condition allowing liquid and sludge disposal. They argued that liquids and sludges could not be classed as 'fill' as per the 1969 Condition 4. The Judge quoted evidence from the County Council Surveyor that a proportion of such materials was acceptable, having regard to final use as agricultural land.

North East Derbyshire District Council quoted the site licence condition requiring the construction of an access road. They argued that this was an engineering operation outside the 1969 permission. The Judge replied that the licensed area only included the disposal site and the access road was ancillary thereto. The County Council was satisfied that planning permission was valid for infilling of the licensed area. The County Council was, further, empowered by Section 6(2(g)) of the Control of Pollution Act to require works to be carried out before waste disposal activities were begun and this may require the carrying out of works, notwithstanding that the licence holder is not entitled as of right to carry them out. Thus, the absence of planning permission to construct the access road was immaterial.

The North East Derbyshire District Council application was dismissed and they were ordered to pay costs of both Cambro and the County Council.

(11) Leigh Interests Ltd v. Stoke-on-Trent City Council
(Ref: APP/5375/A/80/12128 – West Midlands Regional Office)

Leigh appealed against the refusal of the Council to grant an application for the continuance of the use of land already permitted for wastes disposal but without complying with the condition: 'The proposed materials to be tipped shall be inert, non-toxic and non-putrescible.'

The appeal grounds were based upon the contention that the types of material to be authorised for disposal at a site was a site licence matter and not a planning matter.

The Secretary of State agreed with this contention and referred to the advice in Circulars 5/68 and 55/76 which set down the Department's view on planning/licensing.

The appeal was allowed and the condition thereby deleted from the extant consent.

Enforcement provisions

The 1971 Act confers various powers and duties onto the planning authorities so that they are fully equipped to enforce the provisions of the Sections. Planning applications are essentially public documents and planning authorities are charged, amongst other things, with ensuring that 'unneighbourly' development is not permitted. If developments do take place without appropriate consent then the authorities have very real powers to stop continuation of the development and to restore the original status if necessary.

Also, the planning authorities, having given conditional planning permission for developments, must ensure that the conditions are complied with. Again, powers of enforcement are real. (Telling [33] gives a useful guide to planning law and procedure.)

Minerals Bill

At the time of writing, there is a Minerals Bill being discussed in parliament. Two important provisions are in the draft Bill in so far as waste disposal is concerned.

Firstly, the Bill, if it became law, would confer powers upon local authorities to impose conditions of restoration onto existing mineral workings that have no such condition.

Secondly, it would empower local authorities to impose time limits for mineral extraction and restoration.

Both of these provisions recognise the historic problems that have resulted in spent and derelict mineral excavations never being reclaimed. Thus, planning authorities would possess real powers for restoration of mineral workings which, effectively, means importation of waste materials for infilling.

There is, naturally, opposition to the Bill and it will be interesting to see the final Act when passed.

The carriage of wastes

Apart from the general provisions of the Health and Safety at Work (etc.) Act 1974 [34] and the Motor Vehicle (Construction and Use) Regulations 1973 (*et seq.*) [35], there exists specific legislation concerning the carriage of liquid, but not solid, wastes.

Petroleum Act

The Petroleum (Consolidation) Act 1928 [36] and the Petroleum-Spirit (Conveyance by Road) Regulations 1957 [37] as amended in 1958 [38] and 1966 [39] relate to the carriage of prescribed substances that are inflammable. To date, 428 such substances have been named in the various Regulations [37, 38, 39, 40]. The 1928 Act related to petroleum spirits having a flashpoint of less than 73°F (22.8°C) and the Conveyance Regulations specified the type of vehicle and equipment to be used for carriage of such substances in bulk. The Regulations specify minimum construction requirements for vehicles and for tanks/tank trailers and include such items as emergency shut-off valves, fire resistant screening, exhaust pipe location, electrical insulation, tank construction, internal divisions, strengthening and protective frames, manhole covers, etc.

Labelling regulations

The Hazardous Substances (Labelling of Road Tankers) Regulations 1978 [41] came into force on 28 March 1979 and specified, as

Table 1. *Transport Regulations: labelling requirements*

Name of substance	Substance identification number	Emergency action code	Hazard warning
Hazardous waste, liquid, containing acid	7006	2WE	Other hazardous substance
Hazardous waste, solid or sludge, containing acid	7007	2WE	Other hazardous substance
Hazardous waste, liquid, containing alkali	7008	2WE	Other hazardous substance
Hazardous waste, solid or sludge containing alkali	7009	2WE	Other hazardous substance
Hazardous waste, flammable liquid, flashpoint below 23°C	7010	3WE	Other hazardous substance
Hazardous waste, flammable liquid, flashpoint 23°C to 60.5°C	7011	3W	Other hazardous substance
Hazardous waste, flammable, solid or sludge nos*	7012	3WE	Other hazardous substance
Hazardous waste, solid or sludge nos	7014	2X	Other hazardous substance
Hazardous waste, liquid nos	7015	2X	Other hazardous substance
Hazardous waste, solid or sludge, toxic nos	7016	2X	Other hazardous substance
Hazardous waste, liquid, toxic nos	7017	2X	Other hazardous substance
Hazardous waste, solid, containing inorganic cyanides	7018	4X	Other hazardous substance
Hazardous waste, liquid, containing inorganic cyanides	7019	4X	Other hazardous substance
Hazardous waste, solid or sludge, agrochemicals, toxic nos	7020	4WE	Other hazardous substance
Hazardous waste, liquid, agrochemicals, toxic nos	7021	4WE	Other hazardous substance
Hazardous waste containing isocyanates, nos	7022	4WE	Other hazardous substance
Hazardous waste containing organo-lead compounds nos	7023	4WE	Other hazardous substance

*nos – not otherwise specified.
Source: Reference [41].

the title suggests, certain minimum labelling requirements for vehicles carrying bulk liquid hazardous substances including wastes. The Regulations seek to identify the individual substances to which they apply and this is relatively straightforward for anything other than waste products which are, almost by definition, mixtures of substances. The categories serving waste disposal requirements are, hence, vague.

Labelling of the road vehicles involves the displaying of prescribed information on both sides and on the rear of the tank(er). The size, material of construction, colour and location of the display boards is prescribed.

Regulation 4 requires the operator of a road tanker to display:

the emergency action code (Hazchem);
the substance identification number;
the appropriate hazard warning sign;
the telephone number, or other approved text, indicating from where specialised advice can be obtained at all times.

Additionally, there may be displayed:

the name of the prescribed substance and its trade name;
the name of the manufacturer.

Details of the waste classifications in the Regulations are as shown in Table 1.

The 'emergency action code' (Hazchem) comprises a number and a letter or letters being coded to mean:

'1' – water jets;
'2' – water fog;
'3' – foam;
'4' – dry agent.

The first letter indicates the hazards, as shown in Table 2, and if the code is

Table 2. *Transport Regulations: warning signs*

Letter	Danger of violent reaction or explosion	Protective clothing and breathing apparatus	Appropriate measures
P	Yes	Full protective clothing	Dilute
R	No	Full protective clothing	Dilute
S	Yes	Breathing apparatus	Dilute
S (reversed, white on black)	Yes	Breathing apparatus for fire only	Dilute
T	No	Breathing apparatus	Dilute
T (reversed, white on black)	No	Breathing apparatus for fire only	Dilute
W	Yes	Full protective clothing	Contain
X	No	Full protective clothing	Contain

Table 2. (*cont.*)

Letter	Danger of violent reaction or explosion	Protective clothing and breathing apparatus	Appropriate measures
Y	Yes	Breathing apparatus	Contain
Y (reversed)	Yes	Breathing apparatus for fire only	Contain
Z	No	Breathing apparatus	Contain
Z (reversed)	No	Breathing apparatus for fire only	Contain

Source: Reference [41].

followed by the letter 'E' then evacuation of people from the neighbourhood of any incident is to be considered.

The selection of warning signs available is comprised of those shown in Fig. 1. (All of the hazardous waste categories, at present, are represented by the exclamation mark symbol but the other symbols are included for reference to take account of forthcoming changes in legislation – see later.)

The layout of the information is also specified and a typical display board for a hazardous sludge containing acid would be as shown in Fig. 2.

In-cab information

For many years, the waste disposal industry has, through the National Association of Waste Disposal Contractors, been using a system of Hazardous Waste Cards which are carried in the cab of a vehicle. These cards present the driver with certain information about the load, upon which he can base his decisions and actions in the event of an emergency. These cards have been rationalised as the Transport Emergency Cards (Tremcards) and are coded so as to cross-refer to the hazard warning display boards.

The 7007 example (sludge containing acid) is reproduced in Fig. 3 for information.

Discussion

As has been highlighted, the classifications for wastes are vague. For example, the distinction between acidic liquids (7006) and sludge (7007) is more apparent than real. The category 'hazardous waste liquid not otherwise specified' (7015) is quite a potential loophole.

NAWDC has adopted the use of Tremcard No. 7016 for the hazardous (non-toxic) sludges 7014 category and Tremcard No. 7017 for the

Fig. 1. Transport Regulations: warning signs. (Source: Reference [41].)

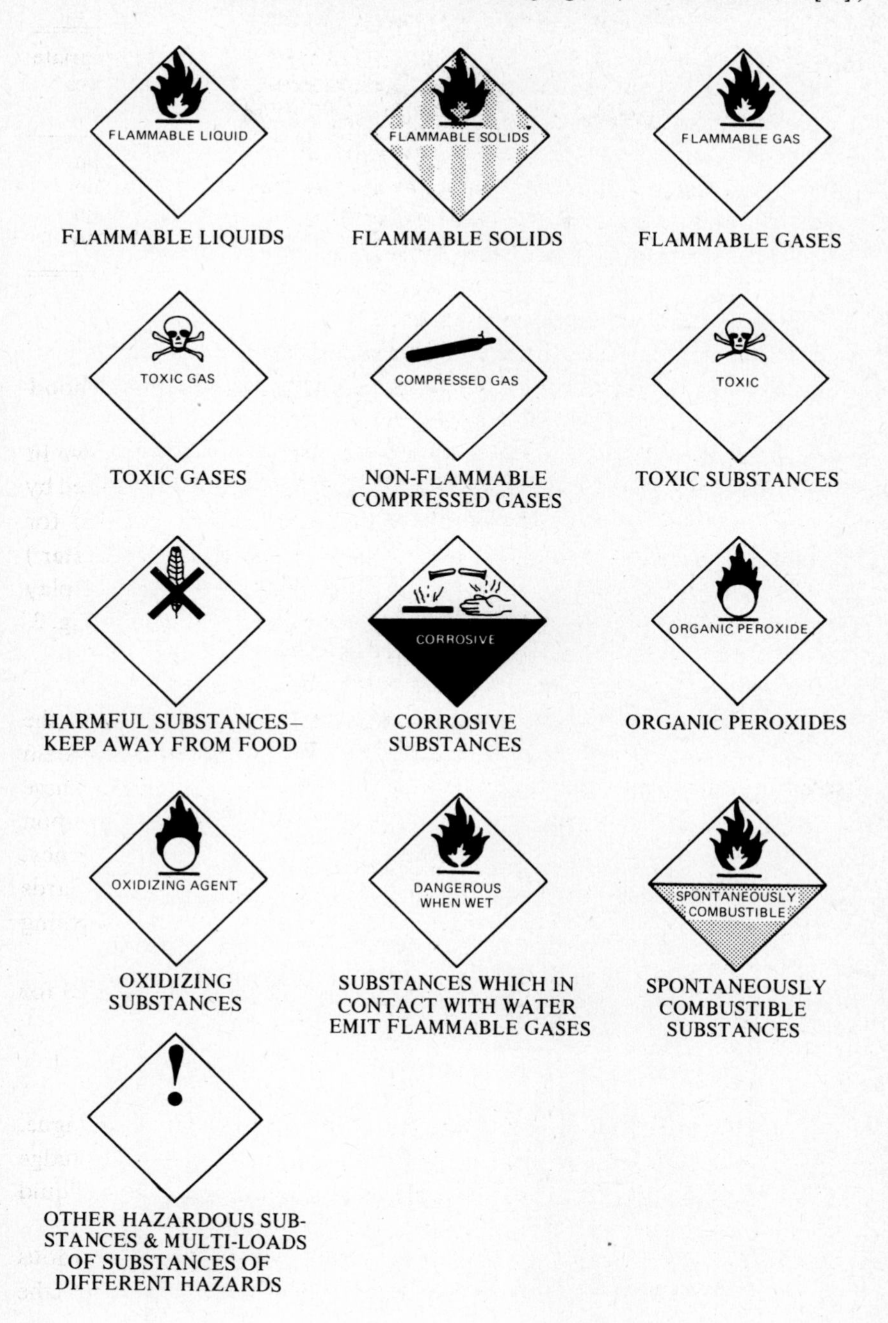

hazardous (non-toxic) liquids 7015 category on the basis that if they are hazardous they should be classed as potentially toxic: 7016 and 7017 refer to toxicity.

If more than one load is scheduled for carriage during a day then the vehicle must carry with it a selection of display boards appropriate to those loads. The boards must then be changed by the driver when loading a different substance. Driver training is obviously essential.

When a road tanker is empty, the operator must either remove or cover the boards such that the only item visible is the telephone number.

Proposed amendments

The Regulations relating to the carriage of dangerous substances, including hazardous wastes, are scheduled for replacement. The second consultative document of the Health and Safety Commission is now published as the Proposals for the Dangerous Substances (Conveyance by Road in Road Tankers and Tank Containers) Regulations 1980 [42].

The Proposals extend to the construction of vehicles as well as the labelling requirements during transit. No specific requirements are prescribed and the generality of proper design, adequate strength, good construction from sound and suitable material and suitability for the purpose to which the vehicle is put suffice. It is envisaged that Codes of Practice will be issued in due course and attention is drawn to the existing NAWDC Code of Practice [43] which has been submitted to the Health and Safety Commission for consideration. The Proposals also cover the fixtures and fittings of vehicles to prevent leaking of contents. The construction,

Fig. 2. Transport Regulations: a typical display sign. (Source: Courtesy Effluent Disposal Ltd.)

Fig. 3. A typical transport emergency card. (Source: Courtesy National Association of Waste Disposal Contractors.)

Cargo **HAZARDOUS WASTE**

Solid or Sludge containing Acid

Nature of Hazard

Corrosive
Contact with solid, liquid or vapour may cause skin burns and severe damage to eyes and air passages
Attacks clothing
Attacks many metals with liberation of hydrogen which is flammable and may form explosive mixture with air
Reaction with combustible substances generates heat and may cause fire or explosion or toxic fumes
May contain dissolved or suspended toxic constituents

Protective Devices

Suitable respiratory protective device
Goggles giving complete protection to eyes
Plastic or synthetic rubber gloves, boots and suit with hood giving complete protection to head, face and neck
Eyewash bottle with clean water

EMERGENCY ACTION

- Stop the engine
- No naked lights. No smoking
- Mark roads and warn other road users
- Keep public away from danger area
- Keep upwind
- Put on protective clothing

Spillage

- Contain leaking liquid with sand or earth, consult an expert
- Shut off leaks if without risk
- Prevent liquid entering sewers, basements and workpits. Vapour may create toxic atmosphere
- If substance has entered a water course or sewer or contaminated soil or vegetation, advise police
- Warn everybody—evacuate if necessary

Fire

- Keep containers cool by spraying with water if exposed to fire
- Extinguish with water spray, dry chemical or foam

First aid

- If substance has got into eyes, immediately wash out for several minutes
- Remove contaminated clothing immediately and wash affected skin with plenty of water
- Seek medical treatment when anyone has symptoms apparently due to inhalation, swallowing, contact with skin or eyes or the fumes produced in a fire
- Even if there are no symptoms, send to a doctor and show this card

Additional information provided by manufacturer or sender

TELEPHONE

WARNING: This information is intended to assist any person involved in recognising the characteristics of this load. It is emphasized that such substances may vary according to conditions; this information, therefore, must be regarded as a general guide.

testing and periodic re-testing of pressure tanks and vacuum tanks are also covered.

There are, in fact, seven Schedules to the Regulations, namely

Schedule 1: Classification of dangerous substances;
Schedule 2: Construction of road tankers;
Schedule 3: Hazard warning panels and labels;
Schedule 4: Hazard warning signs;
Schedule 5: Unloading of petroleum-spirit at a petroleum filling station;
Schedule 6: Revocations;
Schedule 7: Regulations ceasing to have effect in relation to road tankers and tank containers.

Under proposed Regulation 9, the driver will be obliged to carry prescribed information in the cab of the vehicle and it is expected that the Tremcard system will apply for hazardous wastes.

The prescribed information must be in writing and must enable a driver to know the identity of the substance being carried and the nature and dangers to which the substance may give rise and the emergency action to be taken. An important proviso is 'that no information in writing relating to any dangerous substance other than a substance which is being conveyed at that time is available on the vehicle'. This, of course, is simple and sensible for vehicles routinely employed in carrying one substance but for waste disposal vehicles often collecting a variety of different materials during the same working day it would pose many problems if enforced as written: the industry, through NAWDC, is seeking to amend this.

Special provisions will apply to materials with a flashpoint of less than 23°C: namely, no smoking!

The prescribed information for incorporation into the labels for waste products will be changed slightly along the lines shown in Table 3.

The hazard warning signs for the carriage of wastes would, therefore, be more pertinent to the load carried and would reflect the actual or potential hazards thereof, instead of the present ubiquitous coverage by the exclamation mark. Nevertheless, the disposal industry is resisting this change on the grounds of wasted expense for display boards already in use and marked with the ⟨!⟩ symbol. Definitions relating to these hazard warnings will be prescribed and the Proposals are as shown in Table 4.

At the time of writing it is understood that the definitive Regulations are with the Secretary of State for approval and signature. Apparently, the warning exclamation mark has been retained for all waste categories,

Table 3. *Proposed categories for the carriage of dangerous wastes*

Name of substance	Substance identification number	Emergency action code	Hazard warning sign
Hazardous waste, liquid, containing acid	7006	2WE	Corrosive substance
Hazardous waste, solid or sludge, containing acid	7007	2WE	Corrosive substance
Hazardous waste, liquid, containing alkali	7008	2WE	Corrosive substance
Hazardous waste, solid or sludge, containing alkali	7009	2WE	Corrosive substance
Hazardous waste, flammable liquid, flashpoint less than 23°C	7010	3WE	Flammable liquid
Hazardous waste, flammable liquid, flashpoint 23°C to 55°C	7011	3W	Flammable liquid
Hazardous waste, flammable, solid or sludge, nos*	7012	3WE	Flammable liquid
Hazardous waste, solid or sludge, nos	7014	2X	Other hazardous substance
Hazardous waste, liquid, nos	7015	2X	Other hazardous substance
Hazardous waste, solid or sludge, toxic, nos	7016	2X	Toxic substance
Hazardous waste, liquid, toxic, nos	7017	2X	Toxic substance
Hazardous waste, solid, containing inorganic cyanides	7018	4X	Toxic substance
Hazardous waste, liquid, containing inorganic cyanides	7019	4X	Toxic substance
Hazardous waste, solid or sludge, agrochemicals, toxic, nos	7020	4WE	Toxic substance
Hazardous waste, liquid, agrochemicals, toxic, nos	7021	4WE	Toxic substance
Hazardous waste, containing isocyanates, nos	7022	4WE	Toxic substance
Hazardous waste, containing organo–lead compounds nos	7023	4WE	Toxic substance

*nos – not otherwise specified.
Source: Reference [42].

Table 4. *Proposed interpretations of hazardous wastes signs*

Hazard warning	Characteristic properties of the substances
Non-flammable compressed gases	Substances which have a critical temperature below 50 degrees Celsius or which at 50 degrees Celsius have a vapour pressure of more than 3 bar, other than flammable or toxic gases
Toxic gases	Substances which have a critical temperature below 50 degrees Celsius or which at 50 degrees Celsius have a vapour pressure of more than 3 bar, and are toxic, other than flammable gases
Flammable gases	Substances which have a critical temperature below 50 degrees Celsius or which at 50 degrees Celsius have a vapour pressure of more than 3 bar and are flammable
Flammable liquids	Liquids with a flash point of 55 degrees Celsius and below
Flammable solids	Solids which are readily combustible under conditions encountered in conveyance by road or which may cause or contribute to fire through friction
Spontaneously combustible substances	Substances which are liable to spontaneous heating under conditions encountered in conveyance by road or to heating on contact with air being then liable to catch fire
Substances which in contact with water emit flammable gases	Substances which in contact with water are liable to become spontaneously combustible or to give off flammable gases
Oxidizing substances	Substances which, although not themselves necessarily combustible, may, by yielding oxygen, cause or contribute to the combustion of other material, other than organic peroxides
Organic peroxides	Substances having properties characteristic of organic peroxides
Toxic substances	Substances known to be so toxic to man as to afford a hazard to health during conveyance or which, in the absence of adequate data on human toxicity are presumed to be toxic to man
Harmful substances	Substances known to be toxic to man or, in the absence of adequate data on human toxicity, are presumed to be toxic to man, but which are unlikely to afford a serious acute hazard to health during conveyance
Corrosive substances	Substances which, by chemical action, will cause severe damage when in contact with living tissue, or in the case of leakage will materially damage other freight or equipment

Table 4. (*cont.*)

Hazard warning	Characteristic properties of the substances
Other hazardous substances and multi-loads of different hazards	Substances which, although not having any of the characteristic properties set out above, may create a risk to the health and safety of persons under conditions encountered in conveyance by road

Source: Reference [42].

Table 5. *Regulations which would cease to be applicable to wastes*

Title of Regulations	Reference
The Petroleum-Spirit (Conveyance by Road) Regulations 1957	SI 1957/191; amended by SI 1958/962 and SI 1966/1190
The Carbon Disulphide (Conveyance by Road) Regulations 1958	SI 1958/313; amended by SI 1962/2527
The Corrosive Substances (Conveyance by Road) Regulations 1971	SI 1971/618
The Inflammable Liquids (Conveyance by Road) Regulations 1971	SI 1971/1061
The Inflammable Substances (Conveyance by Road) (Labelling) Regulations 1971	SI 1971/1062
The Organic Peroxides (Conveyance by Road) Regulations 1973	SI 1973/2221
The Petroleum (Consolidation) Act 1928 (Conveyance by Road Exemptions) Regulations 1980	SI 1980/ –

Source: Reference [42].

although the Department of Transport may reconsider this concession in due course. Also, the carrying of more than one Tremcard will be permissible so long as only the one card relevant to the vehicle contents is on display at any one time.

Various existing Regulations will cease to have effect if the Proposals are implemented and these will be (Regulation 24(3)) as shown in Table 5, and various other Regulations will be revoked (Regulation 24(1)), as shown in Table 6.

An important aspect of the Proposed Regulations is that vehicle operators will become responsible for prescribed training of drivers to enable them to:

understand the significance of any hazard warning signs or panels;
understand the requirements of the Regulations for displaying such panels;
understand the requirements of the Regulations as to the loading and unloading of vehicles;
understand the risks associated with the substance to be carried,

Table 6. *Regulations which would be revoked*

1 Title of Order or Regulations	2 Reference	3 Extent of revocation
The Petroleum (Inflammable Liquids and Other Dangerous Substances) Order 1947	SR & O 1947/1443	The whole Order
The Hazardous Substances (Labelling of Road Tankers) Regulations 1978	SI 1978/1702	The whole Regulations
The Petroleum (Consolidation) Act 1928 (Enforcement) Regulations 1979	SI 1979/427	In Regulation 2(1)(*a*), the references to: (*a*) The Petroleum (Inflammable Liquids and Other Dangerous Substances) Order 1947; (*b*) The Petroleum (Corrosive Substances) Order 1970; (*c*) The Petroleum (Inflammable Liquids) Order 1971; (*d*) The Petroleum (Organic Peroxides) Order 1973. In the Schedule the entries from that relating to The Petroleum-Spirit (Conveyance by Road) Regulations 1957 to the end of the Schedule

Source: Reference [42].

the precautions to be observed and the action to be taken in the event of an emergency;

understand the operation of the vehicle and, where appropriate, the proper use and operation of any valves, pumps, hoses, etc;

use any protective clothing and where appropriate, respiratory protective equipment;

use the fire-fighting and first-aid equipment carried on the vehicle.

NAWDC have prepared a Code of Practice for the construction of vacuum tankers [43] as used in the waste management industry. The application of the various transport Regulations and the role of the waste disposal contractor has been discussed by Willetts [44, 45].

Proposals for Packaging and Labelling of Dangerous Substances (Amendments) Regulations 1980 [46]

The Health and Safety Executive is seeking to extend the provisions of the original 1978 Packaging and Labelling Regulations [47] so as to include hazardous materials destined for transport as well as merely 'supply'. It has been argued since 1978 that waste materials could not be considered as materials for supply and, hence, should not be subject to the packaging and labelling provisions of the Regulations. The proposed extension to include waste materials is in order to satisfy the 6th Amendment to the EEC Directive which was formally adopted by the UK in 1979 [48].

No consultative document has yet been issued but some indications as to the provisions of the Regulations can be deduced. It is probable that three categories of packaging will be adopted:

(a) of 220 litres or less (e.g. drums, carboys);

(b) single container of more than 220 litres;

(c) an outer container of more than 220 litres (e.g. skip containing other containers).

It would appear that, in the case of (a), each drum, carboy, etc. on a lorry would require its own individual hazard warning label, including substance identification numbers, hazard diamond signs, telephone number of waste producer and emergency action information. Because of the need to include a producer's telephone number, each waste producer would have to secure his own supply of signs.

In the case of (c), a skip would need to be labelled and also the individual containers within the skip!

For the conveyance of raw chemicals and allied products, the proposed Regulations may be sensible. For the waste disposal industry there will be very many difficulties.

Packaged goods

The Health and Safety Executive propose a third set of Regulations applicable to the movement of wastes. They will deal with the provisions necessary for the conveyance of dangerous substances as packaged goods and dangerous substances will, no doubt, be defined so as to include hazardous wastes.

The consultative draft is due in 1981.

The Control of Pollution (Special Waste) Regulations 1980 [49]

These Regulations, made under Section 17 of the Control of Pollution Act 1974 [21], are related to the transport of 'special', that is, highly hazardous wastes. The Regulations are discussed in the section on Part I of the Act elsewhere.

References

[1] White-Hunt, K., 1980-1. 'Domestic Refuse – A Brief History'. 'Part I': *Solid Wastes*, Vol. LXX, No. 11, November 1980, pp. 609–15. 'Part II': *Solid Wastes*, Vol. LXXI, No. 4, April 1981, pp. 159–66. 'Part III': *Solid Wastes*, Vol. LXXI, No. 6, June 1981, pp. 284–92. Institute of (Solid) Wastes Management, London.

[2] Morris, D., 1967. *The Naked Ape*. Cape, London.

[3] Chadwick, E., 1842. *Report on the Sanitary Condition of the Labouring Population of Great Britain*. Edinburgh University Press, 1965.

[4] *1844 Command 572*. Royal Commission on the Health of Towns: 'First Report'. HMSO, London.

[5] *1845 Command 602*. Royal Commission on the Health of Towns: 'Second Report'. HMSO, London.

[6] *Public Health Act 1848*, 11 and 12, Victoria, Ch. 63. HMSO, London.

[7] *Nuisance Removal and Disease Prevention Act 1848*, 11 and 12, Victoria, Ch. 123. HMSO, London.

[8] *Nuisance Removal Act 1855*, 18 and 19, Victoria, Ch. 121. HMSO, London.

[9] *Public Health Act 1858*, 21 and 22, Victoria, Ch. 97. HMSO, London.

[10] *Sanitary Act 1866*, 29 and 30, Victoria, Ch. 90. HMSO, London.

[11] *Public Health Act 1875*, 38 and 39, Victoria, Ch. 55. HMSO, London.

[12] *Public Health Acts Amendment Act 1890*, 53 and 54, Victoria, Ch. 59. HMSO, London.

[13] *Public Health Acts Amendment Act 1907*, 7, Edward VII, Ch. 53. HMSO, London.

[14] *Public Health Act 1936*, 26, George V and VI, Edward VIII, Ch. 49. HMSO, London.

[15] *1st Royal Commission to inquire into the best means of preventing the pollution of rivers*, appointed 1865.

1866 Command 3634: 'First Report – River Thames': Vol. I – Report.
1866 Command 3634 – I: 'First Report' – Vol. II – Minutes.
1867 Command 3835: 'Second Report – River Lee': Vol. I – Report.
1867 Command 3835 – I: 'Second Report' – Vol. II – Minutes.

1867 Command 3850: 'Third Report – Rivers Aire & Calder': Vol. I – Report.
1867 Command 3850 – I: 'Third Report' – Vol. II – Minutes.

[16] *2nd Royal Commission to inquire into the best means of preventing the pollution of rivers*, appointed 1868.
1870 Command 37: 'First Report – Mersey & Ribble Basins': Vol. I – Report.
1870 Command 109: 'First Report' – Vol. II – Minutes.
1870 Command 180: 'Second Report – The ABC process of treating sewage'.
1871 Command 347: 'Third Report – Pollution from woollen manufacture and connected processes': Vol. I – Report.
1871 Command 347 – I: 'Third Report' – Vol. II – Minutes.
1872 Command 603: 'Fourth Report – Rivers of Scotland': Vol. I – Report.
1872 Command 603 – I: 'Fourth Report' – Vol. II – Minutes.
1874 Command 951: 'Fifth Report – Pollution from mining operations and metal manufacture': Vol. I – Report.
1874 Command 951 – I: 'Fifth Report' – Vol. II – Minutes.
1874 Command 1112: 'Sixth Report – Domestic water supply of Great Britain'.

[17] *Rivers (Prevention of Pollution) Act 1876*, 39 and 40, Victoria, Ch. 75. HMSO, London.

[18] *Rivers (Prevention of Pollution) Act 1893*, 56 and 57, Victoria, Ch. 31. HMSO, London.

[19] *Rivers (Prevention of Pollution) (Border Councils) Act 1898*, 61 and 62, Victoria, Ch. 34. HMSO, London.

[20] *Rivers (Prevention of Pollution) Act 1951*, 14 and 15, George VI, Ch. 64. HMSO, London.

[21] *Control of Pollution Act 1974*, Elizabeth II, Ch. 40. HMSO, London.

[22] *Town and Country Planning Act 1971*, Elizabeth II, Ch. 78. HMSO, London.

[23] *SI 1977 No. 289*, Town and Country Planning General Development Order 1977. HMSO, London.

[24] *SI 1980 No. 1946*, Town and Country Planning General Development (Amendment) Order 1980. HMSO, London.

[25] *Housing, Town Planning (etc.) Act 1909*, 9, Edward VII, Ch. 44. HMSO, London.
Housing, Town Planning (etc.) Act 1919, 9 and 10, George V, Ch. 35. HMSO, London.
Town and Country Planning Act 1932, 22 and 23, George V, Ch. 48. HMSO, London.
Town and Country Planning (Interim Development) Act 1943, 6 and 7, George VI, Ch. 29. HMSO, London.
Town and Country Planning Act 1944, 7 and 8, George VI, Ch. 47. HMSO, London.
Town and Country Planning Act 1947, 10 and 11, George VI, Ch. 51. HMSO, London.
Town and Country Planning Act 1953, 1 and 2, Elizabeth II, Ch. 16. HMSO, London.
Town and Country Planning Act 1954, 2 and 3, Elizabeth II, Ch. 72. HMSO, London.
Town and Country Planning Act 1959, 7 and 8, Elizabeth II, Ch. 53. HMSO, London.
Town and Country Planning Act 1962, 10 and 11, Elizabeth II, Ch. 38. HMSO, London.
Town and Country Planning Act 1963, Elizabeth II, Ch. 17. HMSO, London.
Town and Country Planning Act 1968, Elizabeth II, Ch. 72. HMSO, London.

[26] *SI 1963 No. 709*, Town and Country Planning General Development Order 1963. HMSO, London.

SI 1973 No. 31, Town and Country Planning General Development Order 1973. HMSO, London.
SI 1973 No. 273, Town and Country Planning General Development (Amendment) Order 1973. HMSO, London.
SI 1974 No. 418, Town and Country Planning General Development (Amendment) Order 1974. HMSO, London.
SI 1976 No. 301, Town and Country Planning General Development (Amendment) Order 1976. HMSO, London.
[27] *Town and Country Planning (Amendment) Act 1972*, Elizabeth II, Ch. 42. HMSO, London.
Town and Country Amenities Act 1974, Elizabeth II, Ch. 32. HMSO, London.
[28] *Local Government Act 1972*, Elizabeth II, Ch. 70. HMSO, London.
[29] *SI 1980 No. 2010*, Town and Country Planning (Prescription of County Matters) Regulations 1980. HMSO, London.
[30] *Circular 2/81*, Department of the Environment (Welsh Office 2/81), Town and Country Planning: Development Control Functions. HMSO, London.
[31] *Local Government, Planning and Land Act 1980*, Elizabeth II, Ch. 65. HMSO, London.
[32] *SI 1981 No. 369*, Town and Country Planning (Fees for Applications and Deemed Applications) Regulations 1981. HMSO, London.
[33] Telling, A.E., 1963–1977. *Planning Law and Procedure*, 5th Edn, 1977. Butterworth, London.
[34] *Health and Safety at Work (etc.) Act 1974*, Elizabeth II, Ch. 37. HMSO, London.
[35] *SI 1973 No. 24*, The Motor Vehicle (Construction and Use) Regulations 1973. HMSO, London.
SI 1978 No. 1017, The Motor Vehicles (Construction and Use) Regulations 1978. HMSO, London.
[36] *Petroleum (Consolidation) Act 1928*, 18 and 19, George V, Ch. 32. HMSO, London.
[37] *SI 1957 No. 191*, The Petroleum-Spirit (Conveyance by Road) Regulations 1957. HMSO, London.
[38] *SI 1958 No. 962*, Petroleum-Spirit (Conveyance by Road) Regulations 1958. HMSO, London.
[39] *SI 1966 No. 1190*, Petroleum-Spirit (Conveyance by Road) (Amendment) Regulations 1966. HMSO, London.
[40] *SI 1958 No. 313*, The Carbon Disulphide (Conveyance by Road) Regulations 1958. HMSO, London.
SI 1962 No. 2527, The Carbon Disulphide (Conveyance by Road) Regulations 1962. HMSO, London.
SI 1971 No. 618, The Corrosive Substances (Conveyance by Road) Regulations 1971. HMSO, London.
SI 1971 No. 1061, The Inflammable Liquids (Conveyance by Road) Regulations 1971. HMSO, London.
SI 1971 No. 1062, The Inflammable Substances (Conveyance by Road) (Labelling) Regulations 1971. HMSO, London.
SI 1973 No. 2221, The Organic Peroxides (Conveyance by Road) Regulations 1973. HMSO, London.
SI 1974 No. 1735, Radioactive Substances (Carriage by Road) (Great Britain) Regulations 1974. HMSO, London.
SI 1979 No. 427, The Petroleum (Consolidation) Act 1928 (Enforcement) Regulations 1979. HMSO, London.

SI 1980 No. 1100, Petroleum (Consolidation) Act 1928 (Conveyance by Road Regulations Exemptions) Regulations 1980. HMSO, London.

[41] *SI 1978 No. 1702*, Hazardous Substances (Labelling of Road Tankers) Regulations 1978. HMSO, London.

[42] Health and Safety Executive 1980, 'Proposals for the Dangerous Substances (Conveyance by Road in Road Tankers and Tank Containers) Regulations 1980'. HMSO, London.

[43] National Association of Waste Disposal Contractors. *Code of Practice for the Construction of Vacuum Tankers*. March 1981. NAWDC.

[44] Willetts, S.L., 1979 and 1980, 'The Role of the Waste Disposal Contractor', *Safety in Transit Symposium*, Middlesbrough, 26–30 November 1979 and 10–14 November 1980. Redwood International Consultants Ltd, London.

[45] Willetts, S.L., 1981, 'The Safe Transportation and Disposal of Chemical Waste', *Transchem VIII – Eighth Symposium on the Safe Transportation of Hazardous Substances*, 27–28 May 1981. Cleveland Constabulary/Teesside Polytechnic, Middlesbrough, England.

[46] Health & Safety Executive, 1980, 'Proposal for Packaging and Labelling of Dangerous Substances (Amendments) Regulations 1980'. HMSO, London.

[47] *SI 1978 No. 209*, The Packaging and Labelling of Dangerous Substances Regulations 1978. HMSO, London.

[48] *C E C Directive 67/548/EEC:* 6th Amendment – 79/370/EEC. *O J* L88/1, 2 February 1979.

[49] *SI 1980 No. 1709*, The Control of Pollution (Special Waste) Regulations 1980. HMSO, London.

3

Hazardous wastes legislation: pollution and public health

Water pollution

Certain liquid wastes may be suitable for discharge directly to surface (fresh) waters, or to sewers, either in their raw state or after some form of treatment. The potential impact upon aquatic fauna and flora and any subsequent usages of the receiving waters are the governing factors as to whether any particular proposed discharge is 'allowable' at law.

The legislation concerning liquid waste discharges is complex due partly to the dual coverage by common law and by statute law, and partly to the differences existing between the legislation in England and Wales, Scotland and Northern Ireland. In addition, statute law governs water pollution via many subjects such as public health, fisheries protection, radioactive waste disposal, harbours, water resources, as well as water pollution itself.

(A) Civil law

Under civil law the owner of land bordering on a river or stream has certain rights in respect of natural watercourses running through his land. In John Young & Co. v. Bankier Distillery Co. (27 July 1893 [1]) these rights were summarised by Lord MacNaghten as:

> Every riparian proprietor is thus entitled to the water of his stream, in its natural flow, without sensible diminution or increase and without sensible alteration in its character or quality. Any invasion of this right causing actual damage or calculated to found a claim which may ripen into an adverse right entitles the party injured to the intervention of the Court.

Apart from discharges made under statute consent of a water authority (see later), no-one has any right whatever to introduce or direct into a stream anything but the water he has abstracted for domestic or agricultural use, and then only if the returned water is not increased, diminished or changed. Any person introducing effluent into a river without consent is at civil law *prima facie* a wrongdoer.

Of course, if all riparian owners insisted on receiving water in its natural state, major changes would be required in the methods of treating and disposing of polluting discharges. The operation of the civil law was considered by the Armer Committee (Trade Effluents Sub-Committee of the Central Advisory Water Committee) in 1960 [2] which recommended that no change be made. It is left to the good sense of the courts to interpret civil law so as to prevent injunctions suddenly closing down the activities of effluent-producing industries.

Comparative legislation

It is of interest that no definition of the term 'pollution' is given anywhere in the United Kingdom legislation. Various countries have attempted to define it and some examples are quoted:

(i) Swiss Federal Law on the control of water pollution, Section 2(1), of 1955 [3]:

> Measures necessary to control the pollution or other *deterioration* of surface water and groundwater shall be taken so that the *health of man* and of animals *is protected*, that groundwater and springwater is *fit to drink*, that surface water may be treated to render it *fit for consumption* and for *industrial use*, . . ., that fish may live in it, . . ., and that the countryside is not disfigured.

These provisions constitute an indirect definition of water pollution. It is interesting to note that fish are used as an indicator of pollution and they probably serve best as a visual means of assessing the state of a river.

(ii) French Law, 1964 (Law No. 64–1245) [3]: The provisions of Section 1 of this law of 16 December 1964 aimed at

> discharges, drainage, wastes, the storage, whether directly or indirectly, of materials of any kind, and more generally to anything liable to cause or increase *the deterioration in quality* of waters, whether surface water, groundwater, or maritime territorial waters, by changing their physical, chemical, biological or bacteriological characteristics.

(iii) Finnish Water Law, 1961 [3]: Pollution is defined in Section 19 as

> the discharge of dirt, waste, liquids, gas, bark or other materials into watercourses in such a way that, directly or indirectly, a harmful blocking up of the watercourse, a *harmful alteration* in the

> water *quality*, obvious harm to fish, an appreciable decrease in the pleasantness of the surroundings, a *danger to health*, or any other injury to private or public interests, is caused.

This resembles the Swiss definition in terms of interference with the use of water rather than the French in terms of the harmful effects of pollution.

(iv) A more explicit definition is given in Section 1 of the Belgian law of 11 March 1950, making it illegal to discharge into waters anything that

> is capable of harming the waters by making them either *malodorous* or *putrescible*, or harmful to the natural, or cultivated, or reared aquatic fauna and flora, or rendering them unsuitable for the watering of animals, the irrigation of land, or for industrial or domestic use.

Definitions of pollution

In the absence of any corresponding UK definition, the following have been held by the courts to constitute pollution under the civil law [4, 5]:

(1) rendering water unfit for domestic and agricultural purposes;
(2) fouling a river so as to kill or drive away fish;
(3) rendering water unsuitable for sheep washing or cattle drinking;
(4) raising the temperature of water;
(5) adding hard water to a soft water stream;
(6) causing canal water to become offensive;
(7) fouling a stream by discharging sewage or trade waste into it;
(8) throwing noxious refuse into a river.

Apart from the civil law rights of riparian owners, it is statute law that comprises the bulk of the legislation.

(B) Criminal law

As has been seen, the first pertinent legislation concerning liquid wastes was the Rivers (Prevention of Pollution) Act of 1876 [6] amended in 1893 [7] and 1898 [8].

This Act provided the basis for further pollution legislation. Remembering the civil law rights of riparian proprietors to abstract river water for agricultural purposes, and the absence of any specific definition of pollution, some guidelines of limits for upstream discharges became necessary. These were included in the establishment in 1898 of the Royal Commission on Treating and Disposing of Sewage (including any liquid from any factory or manufacturing process). This body remained in existence until 1915 [9] and issued ten Reports under its Chairman, Lord

Iddesleigh. These Reports dealt very fully with the treatment and disposal methods for sewage and advised certain Standards with which effluents should comply before being allowed to discharge into rivers. The 'Royal Commission Standards' taken from the Eighth Report have become accepted as indicating acceptable quality for discharges, though these Standards, or any others, have not been prescribed by statute law in the UK.

Royal Commission Standards

The Eighth Report [9] stated that channel experiments led the Commission to believe that 100 000 cubic centimetres of river water taking up not more than 0.4 grammes of dissolved oxygen in five days is ordinarily free from pollution. This is more conventionally expressed as biological oxygen demand (BOD) of 4 parts per million (ppm) and the limiting 4 ppm served as the foundation for the Commission's Standards. Temperature was noted to be important and 65°F (18.3°C) was adopted as the Standard. The dissolved oxygen uptake test was to be applied where a river received and mingled with a discharge and the quality of the receiving water was defined thus:

Description	BOD
Very clean	1 ppm
Clean	2 ppm
Fairly clean	3 ppm
Doubtful	5 ppm
Bad	10 ppm

It is obvious that river water already at 4 ppm cannot be used for dilution purposes. The Commission used the 2 ppm as the average state of river water under ordinary conditions, and recommended a Standard of 20 ppm be set for effluent discharges. A clean river dilution of eight times would result in a river of 4 ppm BOD. The dilution factor of 8 was safely assumed since 'the great majority of effluents are diluted by more than eight times their volume of river water'.

Similarly, suspended solids (SS) content of effluents was said in the Fifth Report (Command 4278 HMSO 1908 [9]) to be not more than 30 ppm after normal treatment.

Thus the Standards recommended for effluents discharging into rivers were taken as 20/30, or 20 ppm BOD and 30 ppm SS, maxima. Sewage authorities could consistently reach these Standards but, as population and industrial activity increased, discharges into rivers consequently increased and these Standards became harder to attain due to overloading of sewage works and also because increased discharge volumes decrease the useful dilution factor of the receiving waters.

The final Report in 1915 [9] recommended the establishment of the Water Pollution Research Laboratory which continues its activities as the Water Research Centre.

Examination of the 'Standards'

A consideration of the Iddesleigh Eighth Report reveals some interesting contradictions, sometimes in successive paragraphs. Since the Report lays down recommendations and not legislation, these may not be of great import but, for a water authority working to the recommendations, it must afford a certain degree of confusion. On the subject of the receiving water, the Commission had this to report:

Paragraph 15(c): 'Quality and quantity of river water are, however, highly important local conditions of which account should be taken.'

Paragraph 16: 'variations in the quality of the diluting water should not be taken into account'.

Paragraph 17: 'the quality of the diluting water should be assumed to be constant', this being represented as requiring 20 ppm dissolved oxygen in five days.

Paragraph 23: this paragraph insists that smaller than 8:1 dilution requires more stringent standards and greater dilution more relaxed standards. Again, the Commission takes into account the receiving water.

These contradictions as to whether or not the receiving waters should be considered when setting standards for the effluent must confuse their actual applications.

The validity of the five-day BOD test may itself be questioned as it was originally chosen to 'reflect . . . the observed conditions of the streams' (Eighth Report, Paragraph 9(1)). However, a very small sample placed motionless in a stoppered bottle in an incubator and kept in the dark for five days can hardly represent the actual conditions of a stream in full flow.

Somewhere in the history of the legislation and standards the temperature of the BOD test has risen from 18.3°C (65°F) to 20°C as laid down in the Ministry of Housing's Methods of Chemical Analysis as applied to Sewage and Sewage Effluents (1956 [10]) under 'Determination of BOD' (page 55). The temperature quotient for bacterial activity, Q10, is about 2 for every ten-degree-centigrade rise in temperature (Dixon & Webb [11]) and so the actual Standard of 20 ppm at 18.3°C becomes 23.4 ppm at 20°C.

Thus, the 20/30 limit of today is slightly less relaxed than that of the original Commission. This, perhaps, is a very minor point but, taking the previous observations into account, there may be a case for abandoning the Royal Commission Standards as a working policy.

(C) Inland fisheries protection

More recently, the Salmon and Freshwater Fisheries Act 1923 [12] provides in Part 1, Section 8, that:

> No person shall cause or knowingly permit to flow, or put or knowingly permit to be put, into any waters containing fish, or into any tributaries thereof, any liquid or solid matter to such an extent as to cause the waters to be poisonous or injurious to fish or the spawning grounds, spawn or food of fish.

The original penalties were a fine not exceeding £50 for the first offence and not more than £5 every day during which the offence was continued after conviction. A third conviction also rendered the offender liable to imprisonment with or without hard labour (!) for not more than three months in lieu of any fine to which he was liable. The Act was revised as The Salmon and Freshwater Fisheries Act 1972 [13] and both were repealed and replaced by the 1975 Act [14] of the same title.

The 1936 Public Health Act [15], in Section 259, extended the definitions of statutory nuisance so as to include

> (a) any pond, pool, ditch, gutter or watercourse which is so foul or in such a state as to be prejudicial to health or a nuisance; (b) any part of a watercourse, not being a part ordinarily navigated by vessels employed in the carriage of goods by water, which is so choked or silted up as to obstruct or impede the proper flow of water and thereby to cause a nuisance or give rise to conditions prejudicial to health.

Relevant authorities have powers to abate or remove statutory nuisances and this is dealt with elsewhere (see 'Solid Wastes' section).

(D) Pollution prevention

In 1951, the Rivers (Prevention of Pollution) Act for England and Wales came into force [16]. The corresponding Act for Scotland is the Rivers (Prevention of Pollution) (Scotland) Act 1951 [17], differing only in certain aspects of legal procedure. In Northern Ireland, the Rivers Pollution Prevention Act 1876 [6], discussed earlier, remained in force.

Sub-section 1 of Section 2 lays down that a person commits a punishable offence:

> (a) if he causes or knowingly permits to enter a stream any poisonous, noxious or polluting matter; or
> (b) if he causes or knowingly permits to enter a stream any matter so as to tend either directly or in combination with similar acts (whether his own or another's) to impede the proper flow of the water of the stream in a manner leading or likely to lead to a substantial aggravation of pollution due to other causes or of its consequences.

The phrase 'causes or knowingly permits' is worthy of close examination. It has been argued that strict liability by way of the very commission of the act of polluting would suffice for a prosecution, otherwise the phraseology would have been 'knowingly causes or knowingly permits'.

Consider Impress (Worcester) Ltd v. Rees (1971 [18]). The appellants had a fuel oil storage tank on their property which was visible from the highway: the tank had an outlet valve that was not locked and no night watchman was employed. A trespasser opened the unlocked valve one night and let the fuel oil escape into the adjacent river. The appellants were convicted under Section 2(1(a)) of the 1951 Act but they appealed to the High Court and were successful. The Court found that the appellants had not 'caused' oil to enter the river but that their fault lay in not locking the valve, which fault 'was not a cause at all but was merely part of the surrounding circumstances'.

This decision means that the prosecution must not necessarily show fault but must at least demonstrate a chain of causation leading back to the defendant.

Consider Alphacell Ltd v. Woodcock (1972 [19]). The appellants prepared manilla fibres and the effluent treatment system comprised two bankside settlement lagoons each equipped with automatic pumps to prevent overflow into the river. Leaves and ferns, on one occasion, blocked a pump inlet and overflow did occur into the river. The appellants were convicted under Section 2(1(a)) of the 1951 Act but appealed to the Court of Appeal and to the House of Lords: both appeals were unsuccessful. The Lords concluded that it was not necessary for the prosecution to show fault or negligence and that it was inescapable that the company's acts 'caused' the pollution: there was no defence in claiming that the pollution was unforeseeable and that the company took all reasonable care.

Consider Price v. Cromack (1975 [20]). The appellant allowed the construction of two effluent lagoons by a company on his land and one lagoon ruptured causing pollution. The appellant was convicted and appealed. The appeal succeeded on the basis that commission of an offence required some positive act and not merely a passive looking-on. It would seem that prosecution of the company that had constructed and used the effluent lagoons would have been successful but certainly not that of the landowner.

It is of interest that no definition of the terms poisonous, noxious or polluting appear in the Act. It has been said that 'these three words must have separate meanings: "poisonous" implies destruction of life, human or animal; "noxious" is lower in degree and signifies some injury, but not of necessity immediately dangerous to life; "polluting" will include both the other qualities and also what is foul and offensive to the senses' (Lumley's Public Health, 12th Edn [21]).

Section 5 of the 1951 Act gave the then river boards power to prescribe standards for matter considered to be poisonous, noxious or polluting. (This power was taken away subsequently by the 1961 Act [22].)

However, negative help is given in Section 11(3) of the 1951 Act such that 'matter shall not be deemed to be poisonous, noxious or polluting by reason of any effect it may have in discolouring a stream, if the discolouration is innocuous'.

Penalties for the offence under Section 2(1) of the 1951 Act are, on summary conviction, a fine of up to £50 or, on conviction on indictment, a fine of up to £200. Continuing pollution carries heavier penalties.

Where an authority believes that a contravention of Section 2 is likely to occur it may apply to the court under Section 3 of the Act for an order prohibiting the activity in question or permitting it subject to terms designed to remove the grounds of complaint.

The 1951 Act, in Section 7, provided for discharges of trade or sewage effluent to be made to watercourses with the consent of the authority. Pre-existing discharges were exempt so long as they were not substantially altered as respects the composition and nature, temperature, volume and rate of discharge. The Rivers (Prevention of Pollution) Act 1961, in Sections 1–3, brought the pre-1951 discharges under similar 'consent' control.

Thus, no person shall, without the consent of the appropriate water authority, bring into use any outlet for the discharge of trade or sewage effluent to a stream or begin to make any such discharge.

Consent, on application, may be made subject to conditions relating to location and specification of the proposed outlet and to nature, composition, temperature, volume and rate of discharge of the effluent. Consent cannot be unreasonably withheld and the conditions must be reasonable: there is an appeal provision to the Secretary of State. (For amplification see Ministry of Housing and Local Government 1968 [23].)

Each water authority is obliged to keep a register but not obliged to make it publicly available. Nevertheless, some authorities will allow public access to the register.

Contravention of Section 7 of the 1951 Act carries the penalty, on summary conviction, of a fine not exceeding £100 or, on conviction on indictment, a fine not exceeding £200.

Establishment of river authorities

A major organisational change was brought about by the Water Resources Act 1963 [24] which dissolved the 32 river boards and established 29 river authorities in England and Wales, and a Water Resources Board empowered to bring to the notice of the river authorities any inland water that needed to be improved and could be improved through exercising the powers of the Rivers (Prevention of Pollution) Acts 1951 and 1961. Whilst not changing existing legislation concerning prevention of pollution, the 1963 Act does make provision for protection of groundwater, and Section 72 makes it unlawful to

> discharge into any underground strata within a river authority area:
> (a) any trade effluent or sewage effluent, or
> (b) any poisonous, noxious or polluting matter not falling within the preceding paragraph, except with the consent of the river authority . . . and subject to any conditions imposed by the river authority.

These conditions may relate to:

> (a) the nature, composition and volume of the effluent or other matter to be discharged;
> (b) the strata into which it may be discharged;
> (c) measures to be taken for protecting water contained in other underground strata through which any well, borehole or pipe containing the effluent or other matter will pass;
> (d) the provision of facilities for inspection, including the provision, maintenance and use of observation wells and boreholes.

The penalty for unconsented discharge or inability to meet the water authority's standards is, on summary conviction, a fine not exceeding £100 or, on conviction on indictment, a fine not exceeding £200.

The authorities have the right to enter and inspect trade premises for the purposes of taking away samples of effluents passing into:

'(a) any inland water in the river authority area, or

(b) any underground strata in that area'.

The Rivers (Prevention of Pollution) (Scotland) Act of 1965 [25] amended the Act of 1951 of the same title to bring Scottish legislation into line with that for England and Wales.

The Water Acts of 1945 [26] and 1948 [27] empowered the river authorities to make bye-laws for the prevention of pollution of waters, both surface and underground. It is an offence under Section 22 to pollute water to be used for human consumption.

The Clean Rivers (Estuaries and Tidal Waters) Act of 1960 [28] extended the jurisdiction of the river authorities to tidal waters and 'controlled' waters inshore. The Act conferred powers to restrict and control effluent discharges and applies to post-29 September 1960 discharges: existing discharges pre-dating this remain uncontrolled unless and until significantly altered.

(E) Control of Pollution Act 1974 [29]

Part II deals with pollution of water.

Section 31, not yet in force, is destined to replace the pollution provisions of the 1951 and 1961 Rivers (Prevention of Pollution) Acts. It extends the offence to 'relevant waters' which includes streams and the sea up to a three nautical mile limit and specified underground waters.

An offence will be committed if a person causes or knowingly permits:

(a) any poisonous, noxious or polluting matter to enter . . . any relevant waters; or

(b) any matter to enter a stream so as to tend (either directly or in combination with other matter which he or another person causes or permits to enter the stream) to impede the proper flow of the water of the stream in a manner leading or likely to lead to a substantial aggravation of pollution due to other causes or of the consequences of such pollution; or

(c) any solid waste matter to enter a stream or restricted waters.

'Restricted waters' will be prescribed by Regulations and will relate to tidal rivers and boat mooring areas.

An offence is not committed if the entry is authorised by consent and complies with any conditions attached thereto or is attributable to an act or omission which is in accordance with 'good agricultural practice' (Codes of Practice by MAFF are in preparation) or is by way of an emergency in order to avoid danger to the public.

Section 32 creates the offence of discharging trade or sewage effluent into relevant waters without consent. There is a special defence for discharges made by water authorities themselves which is important because prosecutions will be able to be brought under this section without the consent of the Attorney General or the water authority. This is a considerable extension of the powers of prosecution currently extant under Section 11 of the Rivers (Prevention of Pollution) Act 1961.

Section 34 relates to applications for consents to discharge effluents into any controlled waters and provides for consideration and determination within three months, or such longer period as may be agreed in writing. Conditions may be imposed:

(a) as to the places at which the discharges to which the consent relates may be made and as to the design and construction of any outlets for the discharges;

(b) as to the nature, composition, temperature, volume and rate of the discharges and as to the periods during which the discharges may be made;

(c) as to the provision of facilities for taking samples of the matter discharged and in particular as to the provision, maintenance and use of manholes, inspection chambers, observation wells and boreholes in connection with the discharges;

(d) as to the provision, maintenance and testing of meters for measuring the volume and rate of the discharges and apparatus for determining the nature, composition and temperature of the discharges;

(e) as to the keeping of records of the nature, composition, temperature, volume and rate of the discharges and in particular of records of readings of meters and other recording apparatus provided in accordance with any other condition attached to the consent;

(f) as to the making of returns and the giving of other information to the water authority about the nature, composition, temperature, volume and rate of the discharges; and

(g) as to the steps to be taken for preventing the discharges from coming into contact with any specified underground water.

relate to certain discharges for which consent was not previously required either by virtue of the Public Health (Drainage of Trade Premises) Act 1937 or the Public Health Act 1961. Occupiers of premises benefiting from 'prescriptive right' discharges or pre-1937 'agreements' were required to serve notice on the water authority by 31 January 1975 and the authority was then required to translate the notice into an 'actual consent' with imposition of conditions if thought fit.

Regulations have been made (SI 1976 No. 958 [39]) relating to appeals by recipients of 'actual consents'.

It seems that 'agreements' made after July 1938 and by virtue of Section 7 of the 1937 Act which do not expressly allow for termination by a water authority remain unaffected.

Public health

The Public Health Act of 1875 was the first realistic and positive step taken for the prevention of nuisance arising from improper disposal of wastes.

Table 3. *Recent trends in sewage characteristics employed to calculate trade effluent charges*

Water Authority		77/78	78/79	79/80	80/81	81/82
Anglian	*Os*	572	572	572	670	640
	Ss	377	337	337	380	380
Northumbrian	*Os*	417	364	372	384	396
	Ss	193	173	164	172	186
Severn–Trent	*Os*	424	403	379	378	375
	Ss	347	342	332	335	340
Southern	*Os*		397	388	386	376
	Ss		354	365	380	370
South West	*Os*	159	178	155	160	318
	Ss	274	280	301	280	285
Thames	*Os*	320	298	292	274	425
	Ss	238	228	221	206	327
Welsh	*Os*	432	432	432	432	500
	Ss	304	304	304	304	350
Wessex	*Os*	400	400	400	400	395
	Ss	350	350	350	350	355
Yorkshire	*Os*	573	538	523	498	434
	Ss	216	203	200	191	171

Source: Courtesy Confederation of British Industry.

The 1936 Act of the same title required the occupiers of houses and industrial premises suitably to prepare their wastes for collection and further disposal. Local authorities were empowered to secure the abatement of 'statutory nuisances' including 'any accumulation or disposal which is prejudicial to health or a nuisance'.

Thus, a nuisance had to occur before a legal remedy could be sought. The sections of particular interest and import in the 1936 Act are as follows:

Section 92: definitions of statutory nuisances;
Section 93: powers to serve abatement notices;
Section 94: powers of the courts to make nuisance orders if an abatement notice is disregarded;
Section 95: penalty for contravention of a nuisance order and powers of local authority to abate a nuisance;
Section 96: recovery of costs incurred by a local authority abating or preventing the recurrence of a nuisance;
Section 97: proceedings where nuisance is caused by acts or defaults of more than one person;
Section 98: powers to proceed where the cause of nuisance arises outside the local authority (district) area;
Section 99: power of the individual to make a complaint as to statutory nuisance;
Section 100: proceedings taken to the High Court by a local authority for abatement of nuisance.

The Public Health (Recurring Nuisances) Act of 1969 [40] gave local authorities the power to prevent the recurrence of an already abated nuisance if such recurrence was deemed likely.

The 1961 Public Health Act gave the authorities the power to remove any accumulations of rubbish that might be 'seriously detrimental to the amenities of the neighbourhood' (Section 34). In this context, 'rubbish' means 'rubble, waste paper, crockery and metal, and any other kind of refuse (including organic matter)'.

This public health legislation, together with the provisions of the Town and Country Planning Acts of 1947, 1962, 1963 and 1968 and the associated General Development Order of 1963, was the only effective controlling mechanism as regards waste disposal for several years. (See Chapter 2 [25].)

It was not until 1972 that waste disposal practices hit the headlines when the concealment of drums of cyanide wastes on land regularly used by children (near Nuneaton) became a national issue. There was such public concern that The Deposit of Poisonous Waste Act [41] was rushed through Parliament at amazing speed and received royal assent on 30 March 1972.

The Deposit of Poisonous Waste Act 1972 [41]

The pre-amble described:

> An Act to penalise the depositing on land of poisonous, noxious or polluting waste so as to give rise to an environmental hazard, and to make offenders liable for any resultant damage; to require the giving of notices in connection with the removal and deposit of waste; and for connected purposes.

The principal offence

Section 1 created the offence of depositing, or causing or permitting the deposit, of poisonous, noxious or polluting waste in such circumstances as would be liable to give rise to an environmental hazard: 'poisonous, noxious or polluting', as has been seen (see section on liquid waste), has been used before (Rivers (Prevention of Pollution) Acts) but has never been subject to close judicial scrutiny and so no sensible (legal) definition can be evinced.

The term 'environmental hazard' is expanded in sub-sections (3) and (4) of Section 1 so as to mean that a deposit constitutes such a hazard if

> in such a manner, or in such quantity (whether that quantity by itself or cumulatively with other deposits of the same or different substances) as to subject persons or animals to material risk of death, injury or impairment of health, or as to threaten the pollution or contamination (whether on the surface or underground) of any water supply; and where the waste is deposited in containers, this shall not of itself be taken to exclude any risk which might be expected to arise if the waste were not in containers.

This definition is important and deserves close study: a hazard is created by a waste deposit if a material risk is thereby presented to public or animal health or of water pollution taking into account previously deposited substances and disregarding the containers within which the waste may be. The Act goes on to say that the assessment of risk should take into account any measures taken to minimise such risk and the likelihood of tampering by children.

The wording of the Act makes it unclear whether or not an environmental hazard actually has to arise consequent upon the deposit of a waste before an offence is committed. Such a hazard could, of course, arise immediately upon deposit or may not occur until, some time after, the rupture of a container or some subsequent and latent chemical event happens. Indeed, if a load of drums was deposited, it might be construed that several offences could have been committed as each drum ruptures.

The prescribed penalty was, on summary conviction, a fine of up to £400 (increased to £1000 by the Criminal Law Act 1977 [42]) and/or imprisonment for up to six months or, on conviction on indictment, to imprisonment for up to five years and/or an unlimited fine.

Statutory defences were provided for persons acting either under the instructions of an employer or relied upon information given by bona fide persons and having no reason to suspect that such information was false or misleading.

Civil liability

Section 2 of the Act introduced civil liability for any damage that may have been caused by a waste deposited in contravention of Section 1. This, of course, was an important aspect of this legislation.

Notification provisions

Section 3 introduced the notification provisions in that certain notices had to be served on relevant authorities prior to the removal or deposit of certain wastes. The relevant authorities at the time of implementation were the local authority (some 1400) and river authority in whose area the waste arose and the same two authorities in whose area the intended disposal would take place. On 1 April 1974 the recipient authorities became the waste disposal authorities and water authorities for the areas of waste production and disposal. (See Local Government Act 1972 and Water Act 1973 and Control of Pollution Act 1974 [43, 36, 29].)

The prescribed information to be served on the authorities was:

- the address from which the waste was to be removed;
- the address of the site on which the waste was to be deposited;
- the nature and chemical composition;
- the quantity together with details of number, size and description of containers;
- the name of the person responsible for haulage.

Such notice had to be served at least three clear, working days before the intended removal or deposit. Additionally, if the notice was raised by a person other than the person (or his employee) who was to undertake the removal then a copy had to be served on that person.

Regulations made by the Secretaries of State for the Environment, for Wales and for Scotland (SI 1972 No. 1017 [44]) defined those wastes that could be removed and deposited without prior notification and came into force on 3 August 1972. The Regulations provided a Schedule of descriptions of wastes and if a particular waste was included on this Schedule and did not contain any hazardous quantity or hazardous concentration of a poisonous, noxious or polluting substance, then it was exempted from the notification procedure.

The relevant Schedule is reproduced in Fig. 1. The Regulations also gave qualified exemptions for wastes of a prescribed description deposited in prescribed circumstances, such as agrochemical preparations, radioactive substances, emergency deposits and deposits specifically consented or authorised by other enactments.

The offence was created of failing to pre-notify a removal or deposit of waste or providing false information, subject to a maximum penalty, on summary conviction, of a fine up to £400. This was subject to statutory defences.

Notification of receipt

Section 4 required the operator of a tip site not to accept wastes for deposit unless he had received, at least three working days prior to intended deposit, a copy of the notice served under Section 3 upon the relevant authorities. It also required the completion of a certificate of receipt for a waste, giving similar information as the pre-notification document, which had to be served on the same authorities within three days of the deposit.

Failure to notify or the giving of false information was a prescribed offence with a penalty of up to £400 fine, subject to statutory defences.

The notices given under Section 3 became known as Part I notices and those under Section 4 as Part II notices. Although no formal document was prescribed, the Circular 70/72 [45] included a suitable format as an appendix and this was generally adopted. Part I is reproduced in Fig. 2.

'Season tickets'

For regular waste arisings and deposits at the same site, the relevant authorities often came to agreement with waste producers and disposers such that 'season tickets' were implemented. These arrangements were designed such that a producer could notify his anticipated waste production for a period, say one month or three months, on the one

Fig. 1. Description of wastes not requiring notification. (Source: Reference [44].)

SCHEDULE

Regulation 3

DESCRIPTIONS OF WASTE WHICH NEED NOT BE SUBJECT TO SECTION 3 OF THE DEPOSIT OF POISONOUS WASTE ACT 1972 IF IT DOES NOT CONTAIN ANY HAZARDOUS QUANTITY OR HAZARDOUS CONCENTRATION OF ANY POISONOUS, NOXIOUS OR POLLUTING SUBSTANCE.

Class 1 Any waste normally arising in the use of premises for domestic purposes.

Class 2 Any waste normally arising in the use of premises as an office for any purpose, or as a retail shop (that is to say, a building used for the carrying on of any retail trade or retail business wherein the primary purpose is the selling of goods or services by retail).

Class 3 Any other waste, however arising, of which the nature and composition are such that
(*a*) if it arose in the use of premises for domestic purposes, it would fall within Class 1;
(*b*) if it arose in the use of premises as an office or retail shop, it would fall within Class 2.

Class 4 Any waste produced in the course of—
(i) the construction, repair, maintenance or demolition of plant or buildings;
(ii) the laundering or dry cleaning of articles;
(iii) working mines and quarries, or washing mined or quarried material;
(iv) the construction or maintenance of highways, whether or not repairable at the public expense;
(v) the dry cutting, grinding or shaping of metals, or the subjection thereof to other physical or mechanical process;
(vi) the softening, treatment or other processing of water for the purpose of rendering it suitable for (a) human consumption, (b) the preparation of foods or drinks, (c) any manufacturing or cooling process, or (d) boiler feed;
(vii) the treatment of sewage;
(viii) the breeding, rearing or keeping of livestock;
(ix) brewing;
(x) any other fermentation process; or
(xi) the cleansing of intercepting devices designed to prevent the release of oil or grease.

Class 5 Any waste (not being waste in any of the foregoing classes) consisting of one or more of the following items whether mixed with water or not:—
(i) Paper, cellulose, wood (including sawdust and sanderdust), oiled paper, tarred paper, plasterboard;
(ii) Plastics, including thermoplastics in both the finished and raw states, and thermosetting plastics in the finished state;
(iii) Clays, pottery, china, glass, enamels, ceramics, mica, abrasives;
(iv) Iron, steel, aluminium, brass, copper, tin, zinc;
(v) Coal, coke, carbon, graphite, ash, clinker;
(vi) Slags produced in the manufacture of iron, steel, copper or tin or of mixtures of any of those metals;
(vii) Rubber (whether natural or synthetic);
(viii) Electrical fittings, fixtures and appliances;
(ix) Cosmetics;

Fig. 1. (*cont.*)

(x) Sands (including foundry and moulding sands), silica;
(xi) Shot blasting residues, boiler scale, iron oxides, iron hydroxides;
(xii) Cement, concrete, calcium hydroxide, calcium carbonate, calcium sulphate, calcium chloride, magnesium carbonate, magnesium oxide, zinc oxide, aluminium oxide, titanium oxide, copper oxide, sodium chloride;
(xiii) Cork, ebonite, kapok, kieselguhr, diatomaceous earth;
(xiv) Wool, cotton, linen, hemp, sisal, any other natural fibre, hessian, leather, any man-made fibre, string, rope;
(xv) Soap and other stearates;
(xvi) Food, or any waste produced in the course of the preparation, processing or distribution of food;
(xvii) Vegetable matter;
(xviii) Animal carcases, or parts thereof;
(xix) Excavated material in its natural state;
(xx) Any other substance which is a hard solid and is insoluble in water and in any acid.

---

document and so avoid the need to pre-notify each and every intended removal/deposit.

Records

Section 5 required local authorities to maintain records of the notices with which they had been served. These are not public documents.

Implementation provisions

The Act came into force on the date of passing (30 March 1972) other than Sections 3, 4, 5(3) and 5(4) which related to the notification provisions which were introduced on 3 August 1972 by SI 1972 No. 1016 [46].

What were the effects of DOPWA?

Circular 37/72 [47] requested that local authorities embark upon a survey of waste disposal facilities in their area and, together with the notifications subsequently received under DOPWA, these authorities collected a considerable amount of data that were, until then, unknown. The scale and nature of hazardous waste disposal activities in Britain gradually became known and quantified for the first time. The administration of such activities was elevated to a local authority level and the controls afforded were obviously beneficial.

Interestingly, an authority receiving a notice of intended disposal could do nothing other than record the details: authorities were not empowered to 'reject' intentions. Nevertheless, many authorities took their duties seriously and studied the notices and did, often, contact persons notifying an intended waste movement and 'advised' that an offence under Section 1 may be committed if such disposal did take place. Such advice was usually tantamount to a 'rejection' anyway.

As far as waste producers were concerned, the Act made them assess the

Fig. 2. Typical notification form for DOPWA. (Source: Reference [45].)

APPENDIX A

DEPOSIT OF POISONOUS WASTE ACT 1972

This copy for:

Notes on the completion of this form appear overleaf

Serial number

Originator's reference

Date of Notice

Part 1 Notice of intention to remove and deposit waste given under Section 3 *(see note (a))*

A (i) I certify that copies of this Part 1 Notice as hereunder completed have been served on the Authorities mentioned in B and C below

Signed ....

On behalf of ....
(person, firm or authority originating and/or disposing of waste)

(ii) Position held by signatory..................

(iv) Earliest permitted date of removal *(see note (c))*

(iii) His address and telephone no. *(see note (b))* ....

B Removal

(i) Premises from which the waste is to be removed:

(ii) Name and address of the Local Authority for this area:

(iii) Name and address of the River Authority or River Purification Board for this area:

C Deposit

(i) Land where the waste is to be deposited: *(see note (d))*

(ii) Name and address of the Local Authority for this area: *(see note (e))*

(iii) Name and address of the River Authority or River Purification Board for this area: *(see note (e))*

D Description of the waste

(i) Nature (general description):

(ii) Principal components (chemical, biological, etc.):

(iii) If in containers, state number, size and description: *(see note (f))*

(iv) Quantity: *(see note (g))*

E Person, firm or authority who is to undertake removal

nature of their wastes, very often for the first time, and the scale of their disposal needs. It also conferred an implicit responsibility to find out what disposal site and/or method was being used by contractors in relation to their particular waste. Otherwise, a waste producer may well be liable for 'knowingly' permitting an offence albeit the direct action may be carried out by a contractor.

For tip site operators, a new standard of operating was required and local authorities started to insist upon proper fencing, proper covering of wastes, adequate site management, etc. concomitant with the new responsibilities under the Act.

An interesting anomaly existed for the operators of treatment plant, transfer stations and incinerators. The Circular 70/72 made it clear that removal of wastes to such installations was subject to the pre-notification provisions of Section 3. However, the certificate of receipts provided by Section 4 were required only from tip operators who recorded the deposit of a waste onto a tip. That is, Part I's were essential but Part II's possibly not so.

References

[1] *Young & Co. v. Bankier Distillery Co.* House of Lords and Privy Council: Appeal Cases, p. 691. *Lord's Journals*, 27 July 1893.

[2] Ministry of Housing and Local Government, 1960, 'Final Report of the Trade Effluents Sub-Committee of the Central Advisory Water Committee'. HMSO, London.

[3] World Health Organization, 1967. *Control of Water Pollution: A Survey of Existing Legislation.* WHO, Geneva.

[4] Wisdom, A.S., 1956. *The Law on the Pollution of Waters.* Shaw, London.

[5] Wisdom, A.S., 1966. *The Law of Rivers and Watercourses.* Shaw, London.

[6] *Rivers (Prevention of Pollution) Act 1876*, 39 and 40, Victoria, Ch. 75. HMSO, London.

[7] *Rivers (Prevention of Pollution) Act 1893*, 56 and 57, Victoria, Ch. 31. HMSO, London.

[8] *Rivers (Prevention of Pollution) (Border Councils) Act 1898*, 61 and 62, Victoria, Ch. 34. HMSO, London.

[9] 1898–1915 Command Papers. *Royal Commission to inquire and report what methods of treating and disposing of sewage (including any liquid from any factory or manufacturing process) may properly be adopted.*

1901 Command 685: 'Interim (First) Report' – Vol. I – Report.
1902 Command 686: 'Interim (First) Report' – Vol. II – Evidence.
1902 Command 686 – I: 'Interim (First) Report' – Vol. III – Appendices.
1902 Command 1178: 'Second Report'.
1903 Command 1486: 'Third Report' – Vol. I – Report.
1903 Command 1487: 'Third Report' – Vol. II – Evidence.
1904 Command 1893: 'Fourth Report' – Vol. I – Report.
1904 Command 1884: 'Fourth Report' – Vol. II – Evidence.
1904 Command 1885: 'Fourth Report – Contamination of shellfish' – Vol. III.

1904 Command 1886: 'Fourth Report' – Vol. IV, Part I – General Report.
1904 Command 1886 – I: 'Fourth Report' – Vol. IV, Part II – Chemical Report.
1904 Command 1886 – II: 'Fourth Report' – Vol. IV, Part III – Bacteriological Report.
1904 Command 1886 – III: 'Fourth Report' – Vol. IV, Part IV – Engineering and Practical Report.
1904 Command 1886 – IV: 'Fourth Report' – Vol. IV, Part V – Report to the Commission on chemical aspects.
1908 Command 4278: 'Fifth Report' – Vol. I – Report.
1908 Command 4279: 'Fifth Report' – Appendix I.
1908 Command 4280: 'Fifth Report' – Appendix II.
1909 Command 4281: 'Fifth Report' – Appendix III.
1910 Command 4282: 'Fifth Report' – Appendix IV.
1908 Command 4283: 'Fifth Report' – Appendix V.
1908 Command 4284: 'Fifth Report' – Appendix VI.
1908 Command 4285: 'Fifth Report' – Appendix VII.
1908 Command 4286: 'Fifth Report' – Appendix VIII.
1909 Command 4511: 'Sixth Report'.
1911 Command 5543: 'Seventh Report' – Vol. I – Report and Vol. II – Appendices.
1911 Command 5543 – I: 'Seventh Report' – Vol. III – Appendices.
1912 Command 6464: 'Eighth Report' – Vol. I – Report.
1913 Command 6943: 'Eighth Report' – Vol. II – Appendices.
1915 Command 7819: 'Ninth Report' – Vol. I – Report.
1915 Command 7820: 'Ninth Report' – Vol. II – Appendices.
1915 Command 7821: 'Final (tenth) Report'.

[10] Ministry of Housing and Local Government, 1956. *Methods of Chemical Analysis as Applied to Sewage and Sewage Effluents*, 2nd Edn. HMSO, London.

[11] Dixon, M. & E.C. Webb, 1966. *Enzymes*, 2nd Edn. Longmans, Green, London.

[12] *Salmon and Freshwater Fisheries Act 1923*, 13 and 14, George V, Ch. 16. HMSO, London.

[13] *Salmon and Freshwater Fisheries Act 1972*, Elizabeth II, Ch. 37. HMSO, London.

[14] *Salmon and Freshwater Fisheries Act 1975*, Elizabeth II, Ch. 51. HMSO, London.

[15] *Public Health Act 1936*, 26, George V and VI, Edward VIII, Ch. 49. HMSO, London.

[16] *Rivers (Prevention of Pollution) Act 1951*, 14 and 15, George VI, Ch. 64. HMSO, London.

[17] *Rivers (Prevention of Pollution) (Scotland) Act 1951*, 14 and 15, George VI, Ch. 66. HMSO, London.

[18] *Impress (Worcester) Ltd v. Rees* (1971) 2. All England Reports 357.

[19] *Alphacell Ltd v. Woodcock* (1972) 2. All England Reports 475.

[20] *Price v. Cromack* (1975) 2. All England Reports 113.

[21] *Lumley's Public Health Acts*, 12th Edn. Eds. E. Simes, C.E. Scholefield and K.T. Watson. Vol. V, 1955. Butterworth and Shaw, London.

[22] *Rivers (Prevention of Pollution) Act 1961*, 9 and 10, Elizabeth II, Ch. 50. HMSO, London.

[23] Ministry of Housing and Local Government, 1968. *Standards of Effluents to Rivers with Particular Reference to Industrial Effluents.* HMSO, London.

[24] *Water Resources Act 1963*, Elizabeth II, Ch. 38. HMSO, London.

[25] *Rivers (Prevention of Pollution) (Scotland) Act 1965*, 13 and 14, Elizabeth II, Ch. 13. HMSO, London.

[26] *Water Act 1945*, 8 and 9, George VI, Ch. 42. HMSO, London.

[27] *Water Act 1948*, 11 and 12, George VI, Ch. 22. HMSO, London.

[28] *Clean Rivers (Estuaries and Tidal Waters) Act 1960*, 8 and 9, Elizabeth II, Ch. 54. HMSO, London.
[29] *Control of Pollution Act 1974*, Elizabeth II, Ch. 40. HMSO, London.
[30] *Public Health Act 1875*, 38 and 39, Victoria, Ch. 55. HMSO, London.
[31] *Public Health (Drainage of Trade Premises) Act 1937*, 1, Edward VIII and George VI, Ch. 40. HMSO, London.
[32] *Public Health Act 1961*, 9 and 10, Elizabeth II, Ch. 64. HMSO, London.
[33] *Water Resources Act 1968*, Elizabeth II, Ch. 35. HMSO, London.
[34] *Water Resources Act 1971*, Elizabeth II, Ch. 34. HMSO, London.
[35] Jeffries, C., 1981. '3 Big Effects of Trade Effluent Charges', *Water and Waste Treatment*, Nov. 1981.
[36] *Water Act 1973*, Elizabeth II, Ch. 37. HMSO, London.
[37] *SI 1976 No. 957*, The Control of Pollution Act 1974 (Appointed Day) Order 1976. HMSO, London.
[38] *SI 1974 No. 2039 (C.33)*, The Control of Pollution Act 1974 (Commencement No. 1) Order 1974. HMSO, London.
[39] *SI 1976 No. 958*, The Control of Pollution (Discharges into Sewers) Regulations 1976. HMSO, London.
[40] *Public Health (Recurring Nuisances) Act 1969*, Elizabeth II, Ch. 25. HMSO, London.
[41] *Deposit of Poisonous Waste Act 1972*, Elizabeth II, Ch. 21. HMSO, London.
[42] *Criminal Law Act 1977*, Elizabeth II, Ch. 45. HMSO, London.
[43] *Local Government Act 1972*, Elizabeth II, Ch. 70. HMSO, London.
[44] *SI 1972 No. 1017*, The Deposit of Poisonous Waste (Notification of Removal or Deposit) Regulations 1972. HMSO, London.
[45] *Circular 70/72*, Department of the Environment (Welsh Office 149/72). Deposit of Poisonous Waste Act. HMSO, London.
[46] *SI 1972 No. 1016 (C.16)*, The Deposit of Poisonous Waste Act (Commencement Order). HMSO, London.
[47] *Circular 37/72*, Department of the Environment (Welsh Office 86/72, Scottish Development Department letter 26.4.72). Review of Waste Disposal Facilities. HMSO, London.

4

Hazardous wastes legislation: contemporary law

Control of Pollution Act [1] – **Part I**

Part I of the Act is titled: 'Waste on Land'.

Section 1

'It shall be the duty of each disposal authority to ensure that the arrangements made by the authority and other persons for the disposal of wastes are adequate.' This Section is the raison d'être of Part I of the Act but has not yet been implemented: the grounds for not doing so have been quoted as economic. Actually, the cost of implementation would be nil since 'controlled' wastes (see later) are all being disposed in an authorised and adequate manner now, except those individual cases which are actioned by the disposal authorities.

A more insidious reason for holding back on implementation is that this Section would confer responsibility onto the disposal authorities: other Sections confer powers. Power without responsibility can be a very bad thing and it is certainly easier to wield power when there is no concomitant responsibility. The present position very often reflects this situation and dual standards are evident: what a disposal authority may do on its own sites can be different from what it requires to be done on private sites.

Dual standards are further encouraged in the separate provisions for the authorisation or licensing of public authority and private sites.

An interesting point is that the Secretary of State does have powers (Section 97) where an authority is deemed to be in default: detailed cases of overt failures to perform duties should, therefore, be brought to the attention of the Secretary of State.

Section 2

Brought into force on 1 July 1978 in England and Wales (SI 1977 No. 2164 [2]) and 1 September 1978 in Scotland (SI 1978 No. 816 [3]) the

Section requires each disposal authority to carry out an investigation on waste generation and disposal and to draw up a plan for future disposal arrangements.

The 'waste disposal plan'

Sub-section (1) requires each authority:

(a) to carry out an investigation with a view to deciding what arrangements are needed for . . . disposing of controlled wastes;
(b) to decide what arrangements are . . . needed;
(c) to prepare a statement of the arrangements made and proposed to be made;
(d) to carry out . . . further investigations;
(e) to make any modification of the plan . . . and . . . it shall be the duty of the authority to have regard to the effect . . . on the amenities . . . and to the likely cost.

Sub-section (2) requires that the plan should include information as to:

(a) the kinds and quantities of controlled waste . . . situated in its area;
(b) the kinds and quantities . . . brought for disposal into or taken for disposal out of the . . . area;
(c) the kinds and quantities . . . expects to dispose of itself;
(d) the kinds and quantities . . . to be disposed of . . . by persons other than the authority;
(e) the methods by which . . . waste . . . should be disposed of;
(f) the sites and equipment . . . for disposing of . . . waste;
(g) the estimated cost.

Regulations may be made altering (a)–(g) but the Secretary of State has said that there is no such intention (Circular 29/78, Paragraph 6 [4]).

Sub-section (3) requires consultation, in all cases, with the relevant water authority(ies) and collection authorities and, in certain cases, with other prescribed bodies. It also requires that the plan be publicised and that due consideration be taken of representations made by the public.

Sub-section (4) prescribes consultation with appropriate bodies with a view to reclamation of substances from wastes.

Sub-section (5) relates to aspects of plans that may involve neighbouring or other disposal authorities and requires either mutual agreement between authorities or a direction from the Secretary of State in cases of disagreement.

Sub-section (6) requires that the authority publicises its final draft of the plan and submits a copy to the Secretary of State. It is of interest here that the plan is submitted to the Secretary of State purely for information and not for comment, amendment or approval and so differs from the plans submitted under the Town and Country Planning Act 1971 [5] (i.e. 'structure plans', Section 7 (*et seq.*), 'local plans', Section 11 (*et seq.*)). In effect, the disposal authorities have considerable powers in making their plans.

Sub-section (7) provides that the Secretary of State may impose a time limit for preparation of the plans. He has expressed no such intention (Circular 29/78, Paragraph 5) but expected substantial progress within two years. Such progress has, in fact, been singularly lacking and only a handful of authorities have, as yet, prepared their plans.

Guidance documents

There are two *Waste Management Papers* that offer guidance to authorities in respect of the statutory survey and plan: No. 2 – 'Waste Disposal Surveys' [6] and No. 3 – 'Guidelines for the Preparation of a Waste Disposal Plan' [7].

For the purposes of the survey, certain non-'controlled' wastes, such as farm wastes and mine and quarry wastes, are included for the sake of completeness. The categories to be surveyed are:

(a) household and commercial waste (excluding h, j, k or l);
(b) medical, surgical and veterinary waste;
(c) industrial waste;
(d) mine and quarry waste;
(e) radioactive waste;
(f) farm waste;
(g) waste from construction and demolition;
(h) sewage sludge and cesspool and pail closet contents;
(i) old cars, vehicles and trailers;
(j) pulverised household and commercial waste;
(k) screenings from household and commercial waste;
(l) ash from incineration.

The survey must also include details of existing disposal facilities, both treatment plant and landfill sites, and potential future disposal facilities. For existing facilities, the following items must be covered:

Treatment plant

(i) type of plant and location;
(ii) age and remaining useful life;
(iii) tonnage and type of waste handled and spare capacity;

(iv) any restrictions on type of waste handled;
(v) advantages and disadvantages;
(vi) final disposal point for residues;
(vii) transportation methods for wastes and residues;
(viii) origin of waste input.

Landfill sites

(i) description of site, access and location;
(ii) date tipping started;
(iii) type and quantity of wastes being tipped;
(iv) current rate of waste input;
(v) remaining volumetric capacity;
(vi) expected date of completion;
(vii) planning status of site;
(viii) inventory of site plant;
(ix) details of cover material used;
(x) performance rating against Code of Practice (DOE Circular 26/71, W/O Circular 65/71, SDD Circular 27/71 [8]);
(xi) restrictions on types of wastes;
(xii) transportation method for wastes;
(xiii) origin of waste input.

Reference is made to information already existing from the survey of waste disposal sites as requested in the DOE Circular 37/72 (W/O Circular 86/72, SDD letter of 26 April 1972) [9].

Advice is given on the computer logging and coding of the survey information so that data can be stored, manipulated and retrieved by the authorities in a common format.

As regards the use to which the survey data is put, the 'plan' is detailed in *Waste Management Paper No. 3* [7] and the suggested layout is:

(i) introduction;
(ii) summary and conclusions;
(iii) background of waste disposal authority area;
(iv) existing conditions;
(v) likely future conditions;
(vi) objectives;
(vii) evaluation of possible courses of action/recommended courses of action;
(viii) implementation;
(ix) appendix – tables describing existing conditions.

This is the standard format to which disposal authorities should be working.

Sections 3–11, 16 – site licensing

These Sections form the backbone of Part I of the Act and provide for a system of site licensing such that all facilities disposing of controlled wastes have to be properly authorised by the disposal authority.

All were brought into force in England and Wales on 14 June 1976 (SI 1976 No. 731 [11]) and in Scotland on 1 January 1978 (SI 1977 No. 1587 [12]).

The principal offence

Section 3 provides that a person shall not

> deposit controlled waste on any land or cause or knowingly permit controlled waste to be deposited . . . or use any plant or equipment or cause or knowingly permit any plant or equipment to be used for the purpose of disposing of controlled waste . . . unless the land . . . or . . . the site of the plant or equipment is occupied by the holder of a licence,

and that conditions attached thereto are complied with.

The offence carries a penalty of, on summary conviction, a fine of up to £1000 (increased from £400 by the Criminal Law Act 1977 [13]) or, on conviction on indictment, an unlimited fine and/or imprisonment for up to two years. Where an offence is committed and 'the waste in question is . . . poisonous, noxious or polluting; and . . . is likely to give rise to an environmental hazard', then the penalties are increased, on summary conviction, to imprisonment for up to six months and/or a fine of up to £1000 and, on conviction on indictment, to an unlimited fine and/or imprisonment for up to five years.

In Ashcroft v. Cambro Waste Products Ltd (Queen's Bench Division, 30 March 1981 [14]), the term 'knowingly permit' was explored. The disposal authority took Cambro to the court of justice for breach of two licence conditions. The Justices dismissed the case on the grounds that the director involved had no knowledge of the breach since a site foreman was in charge of operations and not the director. The disposal authority appealed to the High Court. The Law Lords decided and agreed that, where a company knowingly permitted the deposit of controlled waste, it was not necessary for the prosecution to prove that the company had knowingly permitted the breach of licence conditions. The Justices were directed to convict.

Defences

Statutory defences are provided being that a person

> took care to inform himself, from persons who were in a position to provide the information, as to whether the deposit . . . would be

> in contravention . . ., and did not know and had no reason to suppose that the information given to him was false or misleading . . .; or, that he acted under instructions from his employer . . .; or, . . . that he took all such steps . . . to ensure that the conditions were complied with; or that the acts . . . were done in an emergency.

Supplementary provisions

Section 4 contains important supplementary provisions, namely that reference to land includes reference to water that covers land above the low-water mark of ordinary spring tides and

> (a) the presence of waste on land gives rise to an environmental hazard if the waste has been deposited in such a manner or in such a quantity (whether that quantity by itself or cumulatively with other deposits of the same or different substances) as to subject persons or animals to a material risk of death, injury or impairment of health or as to threaten the pollution (whether on the surface or underground) or any water supply; and
>
> (b) the fact that waste is deposited in containers shall not of itself be taken to exclude any risk which might be expected to arise if the waste were not in containers.
>
> 6. In the case of any deposit of waste, the degree of risk relevant for the purposes of the preceding sub-section shall be assessed with particular regard:
>
> (a) to the measures, if any, taken by the person depositing the waste, or by the owner or occupier of the land, or by others, for minimising the risk; and
>
> (b) to the likelihood of the waste, or any container in which it is deposited, being tampered with by children or others.

Exemptions

Sub-section (3) requires the Secretary of State to consider exempting cases where deposits are so small or uses of plant and equipment so innocuous or where adequate controls are provided by other enactment. This has been done and the list of exempted cases is prescribed in SI 1976 No. 732 [15] as amended by SI 1977 No. 1185 [16]. (In Scotland, the relevant Regulations are SI 1977 No. 2006 [17].) Circulars 55/76 [18] and

79/77 [19] are relevant. Thus, the following cases relating to disposal of wastes are not subject to the licensing provisions:

(a) waste produced in the course of constructing, improving, repairing or demolishing any building or structure is deposited or disposed of on a site being used, or about to be used, for the construction, improvement or repair of a building or structure, provided always that the deposit or disposal is made by, or with the consent of, the occupier of the site;

(b) waste ash is deposited or disposed of on a site being used, or about to be used, for the construction or improvement or repair of a building or structure, provided always that the deposit or disposal is made by, or with the consent of, the occupier of the site;

(c) waste produced in the course of demolishing a building is deposited on the site of such demolition;

(d) spent railway ballast is deposited on operational land belonging to the British Railways Board;

(e) waste produced in the course of dredging operations for the purpose of land drainage or the maintenance of a watercourse, is deposited along the banks of a watercourse;

(f) waste produced in the course of maintaining any park, sports field, public garden or other recreation ground is, by, or with the consent of, the occupier thereof, deposited or disposed of within the boundaries of the ground in which it is produced;

(g) waste is deposited or disposed of for the sole purpose of research into the effect of waste on the natural environment or, as the case may be, into the performance of plant or equipment designed or adapted to deal with waste;

(h) waste is deposited directly on land for a period not exceeding one month by, or with the consent of, the occupier of the land, other than waste temporarily deposited at a site specifically designed or adapted for the reception of waste with a view to its being disposed of elsewhere;

(i) waste is deposited in a receptacle that has been provided or adapted for the reception of waste with a view to that waste being disposed of elsewhere, provided always that the deposit is made by, or with the consent of, the owner of the receptacle;

(j) waste is disposed of on the site on which it is produced by means of static plant with a disposal capacity of not more than 200 kilogrammes per hour;

(k) waste is disposed of as an integral part of the industrial process that produces it;

(l) waste is deposited by means of a line of pipes:
  (i) on the foreshore, or
  (ii) on any land above high-water mark which is covered by the sea from time to time.

However, exceptions (a)–(j) inclusive do not apply where the waste is poisonous, noxious or polluting and its presence is liable to give rise to an environmental hazard. Again, 'environmental hazard' is prescribed in this respect as:

> if the waste has been deposited or disposed of in such a manner, and in such quantity (whether that quantity by itself or cumulatively with other deposits of the same or different substances) as to subject persons or animals to material risk of death, injury or impairment of health, or as to threaten the pollution or contamination (whether on the surface or underground) of any water supply; and where waste is deposited or disposed of in any receptacle, whether sealed or not, this shall not of itself be taken to exclude any risk which might be expected to arise if the waste were not so deposited or disposed of.

There is a potential loophole in exemption (i) in that no maximum size is specified for the receptacle and so a multi-thousand-gallon storage facility may claim exemption from licensing.

However, recent moves suggest that the DOE are considering an amendment so as to set a maximum volume of 10 m^3 for any one receptacle or 20 m^3 for any group within the same curtilage: such an amendment would, therefore, re-inforce the intention as stated in Circular 55/76 that the exemption is provided for smaller bins, skips and the like and not for vast storage complexes.

Similarly, there is a loophole in exemption (h) in that waste storage/transfer operations in simple yards or compounds may claim exemption from licensing controls by virtue of absence of any specific design or adaptation for the reception of waste. Likewise, it is believed that the DOE is considering amending this category so as to encompass all waste transfer stations within the scope of the licensing provisions.

Site licence applications

Section 5 relates to the mechanics of the application process for a disposal licence. Sub-section (2) provides that a disposal licence shall not be issued for a use of land, plant or equipment for which planning permission is required unless such permission is in force and provides for the making of Regulations concerning concurrent planning and site licence applications. No such Regulations have yet been made.

There is no statutory form on which to make a licence application but such applications must be in writing and must include the following prescribed information (SI 1976 No. 732) [15]:

(a) a map showing its location;
(b) the full address of that location;
(c) a plan showing its layout;
(d) the form of deposit or disposal for which the licence is being sought;
(e) the types and estimated quantities of controlled waste it is proposed to deposit or dispose of, and
(f) details of any planning permission under the Town and Country Planning Act 1971 which has been granted in respect of the use which is the subject of the application.

Where a disposal authority receives an application for a disposal site licence for a use of land, plant or equipment for which planning permission is in force, it is the duty of the authority not to reject the application except on grounds of preventing water pollution or danger to public health. It follows that a disposal licence issued for a site for which planning permission is needed but has not been obtained is void.

Consultations

Before a disposal authority can issue a licence, it must refer its proposal to the relevant water authority and the collection authority and to prescribed persons (SI 1976 No. 732), namely the Health and Safety Executive and, in cases relating to waste disposal via shafts, galleries, wells, boreholes or pipes into fractures, fissures or intergranular pore-spaces in geological formations, the Institute of Geological Sciences. Twenty-one days are allowed for consultation, unless extended in writing by agreement, and a disposal authority must consider, but may disregard, representations from consultees except that if a water authority requests refusal of an application or disagrees with the prospective conditions then either authority may refer the matter to the Secretary of State for a decision. In effect, if a water authority feels strongly enough then it has a virtual power of veto either by its own persuasive argument or by reference to the Secretary as a last resort.

Telling lies!

A person giving false or reckless statements in an application is guilty of an offence with a penalty of a fine of up to £1000 on summary conviction or, on conviction on indictment, to imprisonment for up to two years and/or an unlimited fine.

Licence conditions

Section 6 provides that Regulations can be made specifying conditions which are or are not to be included on licences. No such Regulations have yet been made but sub-section (2) goes on to list various matters to which conditions may relate:

(a) the duration of the licence;

(b) the supervision by the holder of the licence of activities to which the licence relates;

(c) the kinds and quantities of waste which may be dealt with in pursuance of the licence or which may be so dealt with during a specified period, the methods of dealing with them and the recording of information relating to them;

(d) the precautions to be taken on any land to which the licence relates;

(e) the steps to be taken with a view to facilitating compliance with any conditions of such planning permission as is mentioned in sub-section (2) of the preceding section;

(f) the hours during which waste may be dealt with in pursuance of the licence; and

(g) the works to be carried out in connection with the land, plant or equipment to which the licence relates, before the activities authorised by the licence are begun or while they are continuing.

In any event, a disposal authority can impose any condition as 'it sees fit to specify'. However, by analogy to planning law, any condition must satisfy each of three criteria:

(i) it must be reasonable;

(ii) it must not be '*ultra vires*' (i.e. it must relate only to land within the control of the licence holder);

(iii) it must be capable of being enforced by the means envisaged in the condition,

except that a condition may require the carrying out of works notwithstanding that the licence holder is not entitled as of right to carry out the works.

Public registers

Sub-section (4) requires that each disposal authority maintains a public register of all issued licences and affords members of the public reasonable copying facilities. Prescribed information to be kept on the register is (SI 1976 No. 732):

(a) the date of the granting of the disposal licence;
(b) the full name and address of the holder of the licence;
(c) the full name and address of the local representative (if any) of the holder of the licence;
(d) the location of the site to which the licence relates;
(e) the form of deposit or disposal to which the licence relates;
(f) the types of waste of which the deposit or disposal is authorised by the licence, and any limitation as to quantity specified therein; and
(g) the conditions (if any) attached to the issue or variation of the licence.

Determination of applications

Sub-section (5) stipulates a two-month period for determination of licence applications which may be extended by written agreement. In the absence of agreement to extend the time limit, then a non-determination is classed as a deemed refusal.

Modifications/revocations

Section 7 concerns variation and revocation of licences. An authority may modify a licence either on its own initiative or on application from the licence holder. Additionally, it has a duty to modify a licence where it seems that such modification is necessary to prevent water pollution or danger to public health or detriment to the amenities of the locality. Where an authority receives an application for modification of a licence, it may disregard the consultation provisions of Section 5(4) if it is satisfied that a particular consultee would not be affected by a proposed modification.

If an authority is satisfied that a mere modification to a licence would not be sufficient to prevent water pollution or danger to public health or serious detriment to the locality then it has a duty to serve a notice of revocation stating the time such notice takes effect.

Transfer provisions

Section 8 provides a mechanism for transfer of licences and powers of authorities to decline to accept prospective transferees. It also allows for the cancellation of licences by licence holders.

Supervision of licensed activities and sites

Section 9 requires the disposal authority to supervise licensed activities and take all such steps needed:

(a) for the purpose of ensuring that the activities to which the licence relates do not cause pollution of water or danger to public health or become seriously detrimental to the amenities of the locality affected by the activities; and
(b) for the purpose of ensuring that the conditions specified in the licence are complied with.

Authorised officers of a disposal authority are empowered to carry out works by reason of emergency and the authority may recover expenditure so incurred.

Where it appears to an authority that a licence condition is not being complied with then it may:

(a) serve on the licence holder a notice requiring him to comply with the condition before a time specified in the notice; and
(b) if in the opinion of the authority the licence holder has not complied with the condition by that time, serve on him a further notice revoking the licence at a time specified in the further notice.

Section 10 relates to appeal provisions for cases where:

(a) an application for a disposal licence or a modification of a disposal licence is rejected; or
(b) a disposal licence which specifies conditions is issued; or
(c) the conditions specified in a disposal licence are modified; or
(d) a disposal licence is revoked.

It is of interest for cases falling within (a) that an application for a licence is deemed to be rejected upon expiry of the two-month time limit (Section 6(5)) but that no such time limit applies for applications for modification(s) to an existing licence. It might, therefore, be worthwhile for a prospective applicant who is seeking even a simple modification to lodge a completely new licence application!

Appeal provisions

Where an appeal is lodged against a licence modification (case (c)) or revocation (case (d)) then the notice of modification or revocation under Section 7 or 9(4)(b) is ineffective unless such notice includes a statement to the effect that 'it is necessary for the purpose of preventing pollution of water or danger to public health'. But, if on application of the licence holder, the Secretary of State determines that the authority acted unreasonably, then the holder is entitled to recover compensation from the authority.

Regulations (SI 1977 No. 1185 [16]) prescribe that an appeal must be made within six months of a decision, or deemed rejection and must normally include:

> two copies of a statement of the reasons for his appeal and two copies of any or all of the following documents:
> (a) the application, if any, to the disposal authority for a disposal licence or for a modification of such a licence;
> (b) any relevant plans, drawings, particulars and documents submitted to the disposal authority in support of the application;
> (c) any relevant record, consent, determination, notice or other notification made or issued by the disposal authority;
> (d) any relevant planning permission in force under the Town and Country Planning Act 1971;
> (e) all other relevant correspondence with other authorities.

The Secretary of State will decide whether to determine an appeal by written representations or by local inquiry.

Council-owned premises

Section 11 relates to land occupied by disposal authorities and if such land is to be used for waste disposal purposes then a 'resolution' must be passed by the authority, being the equivalent of a licence. Planning consent, if required, must first be obtained. Consultations with statutory bodies, the same as for licences, are obligatory. The conditions attached to a resolution are supposed to mirror those of licences. A water authority may request the Secretary of State to direct a disposal authority to discontinue activities pursuant to a resolution if it considers that such activities are causing or are likely to cause pollution of water.

Power for removal of wastes

Section 16 provides for powers relating to the removal of wastes deposited in breach of licensing provisions. A disposal authority may, by notice, require the removal of waste which it believes to have been deposited in contravention of Section 3(1) within a period of not less than twenty-one days or may require the elimination or reduction of the consequences of a particular deposit or may require both. The recipient of such a notice may appeal within twenty-one days to a magistrates' court against the notice and the court must quash the notice if:

(a) the appellant neither deposited nor caused nor knowingly permitted the deposit of the waste on the land; or
(b) service of the notice on the appellant was not authorised by the preceding sub-section; or
(c) there is a material defect in the notice.

A notice under appeal is ineffective during the time of appeal.

Failure to comply with a notice, unless under appeal, is an offence subject to a fine of up to £400 and a further fine of up to £50 for each day the failure continues. An authority may itself undertake the work envisaged in a notice and recover costs. Additionally, and without service of a notice, an authority may undertake removal of waste, or other works, in order to prevent pollution of water or danger to public health and recover costs incurred in so doing from the occupier of the land or the person who deposited or caused or knowingly permitted the deposit.

Practical assessment

Having discussed the legislative provisions of Sections 3–11 and 16 regarding site licensing, it is germane to discuss their practical application in two sections: firstly, the mechanics of site licensing and, secondly, the decisions of the Secretary of State thus far on licensing appeals. The reader is referred to Willetts [20, 21] for a discussion on the practical effects.

Mechanics of site licensing

Reference must be made here to *Waste Management Paper No. 4* – [10] 'The Licensing of Waste Disposal Sites' and Department of the Environment Circular 55/76 (W/O Circular 76/76) [18].

The stated purposes of the licensing system are:

(a) to ensure that waste treatment and disposal is carried out with no unacceptable risk to the environment and to public health, safety and amenity;
(b) to put at a suitable local level the responsibility for deciding what conditions should be imposed at a given site, so that local circumstances can be taken full account of;

(c) to ensure that changing patterns of waste disposal do not prejudice objective (a) above and equally that those responsible for waste treatment and disposal take proper advantage of technical progress;
(d) to give waste disposers a clear idea of what operating standards are required of them;
(e) as a result of (d) above to secure the provision of sufficient facilities for the treatment and disposal of waste;
(f) to ensure in the interests of proper allocation of national resources, that unnecessarily high disposal standards are not demanded;
(g) to ensure that sufficient information is available to the responsible authorities to enable them to fulfil their statutory duties.

As has been said earlier, there is no statutory form on which to make a licence application. However, most authorities have adopted the format suggested in the Circular and so the information required by each authority is common. Basic details of site ownership, location, size, type, planning status, etc. are required, together with an estimation of the types and quantities of waste expected to be disposed at the facility. 'Difficult' wastes are itemised separately in the application and reference usually made to the coding system suggested by the Department of the Environment and reproduced in Table 1 of Chapter 1.

Application details

The more important part of the application concerns the site working/operational plan and the statement of intent.

For landfill sites the plan must show:
(i) gates and boundary fencing;
(ii) direction of working and phasing of filling/restoration;
(iii) emergency tipping areas;
(iv) location of cover material stocks;
(v) location of primary site road;
(vi) location of site office and other fixed buildings;
(vii) location of tanks or lagoons used for storage of wastes;
(viii) existing contours;
(ix) proposed final contours;
(x) existing drainage facilities;
(xi) location of watercourses, adits, shafts, etc.,

and the 'statement of intent' must include:
(i) site preparation works;

(ii) drainage and outfalls;
(iii) record-keeping provisions;
(iv) litter screens;
(v) lighting equipment;
(vi) maximum width of working face;
(vii) methods of handling difficult wastes;
(viii) pest control measures;
(ix) water sampling and monitoring;
(x) site security;
(xi) final restoration;
(xii) waste sampling provisions;
(xiii) number of operatives;
(xiv) inventory of equipment;
(xv) geological statement.

For treatment plant the requirements are slightly different to take account of the fact that wastes are being 'treated' prior to disposal in mechanical plant.

The guidelines for the consideration of an application include:

(i) The proposed facility:
past history;
position in relation to other development;
site conditions (hydrogeology, aquifers, rainfall, site works, etc.);
treatment plant/transfer stations.

(ii) The waste:
type of waste;
quantity of waste;
mix of waste.

(iii) Alternative means of disposal.

(iv) Interaction with other legislation:
planning;
radioactive substances;
water authority;
controls;
health and safety.

Some 37 model conditions for landfill sites are presented in *Waste Management Paper No. 4* [10] with an additional three where pulverised fuel ash is involved. The disposal authority can pick and choose or adapt and adopt from this list: also, it has unrestricted powers to impose what other conditions it thinks fit.

For treatment plant, twelve model conditions are suggested.

Site licence appeals

A number of appeals has been determined to date and a brief summary is presented here. These decisions are important pointers to the way in which the Department of the Environment and, ultimately, the courts, will seek to administer and interpret the legislative and other provisions for site licensing.

(1) Staffordshire County Council v. Severn–Trent Water Authority (STWA)
(18 May 1978, Ref. WD/JB/2)

Site
Ounsdale Quarry, Wombourne, Wolverhampton. Sand quarry. Eight hectares. Upper Mottled Sandstone of Bunter Series forming part of main Triassic aquifer. Batch pumping station 200 metres to north yielding 11.36 million litres per day.

Dispute
Staffordshire County Council proposed to issue a licence for tipping. STWA referred matter to the Secretary of State (Section 5(4) of the Act).

STWA case
That a licence should be refused on the grounds of:

(i) proximity of site to abstraction wells;
(ii) nature of wastes and leachate composition;
(iii) risk of fly-tipping if site opened.

STWA presented evidence on leachate composition from foundry sand and that leachate would migrate against direction of groundwater flow and in the direction of the wells because of cone of influence from pumping.

Waste Disposal Authority case (WDA)

(i) The site was extremely valuable and there was a lack of landfill sites;
(ii) Contaminated foundry sand could be segregated at source.

Determination

(i) The evidence on groundwater movement was not conclusive.
(ii) The leach tests used for foundry sand were not realistic.
(iii) The waste segregation at source was not adequate.
(iv) There was a lack of relevant evidence from both parties.
(v) There was criticism of the fact that neither party referred to the Department of the Environment document 'Hazardous Wastes in Landfill Sites'.

Conclusions

(i) Total prohibition of the use of the site was not justified so long as phenolic sand, etc. could be guaranteed to be segregated.

(ii) STWA concern about fly-tipping was unfounded as such material could be removed.
(iii) The wells were very close to the site.
(iv) STWA concern over control of waste segregation was justified.
(v) Hydrogeological evidence was inconclusive; therefore, the possibility of migration towards wells had to be considered.

Decision

(i) Staffordshire County Council was directed not to issue a licence.
(ii) It was noted that this decision did not affect a concurrent appeal by the licence applicant.

(2) J.F. Somerset v. West Sussex County Council
(24 May 1978, Ref. WD/APP/HE/27)

Site

Castle Goring Farm. Used for ten years prior to licensing. Two years' life on licence application. Nine months' life remaining.

Appeal

Nineteen conditions appealed against, fourteen of which were drawn from *Waste Management Paper No. 4* and five additional ones:

(i) Working plan requirement. WDA was subsequently satisfied.
(ii) Tarmac access road required. The Inspector said no.
(iii) Public highway to be kept clean. County Surveyor would sweep the road as necessary.
(iv) Site notice. Appellant agreed.
(v) Site office, toilet, etc. The Inspector said a mobile caravan without toilet and without telephone was satisfactory as this was a short-life site.
(vi) Clay bund required. This condition was subsequently met.
(vii) Site roads. The Inspector said there was room for improvement.
(viii) Litter screens. The Inspector said these were necessary if it was windy.
(ix) Fencing. Inspector said this should be at the appellant's discretion and not a licence condition as the site was well away from the main road.
(x) Uncovered waste at 1 August 1977. WDA agreed this was superfluous.
(xi) Boundary marker posts. The Inspector said these were required.
(xii), (xiii), (xiv), (xv) Wastes to be compacted, covered, large items

crushed and no fires. The Inspector said this condition should be enforced.

(xvi) Record keeping. Appellant agreed.

(xvii) Half a metre final cover. Appellant agreed.

(xviii) Tidiness and weed control. The Inspector said this was to be left to the appellant's discretion as he cultivated completed areas.

(xix) Skips and bins to be stored in approved positions. The Inspector said the appellant required his own orderliness and this should not be a condition.

Decision

(i) Conditions (ii), (v), (vii), (ix), (xviii) and (xix) were deleted.

(ii) Condition (x) was deleted as superfluous.

(iii) All other conditions stood.

(3) Wold Newton Parish Council v. Humberside County Council
(17 August 1978, Ref. WD/APP/HE/34)

Site

Parish Council tip. Stone pit used by Parish.

Appeal

Nine conditions on the grounds of impracticability and expense:

(i) Access to be gated.

(ii) Site inspection once/week, records of inspections, unusual deposits to be notified.

(iii) Identification board including types of waste allowed.

(iv) Wooden post and wire mesh fence to be provided.

(v) Hollow containers to be crushed, broken or filled.

(vi) Once per month litter collection, to keep tidy.

(vii) No deliberate fires and any fires to be fought and reported.

(viii) Rubbish to be covered and graded.

(ix) Parish Council not to allow sorting-over of rubbish without WDA approval.

Conclusions

(i) Extra expenditure would be necessary to bring the site up to the standard specified in the licence.

(ii) The Parish Council were over-estimating such expenditure.

(iii) The Parish Council ought to be able to communicate the rules of tipping to likely users without undue effort.

(iv) The present standard of operation was in need of improvement.

(v) The site was an asset to the locality and if it were not in existence the County Council would have to provide an alternative.

Decision

The entire appeal was dismissed as all the conditions were considered reasonable and appropriate.

(4) E & A Metals (Wigan) Ltd. v. Lancashire County Council
(17 August 1978, Ref. WD/APP/HE/42)

Site

Transfer station/compactor at Skelmersdale. Steel-framed, corrugated sheet clad building housing refuse compactor.

Appeal

Against condition relating to fencing and gating requirement, specifically the section requiring gates to be locked when the site was unattended.

Appellant's case

(i) The site was in constant use throughout the day.
(ii) The compactor is key operated with each driver having his personal key: thus, there was no possibility of illegal use.
(iii) The site was in an isolated area.
(iv) It was impractical and unduly expensive to employ an attendant.
(v) It was impossible to force drivers to lock gates after they had left the station.
(vi) Locking of gates did not prevent access to anyone who really wanted to gain access.

WDA case

(i) The site was not isolated.
(ii) Unauthorised and possibly hazardous, waste could be tipped into the charge box and would then be compacted along with authorised waste, unnoticed.
(iii) Fire risk.
(iv) There was a risk of injury to children who might enter the station and hide when a vehicle approached and used the compactor.
(v) Permanent manning would not be suggested.

Conclusions

(i) A site visit revealed evidence of vandalism and, hence, unauthorised access.
(ii) It was held to be impossible to comply with the condition as lorry drivers were outside the direct control of site operators and

the County Council did not wish for permanent supervision. Nevertheless, the locking of gates was a sensible precaution.

Decision

The appeal was dismissed and the condition stood.

(5) Douglas Holton Ltd v. Derbyshire County Council
(17 November 1978, Ref. WD/APP/HE/29)

Site

High Peak Silica Quarry, Buxton. 550 yards × 200 yards with three separate holes, each between 30 and 50 feet deep.

Appeal

Against the condition applicable to two of the three holes requiring tyres and battery cases to be restricted to five-metre layers and covered with incombustible material each day.

WDA's case

Previous fire on the site and on a neighbouring site.

Appellant's case

(i) The previous fire was unconnected with the present tipping practice and the waste that caused the fire was no longer taken.
(ii) The third hole did not have this condition.
(iii) The financial burden of importing extra cover would close the site down.
(iv) The site fulfilled a national need.
(v) Suggested an alternative of ten-metre layers to reduce the requirement for cover material. The County Council contended that ten-metre layers would present difficulties in covering the whole face.

Conclusions

(i) A site visit revealed indiscriminate dumping of tyres and evidence of recent fires.
(ii) The County Council were well aware of the national difficulties if the site was closed down and had drafted a condition that was much more lax than it would otherwise have been.

Decision

The appeal was dismissed and the condition stood.

(6) Low Moor Land Reclamation Ltd v. Derbyshire County Council
(15 January 1979, Ref. WD/APP/HE/23)

Site

Low Moor Farm, Parwich. Area of limestone deposits with several 'sink holes' partially filled with clay and silica sand. Artificial watercourse,

Meerbrook Sough, some eight miles away collects groundwater and is a major public supply. Three excavated sink holes form the site, each 100–150 yards diameter and ten yards deep. The site has been used for waste disposal since 1966.

Appeal

Originally against all 32 licence conditions but was subsequently restricted to one condition, namely

Types and quantities of waste each day:

1. Construction industry – unlimited
2. Solid polymeric scrap – 20 tonnes
3. Asbestos – 20 tonnes

All persons wishing to deposit waste at the site to inform County Surveyor that waste does not contain substances that are poisonous, noxious or polluting.

All persons wishing to deposit asbestos to inform County Surveyor of precautions to be taken.

This condition effectively prohibited liquid disposal.

Appellant's case

(i) Quoted 'Hazardous Wastes in Landfill Sites', saying ultra-cautious approach was unjustified.

(ii) Liquids had been deposited for many years and there was no evidence of pollution of Meerbrook Sough.

(iii) Produced a model giving wastes that could be accepted on the basis of a maximum permissible leachate from the raw waste. Said on-site testing could be used to determine the nature of wastes.

(iv) Contended that leachates would arrive at Meerbrook Sough in seven to ten days and the dilution factor was obviously large as no pollution had yet been detected.

WDA/STWA case

(i) The Limestone Series was not as stated by the appellants and leachate would take seventeen years to reach Meerbrook Sough.

(ii) The aquifer in question was of major present significance and was also scheduled for future development.

(iii) The dilution factor was large, as the appellant stated, but localised pockets of leachate may have existed.

(iv) Doubted whether on-site testing would have been adequate to screen wastes for leachate composition.

Conclusions

(i) Appellant and WDA held opposite views on leachate movement and neither could be proved. Future development of the aquifer was important so must err on the side of the WDA.

(ii) On-site analysis could not be adequate.

(iii) The approach of the WDA and STWA was not ultra-cautious.

Decision

(i) The appeal was dismissed and the condition stood.

(ii) A note was appended by the Secretary of State saying that the appellant may like to put more effort into demonstrating the safety of the site and then apply for an amendment to the licence.

(7) Martell, Hudson & Knight Ltd v. Staffordshire County Council (29 January 1979, Ref. WD/APP/HE/41)

Site

Sand Quarry, Hopwas. Partially worked out sand hole. Tipping took place prior to 1974 but was suspended when the Control of Pollution Act was introduced. A site licence application was submitted and the County Council requested an extension to the time limit. The appellants refused.

Appeal

Against the deemed rejection of application.

WDA case

(i) Heavy lorry movement in the village was unacceptable. Petition had been received from residents.

(ii) The County Council was aware that they could only refuse a licence where valid planning permission existed on grounds of water pollution or danger to public health. STWA said there was no danger of pollution. County Council Highways Department said that lorry movements were of no danger to public health. However, the Highways Committee considered lorry movements to be undesirable. Thus, the County Council did not issue a licence.

(iii) The County Council also questioned whether the original planning permission was valid.

(iv) A local enquiry was requested.

Appellant's case

(i) Contended that the County Council delayed solely because of local opposition and not on technical grounds. Said that the County Council was not permitted and not required to consult the public under the Control of Pollution Act consultation procedure.

(ii) The volume of lorry traffic on local roads was irrelevant to whether a licence should be granted to a waste disposal site or not.

(iii) Opposed a local enquiry; thought it was merely a platform for public opposition.

Conclusions

(i) Planning permission was valid.

(ii) The question of lorry movements did not affect operations of the site and such wider aspects were for planning and not licensing consideration. Since planning permission existed, it was up to the planning authority to take action on this point but, in any event, considerations of site licence application were unaffected.

(iii) A public enquiry could not have adduced any further evidence.

(iv) There was no danger of water pollution.

(v) There was no danger to public health.

Decision

The appeal was upheld and the County Council instructed to issue a licence.

(8) Sussex Tipping Ltd v. West Sussex County Council
(30 May 1979, Ref. WD/APP/HE/65)

Site

Old Mead Road, Littlehampton. Planning permission for land raising had been granted in July 1975. Non-toxic wastes had been used since then. A licence with six conditions was issued in June 1977. The Company requested a modification to allow liquid waste disposal and the licence was duly modified.

In January 1978, the County Council served notice alleging breach of two conditions, one specifying access road to be tarmacked and properly drained and one requiring a wheel washer to be installed. The notice required compliance.

In March 1978, the County Council revoked the licence on the grounds of non-compliance with the January 1978 notice.

Appeals

April 1978: Against the wording of the condition but this was withdrawn when the County Council agreed to re-wording to delete tarmac requirement and just require a reasonable road surface.

April 1978: Against revocation of licence.

Appellant's case

(i) No complaints had been received about any mud being carried onto public highway.

(ii) Road had been made up to a reasonable condition without tarmac.

(iii) The access road was jointly used by others, particularly the Southern Water Authority who had taken considerable amounts of excavation spoil along it and had fouled the public highway.

(iv) A wheel wash system was installed in 1977 but removed after the County Council requested re-siting. However, it was not re-installed in anticipation of mains water being available and new washers to be purchased. April 1978: Refitted a non-mains washer since mains water was not, after all, available at the original siting, with County Council approval.

(v) The access road was outside the operational and licensed site; thus, maintenance of it was subject to a planning condition. It was not the appellant's property and he had no right to repair it – joint usership.

(vi) Contended that revocation of licence under Control of Pollution Act legislation had been improperly used and that Town and Country Planning legislation was appropriate to the state of the road. The latter legislation includes compensation clause: the Control of Pollution Act does not.

WDA case

(i) Revocation was justified at the time but subsequently the appellant had improved the road, etc.

(ii) The County Council therefore wished to withdraw their objection to the appeal, so long as no hearing would be held and no order as to costs would be made against them.

Decision

(i) The appeal was upheld and the County Council was instructed to withdraw the notice of revocation.

(ii) On the question of costs, the Secretary of State noted that he had no powers to award costs unless a hearing under the Local Government Act 1972 [22] was held.

(9) Cook, Lubbock & Co. v. Northants. County Council
(26 June 1979, Ref. WD/APP/HE/63)

Site

Storefield Lodge Farm, Rushton. Former ironstone gullet. Planning application in March 1973 to infill with non-notifiable waste as per list agreed with the former Welland & Nene River Authority was rejected. There was an appeal and the Secretary of State granted permission in June 1975.

A site licence was applied for and the County Council refused.

Appeal

Against the refusal of the County Council to issue a licence.

Discussions with the County Council led to the submission of a list of materials to be deposited. Identical to the list on the original planning application plus 'such other liquid waste and sludge as the disposal authority and the Anglian Water Authority will, from time to time, allow'.

A licence was issued in February 1978 restricting wastes to the original list and not allowing the additional range.

Appellant's case

(i) Planning permission was for a waste disposal site and a variation in material to be deposited did not constitute a material change of use.

(ii) Types of waste to be deposited were subject to licensing control which had superseded planning control.

WDA case

(i) Planning permission was granted, on appeal, for disposal of waste in accordance with the planning application and submission, i.e. the accompanying list.

(ii) The planning authority may need to maintain control over types of waste to be deposited on grounds other than water pollution or public health.

Conclusions

(i) Planning permission was granted prior to implementation of site licensing provisions. Attention was drawn to Circulars 5/68 [23] and 55/76 [18] which set out the principle that planning conditions should not cover items for which control under other legislation exists.

(ii) Concluded that permission did relate only to the list of materials as submitted.

Decision

The appeal was dismissed and the licence condition stood.

(10) Robinson Bros. v. West Midlands County Council (13 August 1979, Ref. WD/APP/HE/77)

Site

Chemical Works, Phoenix Street, West Bromwich. Two boreholes sunk (in 1930 and 1964) for waste disposal and also an incinerator. The County Council issued a licence for the incinerator and for one borehole in August 1978. Consultations with the National Coal Board (NCB) on the licence application elicited the response that the voids belonged to the NCB. One borehole was in use prior to 1938 and the NCB therefore had no grounds for objection but the later one did not have NCB consent.

Appeal

Against the County Council refusal to issue a licence for the second borehole.

Appellant's case

The NCB had not produced evidence that voids in question were left as result of the mining of coal.

WDA case

Suggested that the use of the borehole amounted to use of land, plant or equipment for which planning permission was required and that such permission was not in force. Thus, they could not grant a licence.

Conclusions

Planning permission was not required for the sinking of the boreholes since:

(i) Such sinking was now immune from enforcement action, and

(ii) No material change of use was involved since the original planning permission for the whole site was chemical works without significant variation before or after sinking of the borehole.

(iii) Lawful possession of land was not a pre-requisite for obtaining a disposal licence. There was no need for a person to occupy land before he could apply for a licence. Section 8(1) allowed transfers of licences irrespective of occupation. Even so, the appellant did occupy the land in question and if a licence was granted there could be no action taken under Section 3 of the Control of Pollution Act unless the appellant deposited waste on land outside its occupation.

Decision

The appeal was upheld and the County Council instructed to issue a licence.

(11) J & A Jackson Ltd v. Cheshire County Council
(3 September 1979, Ref. WD/APP/HE/86)

Site
Jacksons Brickworks, Rixton, Warrington.
Appeal
Against the refusal of the County Council to issue a licence on the grounds that planning permission was required but did not exist.
WDA case
The site involved a use of land, plant or equipment for which planning permission was required.
Conclusion
Planning permission was required.
Decision
The appeal was dismissed as Section 5(2) required the County Council not to issue a licence as no valid planning permission existed.
Notes
The Secretary of State invited both parties to consider General Development Order provisions. The County Council made no reply. The appellant said he would seek Court ruling to establish the planning status.

The Secretary of State advised the appellant to regularise planning status either by Court decision or application and issue of required permission and then to re-apply for a site licence.

(12) IMI Ltd v. West Midlands County Council
(3 September 1979, Ref. WD/APP/HE/75)

Site
Kynoch Works, Witton, Birmingham. In-house site for disposal of ash, rubble,sweepings and general factory waste in use before 1948.
Appeal
Against licence condition specifying records to be returned once per month, etc.
Appellant's case

(i) Had no objection to keeping of records in principle but contended that model Department of the Environment conditions, although specifying records to be kept did not specify returns of records.
(ii) The waste deposited was general factory rubbish and records would be of no significance to the County Council.
(iii) The condition was not based on grounds of health, safety or

amenity but because the County Council had a computer and therefore needed to use it.

WDA case

(i) The Department of the Environment model conditions were for guidance only.

(ii) The records were essential to help the County Council in the 'onerous duties imposed on them by the Control of Pollution Act' in terms of their survey and plan.

(iii) A computer was essential for processing survey and records data for meaningful predictions in the plan.

Conclusions

(i) Licensing provisions were essentially intended to ensure waste disposal causes no water pollution and no danger to public health. The County Council was empowered to impose conditions to ensure these ends and the uniqueness of sites was to be taken into account. However, implicit in the legislation is that any condition must be reasonable, i.e. appropriate to the above objectives, and not impose unreasonable burden on the licensee.

(ii) Site visits were the proper control mechanism and not computer processing of data.

Decision

The condition was re-written to provide for a summary of records to be submitted not more frequently than once per year.

(13) R.C. Loosmore v. Dorset County Council
(27 September 1979, Ref. WD/APP/HE/69)

Site

Manor Farm, Charmouth. Pits had been dug in the field, one at a time, to take waste from camp/caravan site and small amounts of farm waste. A pit usually lasted for two seasons and was then covered over and another dug.

Appeal

Against twenty conditions:

(i) Tanks to be bunded.

(ii) Watercourses diverted, culverted or protected.

(iii) Drainage from higher land to be diverted.

(iv) Site roads to be maintained to allow all weather operation.

(v) Site to be manned and supervised.

(vi) Waste to be compacted not later than end of each day.

(vii) Compaction equipment to be employed.

(viii) Depth of layers 2.5 metres maximum.

(ix) Tractor to be used either over-face or up-face.

(x) Faces and flanks to be 1 in 3 gradient maximum.
(xi) Putrescible wastes to be buried each day.
(xii) Hollow articles to be filled, crushed, etc.
(xiii) No waste other than inert into water.
(xiv) No fires.
(xv) Roads to be water-sprayed if dusty.
(xvi) 14 days' notice of commencement and recommencement, etc.
(xvii) Cessations in excess of three months to be notified.
(xviii) Lighting to be available when dark.
(xix) Records to be kept and returns made.
(xx) Width of face not to exceed County Council requirements as specified from time to time.

Appellant's case

Conditions (i), (ii), (iii), (iv), (v), (vii), (viii), (ix), (x), (xi), (xii), (xv), (xvi), (xvii), (xviii), (xix) and (xx) were inapplicable or unnecessary. Condition (xiii) was impossible when it rained and puddles formed. Condition (xiv) and (vi) regarding prohibition of burning and specified method of compaction were objected to as burning had been used as bulk reduction method of operation without compaction for thirty years.

WDA case

(i) Agreed conditions (i), (ii), (iii), (iv), (v) and (xv) were inappropriate.
(ii) Conditions (vi), (vii), (viii), (ix), (x), (xi), (xii) and (xx) were necessary.
(iii) Condition (xiii) was necessary to take into account disposal of polluted water, after possible flooding.
(iv) Conditions (xvi) and (xvii) were required for site inspection purposes.
(v) Condition (xviii) was fulfilled by vehicle headlamps.
(vi) Condition (xix) envisaged a simple summary.
(vii) Condition (xiv) regarding fires was essential due to complaints received by the WDA.
(viii) West Dorset District Council offered one free collection per week from caravan sites and 12.5p per bin from tent sites.

Conclusions

(i) Conditions (i), (ii), (iii), (iv), (v) – not appropriate.
(ii) Conditions (vi), (vii), (ix), (x) – not reasonable to scale of operations but reasonable to expect some compaction in the interests of subsequent stability.
(iii) Condition (viii) – unnecessary as the pits were only 2 metres deep.

(iv) Condition (xi) – putrescible wastes did not need covering.
(v) Condition (xii) – not necessary.
(vi) Condition (xiii) – depositing waste into water was not acceptable and the condition was appropriate for flooding possibility.
(vii) Condition (xiv) – combustion was ineffective at this site as it was below ground level.
(viii) Condition (xv) – not appropriate.
(ix) Conditions (xvi) and (xvii) – not appropriate as the operations were seasonal.
(x) Condition (xviii) – vehicle headlights were sufficient as per the County Council.
(xi) Condition (xix) – a summary was to be kept and sent in annually.
(xii) Condition (xx) – not appropriate as the size of the pit determined it.

Decision

(i) Conditions (i), (ii), (iii), (iv), (v), (vii), (viii), (ix), (x), (xii), (xv), (xvi), (xvii) and (xx) were to be deleted.
(ii) Condition (vi) – re-written to require wastes to be deposited in as compact a layer as is reasonable but without specifying equipment, etc. to be used.
(iii) Condition (xi) – re-written to require burial of putrescible wastes but without specifying tractor required.
(iv) Conditions (xiii), (xiv) and (xvii) – to remain unchanged.
(v) Condition (xviii) – re-written to specify vehicle headlights.
(vi) Condition (xix) – re-written to provide return of summary not more frequently than once per year.

(14) Wright Waste Recycling Ltd v. Salop County Council
(4 December 1979, Ref. WD/APP/HE/71)

Site

Farley Oil Terminal, Much Wenlock. Existing underground storage tank farm formerly used for aviation fuel storage.

Appeal

Against refusal to issue a site licence because of alleged absence of planning permission.

Appellant's case

Rights of existing use were transferred when the site was purchased from the Air Ministry in 1967. Hence, no planning permission was required.

WDA case

Bridgnorth District Council determined a Section 53 application that use of the site for storage was acceptable but a change in use because of oil treatment plant would require planning permission. The County could not, therefore, issue a licence.

Conclusions

No evidence was produced to prove the site had the required planning permission or that it was exempted from requiring any. The Council was, therefore, correct in not issuing a site licence.

Decision

The appeal was dismissed.

(15) Wessex Water Authority v. Somerset County Council
(24 March 1980, Ref. WD/JB/4)

Site

Landshire Lane, Henstridge. A disused limestone quarry, partially filled with liquid tannery and domestic wastes prior to 1971 and with general non-notifiable wastes until 1978 by Haul-Waste Ltd.

Dispute

A draft licence was circulated for consultation on 3 March 1978. In the meantime, Haul-Waste lodged an appeal because no determination had been made. Tipping therefore continued.

Wessex Water Authority requested a condition be included that made Haul-Waste liable if a nearby abstraction point became polluted. The Council replied that such a condition was unreasonable and the subsequent disagreement was referred to the Secretary of State (Section 5(4) of the Act) by the Water Authority.

County Council case

An Institute of Geological Sciences desk study was referred to which indicated migration towards the abstraction area. The cost of moving the abstraction points would be expensive and the establishment of proof of pollution difficult. The types of liquid waste to be tipped were non-hazardous and such an open-ended condition impossible and unreasonable to enforce.

Wessex Water Authority case

Two small abstraction wells were local and produced a quality supply. The Authority claimed that, although no pollution had occurred over the last twenty years, a monitoring borehole and long-term liability should be made conditional.

Determination
In favour of the County Council which was directed to issue a licence. The Secretary of State decided that a monitoring borehole was required but long-term liability was not to be a condition of the licence.

(16) T&S Element Ltd v. West Midlands County Council
(27 May 1980, Ref. WD/APP/HE/72)

Site
Hall Green Road, West Bromwich. Urban derelict sand/gravel quarry, size 350 metres × 250 metres. Previously partially filled. Boundaries have residential properties/factories bordering.

Appeal
A licence existed for construction industry waste, oil/water mixtures and sludges. Oil/water wastes were lagooned and then filled as necessary. Eight conditions disputed in the licence issued on 3 March 1978. Conditions included types and quantities, pH of sludges, a progressive reduction in liquid wastes until no more accepted, fencing and a completion date.

Appellant's case

(i) Many conditions applied to curtail the activities of the site and were not to prevent pollution.
(ii) One occasion of oil pollution was because of overflow to canal. Lagoons now moved to avoid this.
(iii) Fencing was expensive and would have to be phased.
(iv) A life expectancy could not be guaranteed because of uncertainty of obtaining fill materials.
(v) Many local residents had not objected to the site.

County Council case
Conditions reasonable because:

(i) Oil pollution to nearby canal caused by geological connection.
(ii) Fire hazards in the past.
(iii) Children on site because of lack of fencing.
(iv) Smells, flies, etc. complained of by local MP/residents.

Conclusions

(i) The phasing out of liquids was reasonable in view of the likely geology of the site.
(ii) The pH of sludges condition was now accepted by T&S Element.
(iii) Reduction in lagoons was found to be reasonable.
(iv) Fencing to be carried out as requested.
(v) Time limit dismissed but a recommendation was made to agree final levels.

Decision
The appeal was dismissed in relation to six conditions, appeal with respect to one condition withdrawn, one condition deleted (time limit).

(17) Hall Aggregates Ltd v. Berkshire County Council
(25 June 1980, Ref. WD/APP/HE/81)

Site
Smallmead Farm, Burghfield. A worked out sand/gravel quarry of 58½ acres. Dimensions ½ mile long, 300 yards wide and 4 yards deep. Location is between a railway and River Kennet. An adjoining area was already subject to a disposal licence.
Appeal
Objection to a licence raised by Thames Water Authority because the site was the subject of a proposed water storage area. The Council considered that the objection was not intended to prevent pollution and stated its intention to issue a licence: the Water Authority referred it to the Secretary of State and the Company appealed against non-determination during the statutory time limit.
Appellant's case
A licence should be issued because the Water Authority objection was not based on water pollution or danger to public health.
County Council case
A licence could not be issued because of an objection by the Water Authority.
Water Authority case
Supply of water to Reading would be threatened if proposed bankside reserves could not be made. They contended that a licence should not be issued until the outcome of their compulsory purchase/planning appeal in relation to the proposed reserves.
Conclusion
Water Authority case was not based on prevention of water pollution.
Decision
Appeal upheld and County Council instructed to issue licence.

(18) Bywaters (Leyton) Ltd v. Greater London Council
(11 August 1980, Ref. WD/APP/HE/94)

Site
Bent Marshall Depot, Skeltons Lane, London. A transfer station located near commercial premises, houses and a rail viaduct in North East London.
Appeal
The appellant disputed seven conditions of the issued site licence, dated 6

February 1979. The disputed conditions included times of operation, daily tonnage handled, gates, surfaces, wheel cleaning and special storage for asbestos.

Appellant's case

Unnecessarily restrictive for an established longstanding site.

County Council case

Local opposition, public health risk (asbestos and putrescibles), amenity and drainage. A local petition was also made.

Determination

(i) Saturday morning working would create noise – condition upheld.

(ii) No weighbridge on site but tonnages could be calculated from volumes. Proposed input was increased from 100 tonnes to 250 tonnes per day.

(iii) Small amounts of putrescibles permitted.

(iv) Fencing requirement to be subject of compromise.

(v) Drainage and wheel washing had already been complied with.

(vi) Asbestos condition modified to improve storage facility.

(19) Haul-Waste Ltd v. Devon County Council
(18 August 1980, Ref. WD/APP/HE/30)

Site

Higher Kiln Tip, Bampton, Devon. An old hillside limestone quarry used for filling since before 1939. Operated by Haul-Waste Ltd since 1972. Ground is fissured limestone and several springs appear below the quarry.

Wastes currently disposed of consist mainly of liquid industrial with some solids. A long, involved history of the site exists. During 1973–5 the Company sank seven monitoring boreholes around the site and oil pollution was detected in two of them.

Appeal

Appellant appealed against non-determination of licence application within two-month period. The Council resolved to oppose the appeal on the grounds that it would have rejected the application anyway for the purposes of preventing pollution of water.

During the appeal, the appellant applied for and was granted (Tiverton District Council) an 'established use certificate' for disposal of various types of industrial and other wastes.

County Council's case

(i) The established use certificate did not satisfy Section 5(2) of the

Control of Pollution Act that planning permission for the site had to be in force.

(ii) Since 1948, a material change in use had occurred because the superficial site area had increased (Section 22(3(b)) of the Town and Country Planning Act 1971) and because liquids had been deposited only since 1960.

Appellant's case

The established use certificate provided a clear and undisputable statement of the nature, extent and intensity of use.

Conclusions

(i) Section 5(2) of the Control of Pollution Act provides that a licence shall not be issued where planning permission is required unless such permission is in force. Section 5(2) is satisfied, therefore, either if permission is *not* required or if it is both required and in force.

(ii) In this case, no permission was in force and so 5(2) could only be satisfied if none was, therefore, required.

(iii) The Secretary of State was not satisfied that the superficial area of deposit had increased nor that the intensity, nature or extent of use amounted to a material change. Therefore, no planning permission was required.

(iv) The Inspector said that 'dilute and disperse' would seem to take care of the pollution leaking from the quarry and the Secretary of State noted a preference in terminology suggesting 'attenuate and disperse'.

Decision

Appeal upheld. Council directed to issue licence.

(20) B.P. Tyler v. Avon County Council
(4 September 1980, Ref. WD/APP/HE/91)

Site

Conygor Quarry, Avon.

Appeal

Against the refusal of the Council to issue a licence.

County Council case

Planning permission is required and is not in force.

Appellant's case

Various loads of waste had been deposited in the quarry since before 1964 and no enforcement action taken.

Conclusions
The Secretary of State said there was insufficient evidence that the site had been in use prior to 1 July 1948 to support the contention that permission was needed.
Decision
Insufficient evidence to determine the appeal until planning regularised either by specific permission or court ruling that none was required.

(21) B.P. Tyler v. Avon County Council
(15 January 1981, Ref. WD/APP/HE/93)

Site
Gordano Valley, Avon.
Appeal
Against refusal of Council to issue a licence.
County Council's case
Planning permission was required but none was in force. A 1955 permission to tip ash existed but had never been implemented. Material had been deposited at the site but the Council contended that this was merely for construction of a track for agricultural purposes.
Appellant's case
The deposited material was purchased specifically to activate the 1955 permission and was not merely for trackway construction.
Conclusions
The 1955 permission had lapsed and planning permission was required.
Decision
Insufficient evidence to determine the appeal until planning regularised either by specific permission or court ruling that none was required.

Sections 12–15

These Sections deal, principally, with the duties and powers of collection authorities to collect domestic and certain trade wastes and certain provisions relating to Scotland.

There is little point in examining these Sections here as the kinds of wastes with which they are involved and the fact that they relate only to collection and disposal authorities render them inappropriate as subject matter for this text.

Section 17: 'Special waste'

If the Secretary of State considers that controlled waste of any kind is or may be so dangerous or difficult to dispose of that special

> provision in pursuance of this sub-section is required for the disposal of waste of that kind by disposal authorities or other persons, it shall be his duty to make provision by regulations for the disposal of . . . 'special waste'.

This Section was brought into force in England and Wales on 1 January 1976 (SI 1975 No. 2118 [24]) and in Scotland mostly on 18 July 1976 (SI 1976 No. 1080 [25]), but partly on 1 January 1978 (SI 1977 No. 1587 [11]).

Regulations have now been made and were implemented on 16 March 1981 [26] amid quite a lot of furore. Readers are referred to discussions by Willetts [27, 28, 29, 30] for an outline history of the making of the Regulations. Also, the heated debates in the House of Lords and the House of Commons are of interest [31, 32].

The Regulations serve to replace the Deposit of Poisonous Waste Act 1972 [33] and are termed The Control of Pollution (Special Waste) Regulations 1980 (SI 1980 No. 1709 [26]). The 1972 Act was repealed with effect from 16 March 1981 by SI 1981 No. 196(4) [34]. Department of the Environment Circular 4/81 [35] (W/O: 8/81) explains the workings of the Regulations.

Purposes

There are two stated purposes for the Regulations: firstly, to fulfil the UK obligations under the European Communities' Directive on toxic and dangerous waste [36] and, secondly, to make provision for tighter control over the carriage of wastes which can seriously threaten life if disposed of heedlessly. This latter point is important: the Regulations are a transport manifest only, they do not provide for any control over disposal arrangements, which control is said already to exist by virtue of the site licensing provisions of Sections 3–11 of the Control of Pollution Act. Thus, comparison with the Deposit of Poisonous Waste Act 1972 is, perhaps, inappropriate since the 1972 Act sought to control disposal arrangements for poisonous, noxious or polluting wastes. Nevertheless, the provisions of the Regulations bear strong similarities to those of the 1972 Act.

Definitions

The new term 'special waste' is introduced which is not synonymous with 'notifiable'. 'Special' relates to potential hazards presented during transport and 'notifiable' related to characteristics and nature of import to a receiving site, i.e. whether poisonous, noxious or polluting. 'Special' is defined in Regulation 2 in conjunction with Parts I and II of

ig. 1. Schedule 1 of Special Waste Regulations. (Source: Reference 6].)

SCHEDULE 1

Regulation 2

Part I
Listed Substances

Acids and alkalis
Antimony and antimony compounds
Arsenic compounds
Asbestos (all chemical forms)
Barium compounds
Beryllium and beryllium compounds
Biocides and phytopharmaceutical substances
Boron compounds
Cadmium and cadmium compounds
Copper compounds
Heterocyclic organic compounds containing oxygen, nitrogen or sulphur
Hexavalent chromium compounds
Hydrocarbons and their oxygen, nitrogen and sulphur compounds
Inorganic cyanides
Inorganic halogen-containing compounds
Inorganic sulphur-containing compounds
Laboratory chemicals
Lead compounds
Mercury compounds
Nickel and nickel compounds
Organic halogen compounds, excluding inert polymeric materials
Peroxides, chlorates, perchlorates and azides
Pharmaceutical and veterinary compounds
Phosphorus and its compounds
Selenium and selenium compounds
Silver compounds
Tarry materials from refining and tar residues from distilling
Tellurium and tellurium compounds
Thallium and thallium compounds
Vanadium compounds
Zinc compounds

Part II
Meaning of "Dangerous to Life'

1. Waste is to be regarded as dangerous to life for the purposes of these regulations if—
   (*a*) a single dose of not more than five cubic centimetres would be likely to cause death or serious damage to tissue if ingested by a child of 20 kilograms' body weight or
   (*b*) exposure to it for fifteen minutes or less would be likely to cause serious damage to human tissue by inhalation, skin contact or eye contact.

Assessing effect of ingestion

2.—(1) The likely effect of ingestion is to be assessed by the use of reliable toxicity data in the following order of preference:—

Class 1: information about the effect of oral ingestion by children;
Class 2: data derived by extrapolation from information about the effects of oral ingestion by adults;

Fig. 3. Consignment note procedure. (Source: S.L. Willetts, unpublished.)

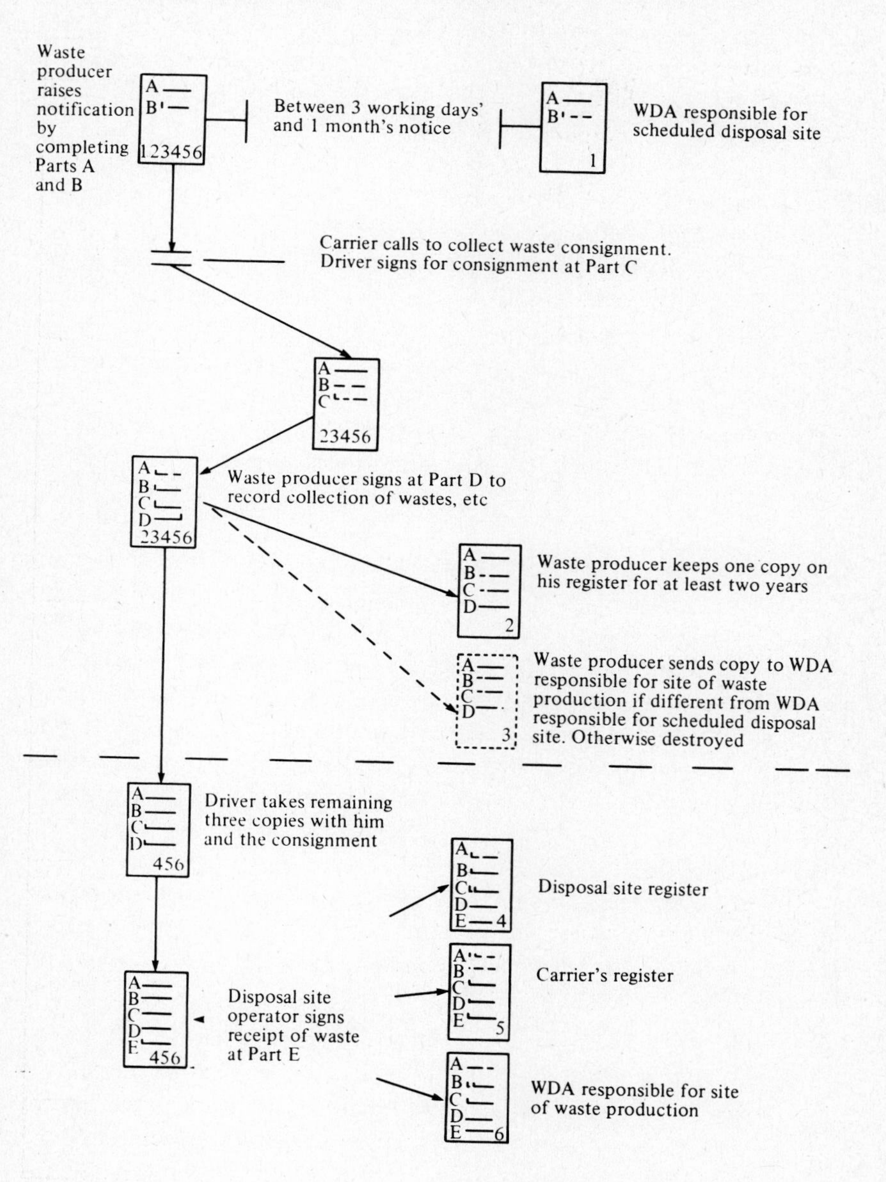

precautionary measures relating to his waste. The producer maintains one copy of this 'certificate of collection' on a statutory register and must send one to the authority responsible for his waste producing premises if, and only if, the waste is scheduled into the area of another authority: otherwise that copy may be destroyed.

Carriage and disposal

The driver carries three copies of the note with the waste consignment and the receiving disposal site clerk completes Box E confirming that the site is licensed to accept the waste as described. The three copies are then distributed and maintained on registers at the site, at the carriers' premises and at the authority responsible for the site of waste production.

Registers of transactions

Registers are maintained by the producer and carrier for a minimum period of two years and by the disposer for the life of the site.

The disposer also has to keep a record of locations of all deposits of 'special wastes' and cross refer such records to the register of consignment notes.

'Season tickets'

The informal 'season ticket' arrangements of the Deposit of Poisonous Waste Act have been formalised under Regulations 9–12 relating to regular waste arisings. Dispensation from the pre-notification provisions as relates to waste producers or from notification of receipt of wastes as relates to waste disposers can be conferred for regularly arising wastes. However, the transport element remains unaltered in that a completed consignment note still must accompany each and every consignment of 'special waste' regardless of whether or not regular arising dispensation has been granted.

The offence

Penalties for non-compliance with the consignment note and record-keeping systems are, on summary conviction, a fine of up to £1000 and, on conviction on indictment, to imprisonment for up to years and an unlimited fine.

Statutory defences are provided in Regulation 16:

> (3) In any proceedings for an offence under these regulations it shall be a defence for the person charged to prove that he took all reasonable precautions and exercised all due diligence to avoid the commission of such an offence by himself or any person under his control.
>
> (4) In any proceedings for an offence under regulation 4, 5 or 6 it shall be a defence for the person charged to prove that he was not

reasonably able to comply with the regulation in question by reason of an emergency and that he took all reasonable steps to ensure that the necessary copies of consignment notes were completed and furnished as soon as practicable after the event.

Radioactive wastes

Regulation 3 provides that the Secretary of State can define certain radioactive wastes as 'special wastes' if they satisfy the 'special waste' criteria in addition to their radioactivity.

The Regulations in practice

An important 'teach-in' was held at Loughborough University on 20 May 1981 [38] to examine the workings of the Section 17 Regulations and the proceedings of that one-day session will be forwarded to the Department of the Environment for consideration at the anniversary. Also, a one-day Symposium to examine the workings of Section 17 and to formulate amendments was held in London on 11 November 1981. The proceedings have been published by Oyez IBC Ltd and the reader is referred to the contribution by Willetts [39] for suggested amendments to the Regulations.

The remaining Sections

Section 18

Section 18 of the Control of Pollution Act empowers the Secretary of State to make Regulations bringing wastes other than 'controlled wastes' under the same provisions as relate to controlled wastes. Mine and quarry wastes and farm wastes are not defined as controlled and the Secretary can, therefore, prescribe these as being 'controlled' in certain situations. The section is in force, but no Regulations have yet been made.

Sub-section (2) provides that a person who deposits, or causes or knowingly permits the deposit, on any land 'any waste other than controlled waste in a case where, if the waste were controlled waste and any disposal licence relating to the land were not in force, would be guilty of an offence under Section 3(3) of this Act shall be guilty of such an offence and punishable accordingly'. That is, a deposit of waste, even if not 'controlled' in circumstances where:

'(a) the waste . . . is poisonous, noxious or polluting; and

(b) its presence . . . is likely to give rise to an environmental hazard; and

(c) it is deposited . . . for such a period that . . . may reasonably be assumed to have abandoned it.'

constitutes an offence with prescribed penalties.

South Yorkshire Metropolitan County Council successfully prosecuted the holder of a disposal licence for tipping piggery waste on his site. Although not a 'controlled' waste, it was held to be likely to cause environmental problems and the licence holder was fined.

Section 19

Section 19 allows disposal authorities to collect information about wastes that are not controlled.

Section 20

Section 20 allows disposal authorities to buy, reclaim or sell wastes.

Section 21

Section 21 allows disposal authorities to generate, from wastes, and sell heat and electricity.

Section 22

Section 22 relates to street cleaning.

Section 23

Section 23 relates to prohibition of car parking to facilitate street cleaning.

Section 24

Section 24 requires disposal authorities to draw up a plan for litter abatement.

Section 25

Section 25 relates to the disposal of wastes underground by the National Coal Board.

Section 26

Section 26 relates to outfall pipes from sewage works.

Section 27

Section 27 prohibits the sorting over of refuse.

Section 28

Section 28 comprises supplementary provisions relating to pipes.

Section 29

Section 29 empowers the Secrétary of State to modify Parts I and II of the Act so as to avoid unnecessary duplication of control.

Section 30

Section 30 includes relevant definitions:

'controlled waste' means household, industrial and commercial waste or any such waste;

'waste' includes:

(a) any substance which constitutes a scrap material or an effluent or other unwanted surplus substance arising from the application of any process; and
(b) any substance or article which requires to be disposed of as being broken, worn out, contaminated or otherwise spoiled.

The meaning of 'industrial waste' has been extended in so far as it relates to site licensing by SI 1976 No. 732 as amended by SI 1977 No. 1185 such that:

For the purposes of sections 3 to 11, 16 and 18(1) and (2), waste of the following descriptions shall be treated as being industrial waste:

(a) waste produced in the course of constructing, improving, repairing or demolishing any building or structure;

(b) waste produced as a result of dredging operations; and

(c) sewage deposited on land, other than
(i) sewage deposited, whether inside or outside the curtilage of a sewage treatment works, as an integral part of the operation of those works; and
(ii) sewage spread on land for agricultural purposes; and
(iii) sewage deposited on land from a sanitary convenience forming part of a moving or stationary vehicle which is being used for the conveyance of passengers; and
(iv) sewage buried on land, being matter taken from a moveable receptacle contained in a sanitary convenience serving a camp site, caravan site, building site, signal box, or other land or premises not being a dwelling-house.

Miscellaneous provisions

Section 91

Section 91 confers powers of rights of entry onto any land or vessel for inspection and examination and for the taking away of samples.

Section 93

Section 93 empowers authorities to demand certain information in writing from any person.

Section 94

Section 94 relates to restrictions on the disclosure of information, gained under the Act, relating to trade secrets.

Regional differences in administration

It is to be noted that each waste disposal authority (WDA) is responsible, of itself, for administering and interpreting the laws relating to waste disposal and, therefore, differences are to be expected. However, these differences may be so large as to warrant investigation and subsequent imposition of common standards.

The desire of central government to devolve responsibilities to the local authorities as a proper basis for democracy is appreciated. However, it must be considered that waste disposal is of regional, in many cases national, import and that central control by way of common standard is essential.

Attention is drawn to the differences in staffing levels of the WDA's and the relationship between the number of staff employed to the amount of waste for which an authority is responsible. The report *Waste Disposal Statistics 1981–82 Estimates* by The Chartered Institute of Public Finance and Accountancy [41] presents various data on the WDA's. It can be seen from that report that the various authorities place different emphases on their waste disposal functions and there is no hint of conformity.

The statistics as presented are worth studying in depth. Additionally, the data shown in Table 1 have been calculated to illustrate some significant differences in the authority manning, pay levels and pollution control expenditure. The data for Cambridgeshire are computed from the 1979–80 'actuals' [40] as the 1981–2 'estimates' are not given. The column for expenditure on the pollution control function is computed from *Waste Disposal Statistics Based on Estimates 1979/80* [42] prepared by the Society of County Treasurers and the County Surveyors' Society.

Some interesting comparisons arise:

Why do Hertfordshire and Shropshire need to have more than one technical/administrative person for each manual worker whereas the Isle of Wight, Northumberland and Wiltshire manage with one technical/administrative person for each five or more manual workers? Are the former two Counties administratively top-heavy?

Why is the average remuneration £12 215 per annum in Gloucestershire but less than half that in Dorset, North Yorkshire and Wiltshire? Is there such a staggering difference in qualifications or training or experience that warrants such a difference in salary levels?

Why does the West Midlands Authority spend 25 pence on pollution control functions for each and every tonne of waste disposed while Greater Manchester spends 3 pence per tonne? Why does Gloucestershire spend 59p per tonne? Is Greater Manchester failing its duty as regards pollution control or is West Midlands over-zealous?

Why do West Midlands and South Yorkshire have very high ratios for expenditure on administration to expenditure on operations? Is the administration framework in these authorities too large and expensive?

Comparing West Midlands with Greater Manchester, both are industrial metropolitan areas with about the same population; why did West Midlands require nearly twice as many people to be responsible for half as much waste and pay them 6% more than Greater Manchester in 1979–80?

It is obvious that such discrepancies in the functioning of the waste disposal authorities warrants very close attention indeed.

Dual standards

It is clear from a comparison of privately-operated and authority-operated disposal facilities that dual standards obtain. There is ample evidence of this and a contributory factor must be the fact that each authority acts as licensee and licensor and, in effect, is its own policeman. This is a wholly bad situation and points, once again, to the need for a national standard of operations and a national means of enforcement.

Quotations from the West Midlands County Council draft waste disposal plan dated June 1980 are informative:

> Failure to do this (secure landfill capacity) will result in the service continuing to stagger from one crisis to the next (Paragraph 9.2.3.10).

> Future landfill sites that are acquired and operated by the County

Table 1. *Review of county council waste management administration*

County Council	Ratio operatives to technical/ admin. staff	Ratio total waste/total employees	Average annual pay/employee £pa	Ratio expenditure on pollution control/amount of waste (p/te)	Ratio expenditure administration/ operational
Avon	1.53	3148	10321	7	0.16
Bedfordshire	4.12	10268	8790	12	0.03
Berkshire	1.26	9848	7387	24	0.20
Buckinghamshire	1.37	6922	8075	12	0.03
Cambridgeshire	1.73	8233	7011	11	0.08
Cheshire	2.50	11224	7878	5	0.06
Cleveland	3.78	5577	7585	22	0.14
Cornwall	4.78	19944	11667	5	0.04
Cumbria	2.14	28182	7414	6	0.21
Derbyshire	2.56	5214	7291	48	0.15
Devon	4.90	5907	8957	—	0.16
Dorset	3.04	5539	6039	13	0.12
Durham	2.43	7514	7268	6	0.09
East Sussex	2.94	6650	7931	5	0.42
Essex	4.13	6610	9202	—	0.16
Gloucestershire	2.57	8642	12215	59	0.15
Hampshire	1.39	5229	8455	4	0.13
Hereford and Worcester	3.17	5756	7353	9	0.21
Hertfordshire	0.11	33395	10800	20	0.23
Humberside	3.10	4046	7465	17	0.11
Isle of Wight	5.00	3865	7237	11	0.22
Kent	2.09	6004	8663	6	0.10
Lancashire	2.33	7795	10159	7	0.18

Leicestershire	3.00	4867	7501	16	0.09
Lincolnshire	3.19	8444	6691	—	0.20
Norfolk	4.89	7059	6970	10	0.09
Northamptonshire	2.40	9706	11856	12	0.18
Northumberland	8.00	9000	6832	—	0.25
North Yorkshire	3.50	4705	1596	2	0.41
Nottinghamshire	4.00	7657	7420	—	0.03
Oxfordshire	1.65	11644	8071	28	0.28
Shropshire	0.96	8444	6871	—	0.26
Somerset	2.64	7604	8881	10	0.21
Staffordshire	2.89	4097	7747	7	0.11
Suffolk	2.20	5542	8500	11	0.09
Surrey	1.87	5426	7090	4	0.45
Warwickshire	1.38	12645	8741	8	0.15
West Sussex	3.89	2878	6944	5	0.09
Wiltshire	5.33	6177	4323	17	0.10
Non-Metropolitan Counties	2.57	6496	7971	10	0.14
Greater Manchester	2.29	3792	8113	3	—
Merseyside	3.60	3400	7116	9	0.14
South Yorkshire	5.13	6530	6343	19	0.43
Tyne & Wear	2.53	3239	8092	—	0.05
West Midlands	2.45	1902	8602	25	0.21
West Yorkshire	2.79	2935	7576	10	0.11
Metropolitan Counties	2.77	3330	7813	9	0.19
Greater London	1.17	3936	8689	8	0.07
English Counties	2.31	4899	8023	9	0.11

Source: S.L. Willetts – unpublished; and partly from Reference [39].

Council will be operated in accordance with the Department of the Environment Report of the Working Party on Refuse Disposal (Paragraph 9.2.5).

Thus it is seen that the County Council's current service staggers from crisis to crisis and that its future facilities carry the promise of being operated to the required standard. There is no parallel mention of private sector operators being in a similar position.

The Department of the Environment has effectively continued to encourage dual standards by granting dispensation from various provisions of the Section 17 Regulations to WDA operations.

There is dispensation for waste disposal authorities in the Section 17 Regulations for the following cases:

(a) Consignment note procedure to be implemented for all 'special' waste transactions except where a waste is collected by a WDA and disposed of in its own area.
(b) Registers of consignment notes are to be kept by all disposal site operators except the WDA's themselves.
(c) Site records of deposits of special wastes shall be referenced to registers of consignment notes for all disposal sites except those operated by the WDA's themselves.

Such dispensation, of course, renders WDA's immune from prosecution as regards offences on those accounts.

Waste disposal plans

The legal status of the waste disposal plans now being prepared under Section 2 of the Control of Pollution Act by the WDA's gives rise to concern.

The Department of the Environment considers that the 'national plan' will comprise the composite of the separate plans from each WDA. This would be acceptable in practice if the waste disposal plans were accorded similar status as the county authority 'structure plans' and were subjected to the same procedure for approval and adoption by central government (the Secretary of State for the Environment). However, the waste disposal plans, when prepared, merely have to be copied to the Secretary of State and there is no mechanism for amendment or rejection by the Minister. This is all the more egregious when it is remembered that each WDA, in preparing its plan, is not obliged to consult with any other WDA, not even neighbouring authorities, unless it exports waste to that other authority. Thus, by definition, the collecting together of the waste disposal plans can only represent an unco-ordinated agglomeration of isolated documents;

there is no possibility of considering this as a realistic national plan.

As to the operation of the plans, when finalised, authorities may seek to implement their content in a similar way to that of 'structure plans' and thereby prevent proper development of private waste disposal facilities 'because such development is not in our 10-year waste disposal plan'. This may be thought an unlikely occurrence bearing in mind the supposedly responsible attitude of most WDA's; however, it must be recognised that the operation of these authorities is subject to local political pressure and, in some cases, inter-authority non-co-operation. WDA's do tend to act as islands. Waste disposal facilites must, of necessity, be considered as nationally available assets and, as such, petty local political interference should be discouraged. Co-operation between authorities and, once again, a national policy input are required.

For example, Essex County Council have written to other waste disposal authorities informing them of its intention to reduce importation of wastes from other authority areas.

Site licensing

It is of interest to observe the, perhaps natural, differences in the site licences that govern the operation of waste treatment and disposal plant compared to those governing operations at a crude landfill site.

A landfill site is basically a dump for rubbish and many sites receive a hazardous waste input. Such sites may require a high initial capital expenditure on acquisition and preparatory site works but the continued operations at the site are extremely crude, simple and cheap. Landfill, for these reasons, would seem to be the cheapest form of waste disposal. There is no requirement for technical input or skilled management.

On the other hand, a waste treatment or recovery or disposal plant or complex has an absolute and fundamental requirement for a technical input and skilled management. In addition to a high capital cost, there is also a high running cost.

It is considerably easier, as a licensing authority, to draft and implement conditions and constraints of operation for a plant than for a dump. This is seen in practice. Licences for plant can contain conditions about pre-sampling and test work; maintenance of on-site laboratory facilities and staff; safety and training requirements; comprehensive instrumentation; fail-safe operation; sampling provisions for processed waste; personnel facilities for washing, eating, etc.; record keeping for waste processing; 24-hour security arrangements; specifications for the product and a host of other rigid, definitive operational constraints. Contrarily, the licences for dumps that receive exactly the same kind of hazardous wastes very often contain none of these strict operational requirements: how can they when the hazardous waste is simply puked out of a tanker into a trench or dumped and buried before anyone can get at it?

For example, the average site licence for landfill facility in the West Midlands County Council area has 25 general conditions or so which have been abstracted from the Department of the Environment *Waste Management Paper No. 4* on site licensing. A waste solidification plant in the same area has a licence with 33 conditions and two appendices relating to monitoring and test work and product specifications with most conditions being specifically drafted to constrain the operation of the facility.

Positive forms of treatment, reclamation and disposal are, in effect, discouraged by the oppressive nature of the licence requirements in comparison to those applicable to dump sites. These unnecessarily strict licences governing the socially desirable treatment and detoxification of wastes are acting as a real disincentive to the advancement of standards in the UK.

Legislative updating

A useful compendium of the Control of Pollution Act is given by Garner [43] and this is regularly updated, on subscription, by the publisher.

Various statutory instruments are extant but have not been specifically discussed here [44] and, similarly, various relevant circulars [45]. These are important of course for the completeness of this discussion.

References

[1] *Control of Pollution Act 1974*, Elizabeth II, Ch. 40. HMSO, London.

[2] *SI 1977 No. 2164 (C.74)*, The Control of Pollution Act 1974 (Commencement No. 11) Order 1977. HMSO, London.

[3] *SI 1978 No. 816 (C.21) (S.70)*, The Control of Pollution Act 1974 (Commencement No. 12) Order 1978. HMSO, London.

[4] *Circular 29/78*, Department of the Environment (Welsh Office, 50/78). The Control of Pollution Act 1974. Part I – 'Waste on Land'. HMSO, London.

[5] *Town and Country Planning Act 1971*, Elizabeth II, Ch. 78. HMSO, London.

[6] *Waste Management Paper No. 2*, DOE, 1976, 'Waste Disposal Surveys'. HMSO, London.

[7] *Waste Management Paper No. 3*, DOE, 1976, 'Guidelines for the Preparation of a Waste Disposal Plan'. HMSO, London.

[8] *Circular 26/71*, Department of the Environment (Welsh Office 65/71, Scottish Development Department 27/71). Report of the Working Party on Refuse Disposal. HMSO, London.

[9] *Circular 37/72*, Department of the Environment (Welsh Office 86/72, Scottish Development Department letter 26.4.72). Review of Waste Disposal Facilities. HMSO, London.
[10] *Waste Management Paper No. 4*, DOE, 1976. 'The Licensing of Waste Disposal Sites'. HMSO, London.
[11] *SI 1976 No. 731 (C.21)*, The Control of Pollution Act 1974 (Commencement No. 5) Order 1976. HMSO, London.
[12] *SI 1977 No. 1587 (C.54) (S.112)*, The Control of Pollution Act 1974 (Commencement No. 10) Order 1977. HMSO, London.
[13] *Criminal Law Act 1977*, Elizabeth II, Ch. 45. HMSO, London.
[14] *Ashcroft v. Cambro Waste Products Ltd.*, Queen's Bench Division, 30 March 1981.
[15] *SI 1976 No. 732*, The Control of Pollution (Licensing of Waste Disposal) Regulations 1976. HMSO, London.
[16] *SI 1977 No. 1185*, The Control of Pollution (Licensing of Waste Disposal) (Amendment) Regulations 1977. HMSO, London.
[17] *SI 1977 No. 2006*, The Control of Pollution (Licensing of Waste Disposal) (Scotland) Regulations 1977. HMSO, London.
[18] *Circular 55/76*, Department of the Environment (Welsh Office 76/76). Control of Pollution Act 1974. Part I – (Waste on Land) Disposal Licences. HMSO, London.
[19] *Circular 79/77*, Department of the Environment (Welsh Office 89/77). Control of Pollution Act 1974. Part I – (Waste on Land) Licensing of Waste Disposal (Amendment) Regulations 1977. HMSO, London.
[20] Willetts, S.L., 1978, 'Site Licensing Rules – O.K.?', *Environmental Pollution Management*, Vol. 8, No. 2, pp. 32–4, March–April 1978.
[21] Willetts, S.L. 1981, 'Aspects of Site Assessment'. *Symposium: The Hazards of Contaminated Land*, 15 May 1981, Geological Society, London.
[22] *Local Government Act 1972*, Elizabeth II, Ch. 70. HMSO, London.
[23] *Circular 5/68*, MHLG (Welsh Office 5/68). 'The Use of Conditions in Planning Permissions.'
[24] *SI 1975 No. 2118 (C.60)*, The Control of Pollution Act 1974 (Commencement No. 4) Order 1975. HMSO, London.
[25] *SI 1976 No. 1080 (C.31) (S.92)*, The Control of Pollution Act 1974 (Commencement No. 7) Order 1976. HMSO, London.
[26] *SI 1980 No. 1709*, The Control of Pollution (Special Waste) Regulations 1980. HMSO, London.
[27] Willetts, S.L., 1979. 'Time for Change in Toxic Waste Controls?', *Surveyor*, 13 September 1979, pp. 41–2.
[28] Willetts, S.L., 1980. 'Toxic Waste Disposal: What happens Before the Cradle?', *Environmental Pollution Management*, Vol. 10, No. 2, pp. 48–52, March–April 1980.
[29] Willetts, S.L., 1980. 'Warning: This New Law Could Damage Your Health', *Surveyor*, 6 November 1980, pp. 22–4.
[30] Willetts, S.L., 1981. 'So What If My Waste Is "Special"', *Environmental Pollution Management*, Vol. 11, No. 2, pp. 45–7, March–April 1980.
[31] *Hansard.* House of Lords. Columns 306–30, 19 January 1981.
[32] *Hansard.* House of Commons. Columns 695–714, 9 February 1981.
[33] *Deposit of Poisonous Waste Act 1972*, Elizabeth II, Ch. 21. HMSO, London.
[34] *SI 1981 No. 196 (C.4)*, The Control of Pollution Act 1974 (Commencement No. 14) Order 1981. HMSO, London.

[35] *Circular 4/81*, Department of the Environment (Welsh Office 8/81). Control of Pollution Act 1974. Control of Pollution (Special Waste) Regulations 1980. HMSO, London.

[36] *CEC Directive 78/319/EEC:* On Toxic and Dangerous Waste, *OJ* L84/43, 31 March 1978.

[37] *Waste Management Paper No. 23*, DOE, 1981, 'Special Wastes: A Technical Memorandum Providing Guidance On Their Definition'. HMSO, London.

[38] Loughborough University of Technology, Centre for Extension Studies. *Proceedings of Symposium: 'Working with Section 17'*, 20 May 1981.

[39] Willetts, S.L. 1981, 'Do the Regulations Require Amendment? – The Contractor Viewpoint'. *Symposium: The Management of Toxic Waste from Cradle to Grave*. Oyez IBC Ltd. London, 11 November, 1981.

[40] The Chartered Institute of Public Finance and Accountancy, London. *Waste Disposal Statistics 1979–80 Actuals*. Statistical Information Service, Ref: 63.81.

[41] The Chartered Institute of Public Finance and Accountancy, London. *Waste Disposal Statistics 1981–82 Estimates*. Statistical Information Service, Ref: 64.82.

[42] The Society of County Treasurers and The County Surveyors' Society, Truro. *Waste Disposal Statistics Based on Estimates 1979/80.*

[43] Garner, J.F., (Ed.), 1976. *Control of Pollution Encyclopaedia* (and updating 'service issues'). Butterworth, London.

[44] *SI 1974 No. 2039 (C.33)*, The Control of Pollution Act 1974 (Commencement No. 1) Order 1974. HMSO, London.
SI 1974 No. 2169 (C.36), The Control of Pollution Act 1974 (Commencement No. 2) Order 1974. HMSO, London.
SI 1975 No. 230 (C.5) (S.18), The Control of Pollution Act 1974 (Commencement No. 3) Order 1975. HMSO, London.
SI 1976 No. 956 (C.28), The Control of Pollution Act 1974 (Commencement No. 6) Order 1976. HMSO, London.
SI 1977 No. 336 (C.12), The Control of Pollution Act 1974 (Commencement No. 8) Order 1977. HMSO, London.
SI 1977 No. 476 (C.17), The Control of Pollution Act 1974 (Commencement No. 9) Order 1977. HMSO, London.
SI 1978 No. 954 (C.23), The Control of Pollution Act 1974 (Commencement No. 13) Order 1978. HMSO, London.
SI 1976 No. 957, The Control of Pollution Act 1974 (Appointed Day) Order 1976. HMSO, London.
SI 1976 No. 958, The Control of Pollution (Discharges into Sewers) Regulations 1976. HMSO, London.
SI 1976 No. 959, The Control of Pollution (Radioactive Waste) Regulations 1976. HMSO, London.
SI 1976 No. 1638, The Control of Pollution (Repeal of Local Enactments) Order 1976. HMSO, London.

[45] *Circular 1/76*, Department of the Environment (Welsh Office 1/76). Control of Pollution Act 1974. Part I – (Waste on Land) Commencement Order No. 4. HMSO, London.
Circular 7/76, Department of the Environment (Welsh Office 9/76). Control of Pollution Act 1974 Commencement Order No. 4. Part IV – Pollution of the Atmosphere. HMSO, London.
Circular 47/78, Department of the Environment (Welsh Office 83/78). Control of Pollution Act 1974. Part I –Waste on Land. HMSO, London.

5

Hazardous wastes legislation: miscellaneous provisions

Dumping at sea

The seas and oceans form a part of the natural water cycle: they comprise a vast living entity and act as a purification sink for all manner of substances. Discharges to rivers, of course, terminate in the seas. That their purification capacity is so vast is beyond doubt but, nevertheless, overloading can happen and has happened. Recognition of potential overload of the natural cleansing properties of the seas has resulted in legislation to govern what and how much is now allowed to be dumped.

In general, the sorts of substances now dumped directly in the sea are those that are too polluting to discharge to inland watercourses, albeit that those watercourses eventually discharge into the sea themselves. As such, the gross polluting nature of such substances must be studied with a view to assessing what, if any, effect may result to marine life: in particular, food chains and harvestable fish. During the early 1960s the Ministry of Agriculture, Fisheries and Food assumed a responsibility for the control of marine pollution and requested that organisations wishing to dispose of wastes at sea should notify their intentions to the Ministry. The 'voluntary consent scheme' apparently worked satisfactorily and in the early 1970s virtually all known sea disposal operations involving industrial wastes and sewage sludges were being notified.

International Conventions

Seas, perforce, do not have or follow international boundaries and it became obvious, as sea dumping increased, that some international codes of practice would be desirable so that marine life would be conserved for the mutual benefit of all interested parties. During 1972 two significant conventions were agreed: The Convention for the Prevention of Marine

Pollution by Dumping from Ships and Aircraft (Oslo [1]) in February, and the Convention on the Prevention of Marine Pollution by Dumping of Wastes and Other Matters (London [2]), in December.

The Oslo Convention

The Oslo Convention applies to those parts of the Atlantic and Arctic Oceans and their dependent seas which lie north of 36°N latitude and between 42°W and 51°E longitude but excluding the Baltic Sea and the Mediterranean Sea and that part of the Atlantic that lies north of 59°N latitude and between 44°W and 42°W longitude.

Thirteen countries signed the Oslo Convention and Article 23 thereof required ratification by seven before it became effective: this occurred on 7 April 1974. To date, Denmark, France, Ireland, The Netherlands, Norway, Portugal, Spain, Sweden and the UK have national legislation that satisfies the requirements of the Convention. These requirements are embodied as:

Annex I: comprising substances that should be prohibited from dumping at sea in other than trace concentrations;

Annex II: comprising substances that can be dumped, but only with special care, and prescribing certain conditions;

Annex III: setting out provisions governing the issue of permits and approvals for sea dumping.

Annex I

The following substances are listed for the purpose of Article 5 of the Convention and should not be dumped at sea:

(1) Organohalogen compounds and compounds which may form such substances in the marine environment, excluding those which are non-toxic, or which are rapidly converted in the sea into substances which are biologically harmless.

(2) Organosilicon compounds and compounds which may form such substances in the marine environment, excluding those which are non-toxic, or which are rapidly converted in the sea into substances which are biologically harmless.

(3) Substances which have been agreed between the Contracting Parties as likely to be carcinogenic under the conditions of disposal.

(4) Mercury and mercury compounds.

(5) Cadmium and cadmium compounds.

(6) Persistent plastics and other persistent synthetic materials which may float or remain in suspension in the sea, and which may seriously interfere with fishing or navigation, reduce amenities or interfere with other legitimate uses of the sea.

Annex II

(1) The following substances and materials requiring special care are listed for the purpose of Article 6 and can be dumped with special care:

(a) arsenic, lead, copper, zinc and their compounds, cyanides and fluorides, and pesticides and their by-products not covered by the provisions of Annex I;

(b) containers, scrap metal, tar-like substances liable to sink to the sea bottom and other bulky wastes which may present a serious obstacle to fishing or navigation;

(c) substances which, though of a non-toxic nature, may become harmful due to the quantities in which they are dumped, which are liable to seriously reduce amenities.

(2) The substances and materials listed under Paragraph (1(b)) above should always be deposited in deep water.

(3) In the issue of permits or approvals for the dumping of large quantities of acids and alkalis, consideration should be given to the possible presence in such wastes of the substances listed in Paragraph (1) above.

(4) When, in the application of the provisions of Annexes II and III, it is considered necessary to deposit waste in deep water, this should be done only when the following two conditions are both fulfilled:

(a) that the depth is not less than 2000 metres;

(b) that the distance from the nearest land is not less than 150 nautical miles.

Annex III

Provisions governing the issue of permits and approvals for the dumping of wastes at sea:

(1) Characteristics of the waste:

(a) amount and composition;

(b) amount of substances and materials to be deposited per day (per week, per month);

(c) form in which it is presented for dumping, i.e. whether as a solid, sludge or liquid;

(d) physical (especially solubility and specific gravity), chemical, bio-chemical (oxygen demand, nutrient production), and biological properties (presence of viruses, bacteria, yeasts, parasites, etc.);

(e) toxicity;

(f) persistence;

(g) accumulation in biological materials or sediments;

(h) chemical and physical changes of the waste after release, including possible formation of new compounds;

(i) probability of production of taints reducing marketability of resources (fish, shellfish, etc.).

(2) Characteristics of dumping site and method of deposit:

(a) geographical position, depth and distance from coast;

(b) location in relation to living resources in adult or juvenile phases;

(c) location in relation to amenity areas;

(d) methods of packing, if any;

(e) initial dilution achieved by proposed method of release;

(f) dispersal, horizontal transport and vertical mixing characteristics;

(g) existence and effects of current and previous discharges and dumping in the area (including accumulative effects).

(3) General considerations and conditions:

(a) interference with shipping, fishing, recreation, mineral extraction, desalination, fish and shellfish culture, areas of special scientific importance and other legitimate use of the sea;

(b) in applying these principles the practical availability of alternative means of disposal or elimination will be taken into consideration.

The London Convention

The London Convention represents a global statement of the provisions of the Oslo Convention. It became effective during 1975 and, similarly, has the three relevant Annexes:

Annex I

Not to be dumped:

(1) Organohalogen compounds.

(2) Mercury and mercury compounds.

(3) Cadmium and cadmium compounds.

(4) Persistent plastics and other persistent synthetic materials, for example netting and ropes, which may float or may remain in suspension in the sea in such a manner as to interfere materially with fishing, navigation or other legitimate uses of the sea.

(5) Crude oil, fuel oil, heavy diesel oil, and lubricating oils, hydraulic fluids, and any mixtures containing any of these, taken on board for the purpose of dumping.

(6) High-level radioactive wastes or other high-level radioactive matter, defined on public health, biological or other grounds, by the competent international body in this field, at present the International Atomic Energy Agency, as unsuitable for dumping at sea.

(7) Materials in whatever form (e.g. solids, liquids, semi-solids, gases, or in a living state) produced for biological and chemical warfare.

(8) The preceding Paragraphs of this Annex do not apply to substances

which are rapidly rendered harmless by physical, chemical or biological processes in the sea provided they do not:

(a) make edible marine organisms unpalatable, or

(b) endanger human health or that of domestic animals.

The consultative procedure provided for under Article 15 should be followed by a Party if there is doubt about the harmlessness of the substance.

(9) This Annex does not apply to wastes or other materials (e.g. sewage sludges and dredged spoils) containing the matters referred to in Paragraphs 1–5 above as trace contaminants. Such wastes shall be subject to the provisions of Annexes II and III as appropriate.

Annex II

The following substances and materials requiring dumping with special care are listed for the purposes of Article 6(1(a)):

(1) Wastes containing significant amounts of the matters listed below:

arsenic, lead, copper, zinc } and their compounds

organosilicon compounds

cyanides

fluorides

pesticides and their by-products not covered in Annex I.

(2) In the issue of permits for the dumping of large quantities of acids and alkalis, consideration shall be given to the possible presence in such wastes of the substances listed in Paragraph (1) and to the following additional substances:

beryllium, chromium, nickel, vanadium } and their compounds.

(3) Containers, scrap metal and other bulky wastes liable to sink to the bottom which may present a serious obstacle to fishing or navigation.

(4) Radioactive wastes or other radioactive matter not included in Annex I. In the issue of permits for the dumping of this matter, the Contracting Parties should take full account of the recommendation of the competent international body in this field, at present the International Atomic Energy Agency.

Annex III

Provisions to be considered in establishing criteria governing the issue of permits for the dumping of matter at sea, taking into account Article 4(2) include:

(1) Characteristics and composition of the matter:

(a) total amount and average composition of matter dumped (e.g. per year);

(b) form (e.g. solid, sludge, liquid or gaseous);

(c) properties: physical (e.g. solubility and density), chemical and biochemical (e.g. oxygen demand, nutrients), and biological (e.g. presence of viruses, bacteria, yeasts, parasites);

(d) toxicity;

(e) persistence: physical, chemical and biological;

(f) accumulation and bio-transformation in biological materials or sediments;

(g) susceptibility to physical, chemical and biochemical changes and interaction in the aquatic environment with other dissolved organic and inorganic materials;

(h) probability of production of taints or other changes reducing marketability of resources (fish, shellfish, etc.).

(2) Characteristics of dumping site and method of deposit:

(a) location (e.g. co-ordinates of the dumping area, depth and distance from the coast), location in relation to other areas (e.g. amenity areas, spawning, nursery and fishing areas and exploitable resources);

(b) rate of disposal per specific period (e.g. quantity per day, per week, per month);

(c) methods of packaging and containment, if any;

(d) initial dilution achieved by proposed method of release;

(e) dispersal characteristics (e.g. effects of currents, tides and wind on horizontal transport and vertical mixing);

(f) water characteristics (e.g. temperature, pH, salinity, stratification, oxygen indices of pollution – dissolved oxygen (DO), chemical oxygen demand (COD), biochemical oxygen demand (BOD) – nitrogen present in organic and mineral form including ammonia, suspended matter, other nutrients and productivity);

(g) bottom characteristics (e.g. topography, geochemical and geological characteristics and biological productivity);

(h) existence and effects of other dumpings which have been made in the dumping area (e.g. heavy metal background reading and organic carbon content);

(i) in issuing a permit for dumping, Contracting Parties should consider whether an adequate scientific basis exists for assessing the consequences of such dumping, as outlined in this Annex, taking into account seasonal variations.

(3) General considerations and conditions:

(a) possible effects on amenities (e.g. presence of floating or stranded material, turbidity, objectionable odour, discoloration and foaming);

(b) possible effects on marine life, fish and shellfish culture, fish stocks and fisheries, seaweed harvesting and culture;

(c) possible effects on other uses of the sea (e.g. impairment of water quality for industrial use, underwater corrosion of structures, interference with ship operations from floating materials, interference with fishing or navigation through deposit of waste or solid objects on the sea floor and protection of areas of special importance for scientific or conservation purposes);

(d) the practical availability of alternative land based methods of treatment, disposal or elimination, or of treatment to render the matter less harmful for dumping at sea.

Broadly, for both Oslo and London Conventions, Annex I (prohibited) categories exhibit or possess toxicity, persistency and bio-accumulation characteristics simultaneously. Annex II categories exhibit one or two of these characteristics only.

The Oslo Annexes are subject to review every five years. The 1979 review recommended no changes be made except that polydimethyl siloxanes could be included on the register of non-toxic organosilicon compounds exempt from Annex I.

The London Annexes are not subject to such a procedure and, instead, proposals have been made that annual plenary sessions should agree any changes in principle for subsequent formalisation on a three-year cycle. The London Convention has agreed in principle to include refined petroleum products and distillation residues in Annex I (prohibited) and substances which, although of a non-toxic nature, may become harmful due to the quantities in which they might be dumped in Annex II. For future consideration will be the inclusion of lead and organic pesticides in Annex I.

The definition of 'significant' for the purposes of Annex II will remain as 0.1% dry weight.

The Dumping at Sea Act 1974 [3]

This attractively descriptive Act is the UK satisfaction of these Conventions. The previous voluntary consent scheme was formalised by this Act that is administered by MAFF for England and Wales and the Department of Agriculture and Fisheries for Scotland and the Department of the Environment for Northern Ireland. There is a provision for licensing of dumping of wastes at sea and the licensing authority is required by Section 2(1) to

> have regard to the need to protect the marine environment, and the living resources which it supports, from any adverse consequences of dumping the substances or articles to which the licence, if granted, will relate, and the Authority shall include such conditions in a licence as appears to the Authority to be necessary or expedient for the protection of that environment and those resources from any such consequences.

The principal offence

The Dumping at Sea Act received royal assent on 27 June 1974 and is prefaced as: 'An Act to control dumping in the sea': the prime provision is (Section 1):

> no person, except in pursuance of a licence granted under Section 2 below and in accordance with the terms of that licence –
> (a) shall dump substances or articles in United Kingdom waters; or
> (b) shall dump substances or articles in the sea outside U.K. waters from a British ship, aircraft, hovercraft or marine structure; or
> (c) shall load substances or articles on to a ship, aircraft, hovercraft or marine structure in the U.K. or U.K. waters for dumping in the sea, whether in U.K. waters or not; or
> (d) shall cause or permit substances or articles to be dumped or loaded as mentioned in paragraphs (a) to (c) above.

Sub-section (d) is of import legally as it means that a person other than the actual 'dumper' can be proceeded against if it can be shown that such a person 'permitted' the illegal loading or dumping. That could well be the *producer* of a waste, of course, and not necessarily the disposal contractor. Thus, it is incumbent upon waste producers who contract the sea disposal method to satisfy themselves that licence conditions applicable to the contractor are being met.

There are certain exceptions to the dumping prohibition, to accommodate incidental discharges and the like during normal marine operations, to cater for emergencies, and to allow for the casting of lobster pots!

A person guilty of an offence is liable on summary conviction to a fine of not more than £1000 (£400 in the Dumping at Sea Act but increased by virtue of Section 28(2) of the Criminal Law Act 1977 [4]) and/or to imprisonment for up to six months or, on conviction on indictment, to imprisonment for up to five years and/or an unlimited fine.

Licensing provisions

Section 2 of the Act relates to licensing and confers various powers upon the relevant licensing authority as regards conditions to impose, fees to charge, fees for testing work if necessary, and modification, transfer and revocation provisions.

An applicant may make representations (Section 3) if a licence application is refused or is granted subject to conditions or a fee is requested or a notice of modification or revocation of a licence is received. Written representations are to be served upon the licensing authority who will constitute a competent committee and an opportunity will be afforded for oral evidence subject to various time limits. Only an applicant can make such representations and environmental groups and the like are not permitted so to do. Incidentally, no representations under Section 3 have yet been made.

Section 4 requires

> A licensing authority shall compile and keep available for public inspection free of charge at reasonable hours the notifiable particulars of any dumping licensed by them . . . and shall furnish a copy of any such notifiable particulars to any person on payment of such reasonable sum as the authority may . . . determine.

The 'notifiable' particulars are those that the UK government is obliged to notify to the international organisations and include such details as the quantity and nature of the waste, its physical state, its chemical composition, the production process from which it was produced, the location and depth of the dumping ground and the biotoxicity test data. The sea disposal registers of MAFF are held at Great Westminster House, Horseferry Road, London SW1P 2AE.

Section 5 relates to the establishment of enforcement officers and inspectors with powers to enter onto land, vessels, etc. and to search, examine, take samples and copy documents.

Section 6 allows for reciprocity as regards any prescribed (by order of the Minister and Secretary of State) procedure of the London, Oslo or any other designated Convention with other signatory governments. No order has yet been made to implement these arrangements.

Section 7 removes liability for actions taken by enforcement officers (British and foreign) in carrying out their duties under the Act and creates the second offence in this Act, that of failure to comply with a reasonable request of an enforcement officer or of preventing others from so complying, or assaulting the officer.

Subsequent sections relate to miscellaneous provisions, definitions, interpretation, etc.

Practical aspects

So much for the legislation itself. How does it operate in practice? For a producer of a waste, the initial step is to locate a reputable disposal contractor offering sea dumping facilities. Individually, or jointly, an

Fig. 1. Application form for a licence to dump waste at sea. (Source: Ministry of Agriculture, Fisheries and Food.)

REFERENCE: MAFF/DAS	FOR OFFICIAL USE : BLR/

1. (a) Name and address of applicant: Phone No:	TECHNICAL CONTACT 1 (b) Name: Phone No:
2. (a) Name and address of producer of waste and place of production if other than in 1 (a) above Phone No:	TECHNICAL CONTACT 2 (b) Name: Phone No:
3. (a) Name and address of sea disposal contractor or shipper if other than in 1 (a) above: Phone No:	TECHNICAL CONTACT 3 (b) Name: Phone No:

4. Category of waste and annual quantity for dumping:	QUANTITY			UNIT (Please state)
(a) Dredging spoil	Capital dredgings	Maintenance dredgings		
(b) Sewage sludge	Type of treatment	Wet Weight	Dry Weight	
(c) Other sludge				
(d) Liquid waste				
(e) Solid waste				
	(net)		(gross)	

application is then made on form DU/1 (Fig. 1) to MAFF, in triplicate, on which the prescribed information is given. A licence must be sought for each separate waste proposed for sea disposal. The analytical requirements are quite specific and information about the method and location of proposed dumping is required. The application fee is currently £45.

Fig. 1. (*cont.*)

5. Nature of waste if other than dredging spoil or sewage sludge:

(a) Arising from manufacture of:

(b) Procedure giving rise to the waste (eg fermentation, electro-plating, dipping, distillation, kiering, tanning, purification of fine chemicals by crystallisation):

(c) Is the waste to be treated before dumping?

(d) (i) If so, please state whether by settlement, filtration, neutralisation or biological treatment:

(ii) or by any other process, namely:

6. (a) Are samples NOW available for testing if required?

(b) If not, when can samples be made available?

7. Are you in possession of a valid consent issued by the Department of Trade under Section 34 of the Coast Protection Act 1949? If so, please give ...

(a) Its reference number:

(b) Its date of issue:

8. (a) Proposed frequency of DUMPING of the waste, i.e. daily, weekly etc:

(b) Quantity for disposal on each occasion:

9. (a) Dumping area. For wastes ex the UK, is the preferred area:

APPROX. LOCATION (e.g. name of sea, ocean, estuary, lightship etc.)

(i) off the European Continental shelf (i.e. a deep-water dump)?

(ii on the European Continental Shelf?

(b) Proposed port of loading:

(c) Co-ordinates of preferred dumping area:

10. For liquid or sludge-type wastes not in containers, please state where known:

(a) Name of dumping vessel

(b) Proposed speed of dumping vessel

(c) Proposed rate of discharge

(d) Whether discharge is to be over ship's side, bottom discharge, or into wake of vessel.

On the basis of the information provided on DU/1, MAFF will either process the application or may require biotoxicity testing on a submitted sample. Usually a one-gallon (4.5-litre) sample is required and the fee for the test work is currently £450. The MAFF Fisheries laboratories at Burnham-on-Crouch, Essex undertake this assay work.

Then, a licence may be refused or granted subject to certain conditions.

Fig. 1. (*cont.*)

11. Physical properties Please state:

(a) Density (not required for sewage sludge)		(f) Total solids (mg/litre)	
(b) pH		(g) Whether the waste is subject to the Radioactive Substances Act, 1960	
(c) 5-day BOD value or COD/PV value or total organic carbon		(h) If so, whether an application for disposal has been made to the Department of the Environment	
(d) Suspended solids (mg/litre)		(i) Details of any references linking up with that application	
(e) Dissolved solids (mg/litre)			

12. Chemical composition. Please complete the following list stating units of concentration and whether the analysis is:
(i) on a dry weight or wet weight basis* and
(ii) on a volume/volume, weight/weight or weight/volume basis*
For negligible quantities please indicate value as less than a number of parts per million:

Mercury		Phenolic substances	
Cadmium		Fluoride	
Arsenic		Chlorine (not chloride)	
Lead		Organohalogens, state which:	
Copper		Pesticides (other than those given above)	
Zinc			
Beryllium		Organosilicon compounds	
Chromium		Antibiotics	
Nickel		Vitamins	
Vanadium		Oil/grease, persistent	
Iron		Oil/grease, non persistent	
Ammonia		Animal oil/fat	
Cyanide		Vegetable oil/fat	

Further information necessary to give a complete description of the waste, eg other constituents and details of the form of constituents where relevant.

*Delete as necessary

The licence document, if granted, will usually prescribe an expiry date, normally one year, a maximum amount licensed for disposal, the method and rate of disposal, the co-ordinates of the disposal area and conditions specifying the making of certified returns.

For deep sea disposals, it will be required that a waste is solid, or rendered solid by absorption into an inert material, packaged so that it will not be released during descent through the thermocline and so reaches the sea bed intact, is disposed at least 150 nautical miles from the nearest coast, away from shipping and submarine cables and in at least 2000 metres of water, and the area of disposal itself may be specified.

Fig. 1. (*cont.*)

13. Containers. If to be used please state:

(a) Type ______________________

(b) Capacity ______________________

(c) Method of sealing (i.e. bung or lid, clipped or welded): ______________________

(d) Whether container is to be cased in concrete: ______________________

(e) Form of any additional ballast necessary to ensure sinking:

NOTES.

(a) Under Section 2 of the Dumping at Sea Act 1974 any person who for the purpose of procuring the grant of a licence knowingly or recklessly makes a false statement is guilty of an offence and liable on summary conviction to a fine not exceeding £1000

(b) You are requested to allow adequate time for this application to be processed, preferably not less than eight weeks before a licence is required.

(c) When this application is completed it should be sent to the Ministry of Agriculture, Fisheries and Food, Fisheries Division, Marine Pollution Branch, Great Westminster House, Horseferry Road, London, SW1P 2AE, together with the fee payable.

DECLARATION.

I certify that to the best of my knowledge and belief the information given in this form is a correct description of the waste for which this application is made.

Signature Position Date

FOR OFFICIAL USE

RECOMMENDATION:

AREA AGREED:

For dumping within the continental shelf area, the method of discharge will relate to sludges/liquids and not to drummed wastes. Discharge may be either via bottom sea cocks or via rear discharge pipes lowered into the wake of the vessel. Dispersion of the waste into the sea is necessary and so discharge will be made from the moving vessel and the minimum sailing speed may well be specified on a licence.

Registers

Two registers are maintained: one for harbour and river dredgings and colliery spoil and one for all other wastes. The latter register records details of the date of issue of the licence, the port of departure, a description of the waste, its physical form, the licensed quantity, the duration of the licence, the frequency of dumping and the dumping site. The licensee is identified by a code.

During 1979 some 8 800 000 tonnes of sewage sludge was licensed for dumping and 1 150 000 tonnes of industrial wastes including latex, detergents, pharmaceutical washings, acids, ammonia salts, paper sludge, grinding sludge, alkalis, phenols, hydroxide sludges, citrates, plating sludges, sulphidic scrubber liquors, food processing wastes, pig manure, arsenic trisulphide, antimony trichloride, sheepskin processing waste, cyanide-bearing wastes, tip leachates, barium sulphate, yeast waste and various organic wastes.

Ports of departure include Avonmouth (Shirehampton Quay), Barrow, Beckstone in Crossness, Birkenhead (Egerton Dock), Blyth, Bristol, Cleveland, Colchester, Davy Hulme, Exeter, Faversham, Goole, Hartlepool, London (Belvedere), Liverpool (Bromborough Dock), Milford Haven, Newcastle, Newport, Port Clarence, Portishead, Plymouth, Runcorn, Salford, Sharpness, Southampton (Wollaston and Fawley), Swansea, Truro, and Tyne, Tees and Billingham.

Principal dumping grounds

Dumping grounds for industrial wastes are defined by the following co-ordinate boxes:

Atlantic deep water – key 11

45° 04′ N 15° 22′ W
43° 50′ N 15° 02′ W
42° 07′ N 17° 14′ W
42° 20′ N 17° 34′ W

Avonmouth – key 8 and 8(a)
1 mile radius from 51° 24′ 5″ N 04° 04′ W or,
in bad weather, 51° 22′ 5″ N 03° 56′ W

Hartland Point – key 7
Circle of 2 miles radius centred on 51° 04′ 48″ N 04° 32′ 18″ W

Liverpool Bay – key 9
Between 53° 30′ 5″ N and 53° 33′ 5″ N
and 03° 34′ 0″ W and 03° 36′ 5″ W

East Coast (off Spurn Head) – key 3
Between 53° 30′ N and 53° 35′ N
and 00° 30′ E and 00° 35′ E

Nab – key 5
50° 36′ 57″ N 00° 56′ 09″ W
50° 36′ 06″ N 00° 55′ 00″ W
50° 34′ 04″ N 00° 58′ 33″ W
50° 35′ 00″ N 00° 59′ 48″ W

Atlantic Ocean – key P
51° N 17°W
51° 30′ N 19° 30′ W
subject to 2000 metres minimum depth and 150 nautical miles from coast.

South Falls – key 4
51° 35′ N 01° 58′ E
51° 35′ N 02° 00′ E
51° 30′ N 02° 00′ E
51° 30′ N 01° 57′ E

Bristol Channel – key 10
51° 26′ 6″ N 02° 57′ 6″ W
51° 26′ 2″ N 02° 59′ 0″ W
51° 26′ 6″ N 02° 59′ 0″ W
51° 27′ 0″ N 02° 57′ 0″ W

Tees – key 2
54° 44′ N 01° 00′ W
54° 51′ N 01° 00′ W
54° 51′ N 00° 48′ W
54° 44′ N 00° 48′ W

Tyne – key 1

55° 07′ N	01° 20′ W
55° 10′ N	01° 21′ 8″ W
55° 07′ N	01° 16′ W
55° 10′ N	01° 17′ 8″ W

English Channel – key 6

Between	50° 20′ N	50° 30′ N
and	01° 55′ W	00° 40′ W

The key numbers identify these on the locations map shown in Fig. 2. Additionally, the Atlantic Ocean has special areas for the dumping of certain wastes:

(i) Cyanide, antimony chloride, arsenic trisulphide:
main area: key P,
bad weather area: key Q,
extreme emergency area: key R, key T, key V;

(ii) Antimony chloride alternative: key U;

(iii) Radioactive wastes ex. Sharpness. Embedded in concrete excluding buoyant material: key S;

(iv) Scotland: administered by DAFS: key A & O.

Sewage sludge and dredging spoil is dumped at other areas:

North Sea (off Blyth)	key B
North Sea	key C
North Sea (off Newcastle)	key D
Leith – St. Abbey Mead	key E
North Sea (North Shields)	key F
North Sea (Harwich)	key G
Thames Estuary (Barrow Deep)	key H
Needles	key I
Lyme Bay	key J
English Channel	key K
English Channel	key L
Irish sea (Milford Haven)	key M
Irish Sea (Liverpool Bay)	key N

Practical enforcement

It will be seen from this discussion that the potential controls on sea dumping are rigorous. But what of practice? MAFF have a complement of inspectors who do sample road vehicles delivering wastes to docks and do accompany sea vessels to observe actual dumping practice. During oral evidence to the House of Lords, MAFF estimated that no more than one in 200 dumpings are sampled for monitoring.

Is it possible for a vessel to dump its load before actually reaching the

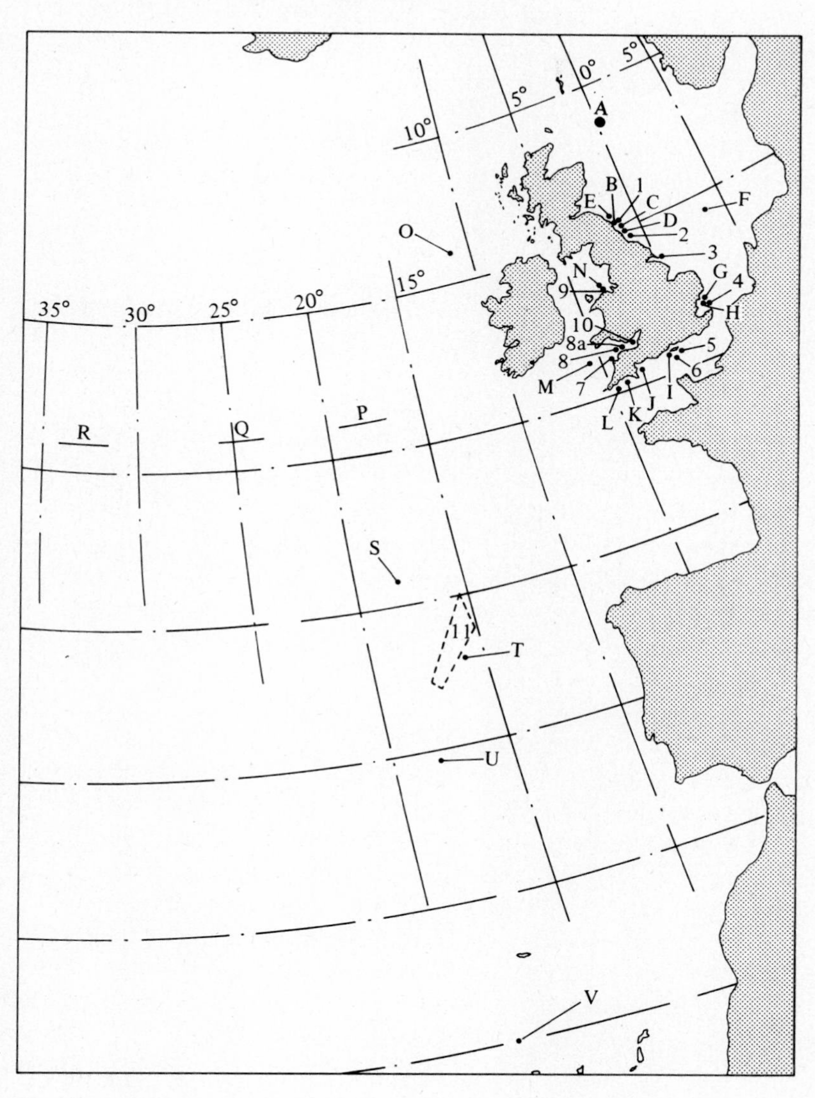

Fig. 2. Location of approved sea dumping grounds. (Source: S.L. Willetts – unpublished survey from Ministry of Agriculture, Fisheries and Food, Public Register.)

specified dumping area? Personal contact with a number of harbourmasters has established that details of all ship movements, with times, are usually logged by the harbourmaster's office. Some harbourmasters would make such records available, upon request, to members of the public but some would definitely not. As a matter of record, MAFF do not consult the harbourmaster's logs to check on sailing times. Thus, it would appear that there is no monitoring of whether ships actually spend the required amount of sailing time that would be necessary to comply with dumping licence conditions.

Other Conventions

Various other Conventions are in force which relate to pollution of the seas by materials other than deliberately dumped wastes. For completeness these are referred to here [5] but are not discussed.

Other provisions

Certain Acts other than the major ones discussed in this chapter can have an impact on waste disposal.

Litter

The Litter Act 1958 [6] and the Dangerous Litter Act 1971 [7] are penal measures creating the offence of depositing or leaving 'any thing whatsoever in such circumstances as to cause, contribute to, or tend to lead to, the defacement by litter of any place in the open air'. The setting of the penalty of a fine of up to £100 is to take into account the 'nature of the litter and any resulting risk (in the circumstances of the offence) of injury to persons or animals or of damage to property'.

Bulky refuse

The Civic Amenities Act 1967 [8] and the replacing Refuse Disposal (Amenity) Act 1978 [9] required local authorities to provide areas where householders could deposit 'bulky' refuse not normally collected by the authority. Additionally, the Acts amended the Public Health Act 1961 [10] so that local authorities were empowered to remove any rubbish from any land in the open air and to recover the costs of so doing.

An offence was created if any person abandoned on land in the open air anything whatsoever without lawful authority subject to a penalty of up to £100 fine or, for second and subsequent convictions, a fine of up to £200 and/or imprisonment for up to three months.

Mines and quarries

The Mines and Quarries Act 1954 [11], as amended by the Mines and Quarries (Tips) Act 1969 [12], defines, for the purposes of Part III of the Public Health Act 1936 [13], a statutory nuisance so as to include:

> a quarry (whether in course of being worked or not) which:
> (i) is not provided with an efficient and properly maintained barrier so designed and constructed as to prevent any person from accidentally falling into the quarry; and
> (ii) by reason of its accessibility from a highway or a place of public resort constitutes a danger to members of the public.

There are also provisions as to the safety of mines and quarries used for the purposes of tipping of mine and quarry waste. Such wastes are exempt from the provisions of the Control of Pollution Act 1974 [14] and this separate Mines and Quarries legislation is, therefore, regarded as sufficient.

Health and Safety at Work (etc.) Act 1974 [15]

Although included here as 'miscellaneous', the Health and Safety at Work Act is, of course, a significant piece of modern legislation as regards public safety and protection. The Act is an enabling measure superimposed over existing legislation such as the Offices, Shops and Railway Premises Act [16] and the Mines and Quarries Act [11]. It extends legislative protection for health, safety and welfare to people at work and to members of the public affected by work activities.

Basically, employers are required to ensure the health, safety and welfare at work of all their employees and to include maintenance of safe plant and systems of work; absence of risk in the use, handling, storage and transport of materials and substances; provision of welfare facilities; training and supervision; maintenance of a safe place of work, etc.

Many Codes of Practice are now in preparation to ensure that employers and employees, so far as is reasonably practicable, maintain safety in their operations.

The Act also, of course, established the now familiar Health and Safety Commission and Executive with wide powers to amend existing laws and to make new Regulations relating to:

> the manufacture, supply or use of any plant;
> the manufacture, supply, keeping or use of any substance;
> the carrying on of any process or the carrying out of any operation.

These powers, and the others listed in Schedule 3 to the Act, are very wide-ranging indeed.

The Act is so important, it is worth reciting the pre-amble:

> An Act to make further provision for securing the health, safety and welfare of persons at work, for protecting others against risks to health or safety in connection with the activities of persons at work, for controlling the keeping and use of dangerous substances, and for controlling certain emissions into the atmosphere.

Central government controls over local authorities [17]

In September 1979 the Secretary of State for the Environment presented a White Paper to parliament which reinforced the determination of the government 'to reduce substantially the number of bureaucratic controls over local government activities'.

Some 300 controls were outlined as being subject to repeal and a number

Table 1. *Repeals and relaxations in central government controls in waste management*

Reference	Nature of provision
Control of Pollution Act 1974	
Section 5(2)	Regulations allowing licence applications to be considered pending receipt of planning permission
Section 5(4(a))	Prescription of bodies to be consulted on proposed issue of disposal licence
Section 6(1)	Prescription of conditions for disposal licences and resolutions
Section 6(4(a))	Prescription of details for register of licences
Section 11(3(c))	Prescription of bodies to be consulted on proposed resolution covering a disposal site operated by the authority
Section 13(7)	Regulations on receptacles for controlled waste
Section 28(1(a))	Prescription of form of map of waste collection pipes
Section 79(5), (6), (7)	Approval of disclosure of information
Section 81	Appeal against notice requiring information and power to make regulations as to such appeals
Refuse Disposal (Amenity) Act 1978	
Section 6(2)	Prescription of notices in respect of removal of refuse

Source: Reference [17].

were outlined as being subject to relaxation. Of relevance to waste disposal activities are the repeals or relaxations as shown in Table 1.

Waste Management Papers [18]

This series of papers is directly related to the issues of waste management and includes both general guidelines and specific advice. The list so far is:

(1) *Waste Management Paper No. 1* – 'Reclamation, Treatment and Disposal of Wastes – An Evaluation of Options'.
(2) *Waste Management Paper No. 2* – 'Waste Disposal Surveys'.
(3) *Waste Management Paper No. 3* – 'Guidelines for the Preparation of a Waste Disposal Plan'.
(4) *Waste Management Paper No. 4* – 'The Licensing of Waste Disposal Sites'.
(5) *Waste Management Paper No. 5* – 'The Relationship between Waste Disposal Authorities and Private Industry'.
(6) *Waste Management Paper No. 6* – 'Polychlorinated Biphenyl (PCB) Wastes – A Technical Memorandum of Reclamation, Treatment and Disposal including a Code of Practice'.
(7) *Waste Management Paper No. 7* – 'Mineral Oil Wastes – A Technical Memorandum on Arisings, Treatment and Disposal including a Code of Practice'.
(8) *Waste Management Paper No. 8* – 'Heat-treatment Cyanide Wastes – A Technical Memorandum on Arisings, Treatment and Disposal including a Code of Practice'.
(9) *Waste Management Paper No. 9* – 'Halogenated Hydrocarbon Solvent Wastes from Cleaning Processes – A Technical Memorandum on Reclamation and Disposal including a Code of Practice'.
(10) *Waste Management Paper No. 10* – 'Local Authority Waste Disposal Statistics 1974/75'.
(11) *Waste Management Paper No. 11* – 'Metal Finishing Wastes – A Technical Memorandum on Arisings, Treatment and Disposal including a Code of Practice'.
(12) *Waste Management Paper No. 12* – 'Mercury-bearing Wastes – A Technical Memorandum on Storage, Handling, Treatment and Recovery of Mercury including a Code of Practice'.
(13) *Waste Management Paper No. 13* – 'Tarry and Distillation Wastes and Other Chemical Based Residues – A Technical Memorandum

on Arisings, Treatment and Disposal including a Code of Practice'.

(14) *Waste Management Paper No. 14* – 'Solvent Wastes (excluding Halogenated Hydrocarbons) – A Technical Memorandum on Reclamation and Disposal including a Code of Practice'.

(15) *Waste Management Paper No. 15* – Halogenated Organic Wastes – A Technical Memorandum on Arisings, Treatment and Disposal including a Code of Practice'.

(16) *Waste Management Paper No. 16* – 'Wood-preserving Wastes – A Technical Memorandum on Arisings, Treatment and Disposal including a Code of Practice'.

(17) *Waste Management Paper No. 17* – 'Wastes from Tanning, Leather Dressing and Fellmongering – A Technical Memorandum on Recovery, Treatment and Disposal including a Code of Practice'.

(18) *Waste Management Paper No. 18* – 'Asbestos Waste – A Technical Memorandum on Arisings and Disposal including a Code of Practice'.

(19) *Waste Management Paper No. 19* – 'Wastes from the Manufacture of Pharmaceuticals, Toiletries and Cosmetics – A Technical Memorandum on Arisings, Treatment and Disposal including a Code of Practice'.

(20) *Waste Management Paper No. 20* – 'Arsenic-bearing Wastes – A Technical Memorandum on Recovery, Treatment and Disposal including a Code of Practice'.

(21) *Waste Management Paper No. 21* – 'Pesticide Wastes – A Technical Memorandum on Arisings and Disposal including a Code of Practice'.

(22) *Waste Management Paper No. 22* – 'Local Authority Waste Disposal Statistics 1974/75 to 1977/78'.

(23) *Waste Management Paper No. 23* – 'Special Wastes – A Technical Memorandum Providing Guidance on their Definition'.

Royal Commission on Environmental Pollution [19]

The Royal Commission was appointed 'to advise on matters, both national and international, concerning the pollution of the environment; on the adequacy of research in this field; and the future possibilities of danger to the environment'. To date, the Commissioners have presented seven reports:

1st Report 'First Report' Cmnd 4585, 1971

2nd Report 'Three Issues in Industrial Pollution' Cmnd 4894, 1972

3rd Report 'Pollution in Some British Estuaries and Coastal Waters' Cmnd 5054, 1972

4th Report 'Pollution Control: Progress and Problems' Cmnd 5780, 1974

5th Report 'Air Pollution Control: an Integrated Approach' Cmnd 6371, 1976

6th Report 'Nuclear Power and the Environment' Cmnd 6618, 1976

7th Report 'Agriculture and Pollution' Cmnd 7644, 1979

Pollution papers [20]

This series represents the publication, for the Department of the Environment, by various bodies, of their work on certain aspects of pollution control.

To date, seventeen have been published:

(1) 'The Monitoring of the Environment in the United Kingdom' (1974).
(2) 'Lead in the Environment and its Significance to Man' (1974).
(3) 'The Non-Agricultural Uses of Pesticides in Great Britain' (1974).
(4) 'Controlling Pollution' (1975).
(5) 'Chlorofluorocarbons and their Effect on Stratospheric Ozone' (1976).
(6) 'The Separation of Oil from Water for North Sea Oil Operations' (1976).
(7) 'Effects of Airborne Sulphur Compounds on Forests and Freshwaters' (1976).
(8) 'Accidental Oil Pollution of the Sea' (1976).
(9) 'Pollution Control in Great Britain: How it Works' (2nd Edn) (1978).
(10) 'Environmental Mercury and Man' (1976).
(11) 'Environmental Standards' (1977).
(12) 'Lead in Drinking Water' (1977).
(13) 'Tripartite Agreement on Stratospheric Monitoring between France, the United Kingdom and the United States of America': Joint Annual Report 1976–7 (1977).

(14) 'Lead Pollution in Birmingham' (1978).
(15) 'Chlorofluorocarbons and their Effect on Stratospheric Ozone' (Second Report (1979)).
(16) 'The United Kingdom Environment 1979: Progress of Pollution Control' (1979).
(17) 'Cadmium in the Environment and its Significance to Man' (1980).

Radioactive wastes

There is a history of legislation concerning the use, transport and disposal of radioactive substances.

Radioactive Substances Act 1948 [21]

Part of this Act, relating to safety Regulations, specifically included the disposal of wastes but no Regulations have been made.

Regulations governing the packaging of radioactive materials for transit have been made:

Radioactive Substances (Carriage by Road) (Great Britain) Regulations 1974 [22];

Radioactive Substances (Road Transport Workers) (Great Britain) Regulations 1970 [23], 1975 [24].

Atomic Energy Authority Act 1954 [25]

This Act established the UKAEA and provided for control over the disposal of its radioactive wastes. Authorisation for a particular disposal had to be obtained from the Secretary of State for the Environment and the Minister of Agriculture, Fisheries and Food and so controls were removed from local authority to a national level: however, the local and water authorities were to be consulted by the Secretary prior to any authorisation for disposal.

Nuclear Installations (Licensing and Insurance) Act 1959 [26]

This Act brought under control the licensed nuclear sites, other than the UKAEA, such as nuclear power stations, and applied the same provisions relating to waste disposal.

Radioactive Substances Act 1960 [27]

This Act came into force on 1 December 1963 and replaced the waste-related provisions of the Atomic Energy Authority Act 1954.

The Act gives permanent effect to the temporary controls of the 1954 and 1959 Acts relating to the UKAEA and to all licensed nuclear sites. All other users of radioactive materials (hospitals, universities, research laboratories) were obliged to register with the Secretary of State.

Thus, the disposals of all radioactive wastes (other than certain defined exceptions such as Crown establishments and insignificant quantities) are now authorised centrally, as is the accumulation of radioactive wastes to form a disposal consignment at a registered premises. Authorisations specify the type of waste, means of disposal, conditions which must be observed and any measurements on wastes or on the environment which may be necessary.

Actual disposals of radioactive wastes are carried out at two facilities provided by the Ministers under the Act. Low-level wastes can be buried at a site operated by British Nuclear Fuels Ltd at the former Royal Ordnance Factory, Drigg in Cumbria. Intermediate-level wastes are handled at the UKAEA facility at Harwell in Oxfordshire where they are usually packaged for deep sea dumping: certain items are incinerated and the ash packaged for deep sea disposal.

Nuclear Installations Acts 1965 [28] and 1969 [29]

High-level radioactive wastes are not 'disposed of' – they are stored. These Acts provide the main statutory control over the processing and storage of such wastes and sites or facilities proposed for these operations have to be licensed by the Secretary of State for Energy.

The Nuclear Installations Inspectorate are responsible for the monitoring of radioactive waste disposal and for ensuring safety of operations.

Incineration of wastes

Certain processes that give rise to particularly noxious or offensive emissions are regulated by the Health and Safety Executive through Her Majesty's Alkali and Clean Air Inspectorate.

The first process to be singled out was the production of alkali from salt which gave rise to emissions of hydrochloric acid to the atmosphere. The Alkali Act of 1863 [30] was the first governing statute and various additions were made and eventually consolidated in the Alkali (etc.) Works Regulation Act 1906 [31]. Over 60 processes have been added by Order and incineration of wastes is one. Certain provisions have been superseded by the Health and Safety at Work (etc.) Act 1974 and the Control of Pollution Act 1974.

Chemical incineration works are defined as

> works for the destruction by burning of wastes produced in the course of organic chemical reactions which occur during the

manufacture of materials for the fabrication of plastics and fibres, and works for the destruction by burning of chemical wastes containing combined chlorine, fluorine, nitrogen, phosphorus or sulphur.

As such, registered incinerators must ensure that the 'best practicable means' are employed for the preventing of the escape of noxious or offensive gases, including smoke, grit and dust, and for the preventing of their discharge to atmosphere. 'Best practicable means' is not defined but is used with respect to the prevention of the escape of gases by reference to the provision and maintenance of appliances designed to prevent such escape and to the manner in which those appliances are used and to the proper supervision, by the owner, of any operation in which such gases are evolved.

'Noxious or offensive' is not defined but it must be said that the Health and Safety Executive is accountable to parliament for the adequacy of measures taken by the Commission for the Control of Emissions to the Atmosphere from Prescribed Classes of Premises.

The Public Health Act 1936 includes 'any dust or effluvia caused by any trade, business, manufacture or process and being prejudicial to the health of or a nuisance to the inhabitants of the district' as being a statutory nuisance and local authorities are charged with the duty to inspect their districts for nuisances and to require their abatement if so found.

The Clean Air Acts of 1956 [32] and 1968 [33] replaced and extended the Public Health Act provisions relating to smoke such that 'dark smoke' may not be emitted from any building. The Secretary of State may relax this prohibition in cases where dark smoke emission is unavoidable.

The Health and Safety Executive is presently considering the discharge limits from chemical waste incinerators and likely standards will probably include:

emissions to be free from visible smoke and, in any case, not more than Shade 1 of the Ringelmann Scale (British Standard: BS 2742, 1969 [34]);

emissions to be substantially free from persistent mist or fume;

not more than 0.46 grammes hydrogen chloride per cubic metre;

not more than 5 parts per million by volume of hydrogen sulphide;

particulate emissions depending upon local conditions;

specific limitations, where appropriate, on heavy metals, halogens and their compounds, and oxides of sulphur and nitrogen.

Also, the Health and Safety Executive propose operational constraints on the receipt, handling and storage of chemical wastes and on the design and performance of the incinerators themselves and flue gas treatment apparatus.

The height of any chimney will be specified.

Marine incineration is designed partly to overcome the regulations on gaseous emissions. No specific legislation exists to control such incineration but the London Convention on the dumping of wastes at sea does impinge. The Inter-Governmental Maritime Consultative Organisation (IMCO) has also made recommendations which have been adopted. The Ministry of Agriculture, Fisheries and Food, as an extension to their duties under the Dumping at Sea Act 1974, certify and licence the technical acceptance for incineration of consignments of incinerable wastes leaving UK ports.

The Oslo Commission Standing Advisory Committee for Scientific Advice is currently considering whether a legal instrument should be adopted to control marine incineration.

European legislation

For better or for worse, the UK is a member of the European Community. As such, the Directives of that Community are imposed upon the UK government and must be implemented.

Those Directives relevant to waste management are listed in Table 2, in chronological order of adoption.

Table 2. *CEC Directives concerned with waste management*

Title and reference	Date of adoption	Official journal reference
Council Directive on the approximation of Member States' legislation relating to the classification, packaging and labelling of dangerous substances (67/548/EEC)	27.6.67	*OJ* L 196/1, 16.8.67
Amended: Council Directive 69/81/EEC	13.3.69	*OJ* L 68/1, 19.3.69
Council Directive 70/189/EEC	6.3.70	*OJ* L 59/33, 14.3.70
Council Directive 71/144/EEC	22.3.71	*OJ* L 274/15, 29.3.71
Council Directive 73/146/EEC	24.6.75	*OJ* L 183/22, 14.7.75
Commission Directive 76/907/EEC	14.7.76	*OJ* L 360/1, 30.12.76
		OJ L 28/32, 2.2.79

Table 2. (*cont.*)

Title and reference	Date of adoption	Official journal reference
Commission Directive 79/370/EEC	30.1.79	*OJ* L 88/1, 7.4.79
Council Directive 79/831/EEC	18.9.79	*OJ* L 259/10, 15.10.79
Council Decision including the Convention for the prevention of marine pollution from land-based sources (75/437/EEC)	3.3.75	*OJ* L 194/5, 25.7.75
Council Recommendation to the Member States regarding cost allocation and action by public authorities on environmental matters (applying the polluter pays principle) (75/436/EEC)	3.3.75	*OJ* L 194/1, 25.7.75
Council Directive on the disposal of waste oils (75/439/EEC)	16.6.75	*OJ* L 194/23, 25.7.75
Council Directive on waste (framework directive) (75/442/EEC)	15.7.75	*OJ* L 194/39, 25.7.75
Council Directive on the disposal of PCB's and PCT's (76/403/EEC)	5.4.76	*OJ* L 108/41, 26.4.76
Council Directive on pollution caused by certain dangerous substances discharged into the aquatic environment of the Community (76/464/EEC)	4.5.76	*OJ* L 129/32, 18.5.76
Council Directive on waste from the titanium dioxide industry (78/176/EEC)	20.2.78	*OJ* L 54/19, 25.2.78
Council Directive on toxic and dangerous waste (78/319/EEC)	20.3.78	*OJ* L 84/43, 31.3.78
Council Directive on the protection of groundwater against pollution caused by certain dangerous substances (80/68/EEC)	17.12.79	*OJ* L 20/43, 26.1.80
Council Decision adopting a programme on the management and storage of radioactive waste (1980–4) (80/343/EURATOM)	18.3.80	*OJ* 78/22, 25.3.80

Source: Reference [35].

There are four Directives of major importance as regards disposal of chemical wastes and these are discussed:

(A) On waste (75/442/EEC)

Article 3 states: 'Member States shall take appropriate steps to encourage the prevention, recycling and processing of waste, the extraction of raw materials and possibly of energy therefrom and any other process for the re-use of waste.'

Article 4 states: 'Member States shall take the necessary measures to ensure that waste is disposed of without endangering human health and without harming the environment.'

Article 8 states: 'any installation or undertaking treating, storing or tipping waste on behalf of third parties must obtain a permit from the competent authority'.

(B) On the disposal of waste oils (75/439/EEC)

Article 3 states: 'Member States shall take the necessary measures to ensure that, as far as possible, the disposal of waste oils is carried out by recycling.'

Article 4 states:

> Member States shall take the necessary measures to ensure the prohibition of:
> (i) any discharge of waste oils into internal surface waters, ground waters, coastal waters and drainage systems;
> (ii) any deposit and/or discharge of waste oils harmful to the soil and any uncontrolled discharge of residues resulting from the processing of waste oils;
> (iii) any processing of waste oils causing air pollution which exceeds the level prescribed by existing provisions.

(C) On toxic and dangerous waste (78/319/EEC)

Article 5 states: 'Member States shall take the necessary measures to ensure that toxic and dangerous waste is disposed of without endangering human health and without harming the environment.'

Article 7 states

> Member States shall take the necessary steps to ensure that:
> – toxic and dangerous waste is, where necessary, kept separate from other matter and residues when being collected, transported, stored or deposited;
> – the packaging of toxic and dangerous waste is appropriately labelled, indicating in particular the nature, composition and quantity of the waste;
> – such toxic and dangerous waste is recorded and identified in respect of each site where it is or has been deposited.

Article 9 states: 'Installations, establishments or undertakings which carry out the storage/treatment and/or deposit of toxic and dangerous waste must obtain a permit from the competent authorities.'

Article 12 states: 'The competent authorities shall draw up and keep up to date plans for the disposal of toxic and dangerous waste.'

Article 14 states:

> 1. Any installation, establishment or undertaking which produces, holds and/or disposes of toxic and dangerous waste shall:
> – keep a record . . .
> – make this information available . . .
> 2. When toxic and dangerous waste is transported . . . it shall be accompanied by an identification form . . .
> 3. Documentary evidence that the disposal operations have been carried out shall be kept for as long as the Member States deem necessary.

The Control of Pollution (Special Waste) Regulations 1980 [36] made under Section 17 of the Control of Pollution Act 1974 were designed to fulfil the outstanding obligation, as seen by the Department of the Environment, under this Directive. The site licensing provisions of Sections 3–11 of the 1974 Act govern the disposal site arrangements and the Section 17 Regulations relate to the transportation element. However, the discussion of the Section 17 Regulations will show that the term 'special waste' to which they apply does not satisfy the broader environmental objectives of this Directive but, rather, concentrates on risk to human health.

The European Commission sponsored a conference in Copenhagen in March 1981 and this resulted in a commitment to make detailed amendments to the directive on toxic and dangerous wastes. The Commission plans to set new and common criteria for the establishment of future waste disposal facilities and to tighten, what it sees as, the current lax control over landfill sites. Indeed, the Commission has said that it regards landfill as a temporary and dangerous second choice (*European Chemical News*, 13 April 1981, p. 19).

(D) On the disposal of PCB's and PCT's (76/403/EEC)

Polychlorinated biphenyls and polychlorinated terphenyls are extremely persistent organic chemicals that are highly toxic. They are mainly used in the electricity generation and distribution industry and their very nature necessitates that they be subject to specific arrangements for control of their disposal.

The Directive requires Member States to prohibit uncontrolled discharge, dumping and tipping of the chemicals or of objects or equipment containing the chemicals. It also requires that Member States ensure that redundant equipment containing PCB's is disposed of and not merely stored indefinitely.

Destruction by high-temperature incineration is the usual method for disposal.

Environmental impact assessment

On 18 June 1980 the Commission of the European Communities put forward its proposal for a Directive concerning the assessment of the environmental effects of certain public and private projects.

The draft Directive [37] proposes:

(a) a general requirement to ensure the advance evaluation of the environmental consequences of any development likely to have a significant impact on the environment;

(b) a compulsory assessment, in a broadly prescribed manner, of the environmental consequences of certain kinds of major development projects prescribed in the Directive;

(c) assessment, at the Member State's discretion, of the environmental consequences of certain other listed developments;

(d) a general requirement to consult appropriate environmental bodies and to inform the public of the issues.

Article 4(1) requires development projects as listed (in the Annex I) to be subjected to the prescribed assessment. Article 4(2) requires projects as listed (in the Annex II) to be subjected to the prescribed assessment 'whenever their characteristics so require'. The 'competent authority' is responsible for setting thresholds and criteria by which 'significant impact on the environment' is to be judged and hence whether or not a particular envisaged development is to be subjected to environmental impact assessment.

Annex I details those development projects that must be subjected to assessment when significant impact on the environment is likely:

Annex I – development projects referred to in Article 4(1)

(1) Extractive industry:
extraction and briquetting of solid fuels;
extraction of bituminous shale;
extraction of ores containing fissionable and fertile materials;
extraction and preparation of metalliferous ores.

(2) Energy industry:
coke ovens;
petroleum refining;
production and processing of fissionable and fertile materials;
generation of electricity from nuclear energy;
coal gasification plants;
disposal facilities for radioactive waste.

(3) Production and preliminary processing of metals:
iron and steel industry, excluding integrated coke ovens;
cold rolling of steel;
production and primary processing of non-ferrous and ferro-alloys.

(4) Manufacture of non-metallic mineral products:
manufacture of cement;
manufacture of asbestos–cement products;
manufacture of blue asbestos.

(5) Chemical industry:
petrochemical complexes for the production of olefins, olefine derivatives, bulk monomers and polymers;
chemical complexes for the production of organic basic intermediates;
complexes for the production of basic inorganic chemicals.

(6) Metal manufacture:
foundries;
forging;
treatment and coating of metals;
manufacture of aeroplane and helicopter engines.

(7) Food industry:
slaughter houses;
manufacture and refining of sugar;
manufacture of starch and starch products.

(8) Processing of rubber:
factories for the primary production of rubber;
manufacture of rubber tyres.

(9) Building and civil engineering:
construction of motorways;
intercity railways, including high-speed tracks;
airports;
commercial harbours;

construction of waterways for inland navigation;
permanent motor and motorcycle racing tracks;
installation of surface pipelines for long distance transport.

Annex II – details those development projects for which an assessment will be required whenever their characteristics so require

(1) Agriculture:
projects of land reform;
projects for cultivating natural areas and abandoned land;
water management projects for agriculture (drainage, irrigation);
intensive livestock-rearing units;
major changes in management plans for important forest areas.

(2) Extractive industry:
extraction of petroleum;
extraction and purifying of natural gas;
other deep drillings;
extraction of minerals other than metalliferous and energy-producing minerals.

(3) Energy industry:
research plants for the production and processing of fissionable and fertile material;
production and distribution of electricity, gas, steam and hot water (except the production of electricity from nuclear energy);
storage of natural gas.

(4) Production and preliminary processing of metals:
manufacture of steel tubes;
drawing and cold folding of steel.

(5) Manufacture of glass fibres, glass wood and silicate wool.

(6) Chemical industry:
production and treatment of intermediate products and fine chemicals;
production of pesticides and pharmaceutical products, paint and varnishes, elastomers and peroxides;
storage facilities for petroleum, petrochemical and chemical products.

(7) Metal manufacture:
stamping, pressing;
secondary transformation treatment and coating of metals;

boilermaking, manufacture of reservoirs, tanks and other sheet-metal containers;
manufacture and assembly of motor vehicles (including road tractors) and manufacture of motor vehicle engines;
manufacture of other means of transport.

(8) Food industry:
manufacture of vegetable and animal oils and fats;
processing and conserving of meat;
manufacture of dairy products;
brewing and malting;
fish-meal and fish-oil factories.

(9) Textile, leather, wood, paper industry:
wool washing and degreasing factories;
tanning and dressing factories;
manufacture of veneer and plywood;
manufacture of fibre board and of particle board;
manufacture of pulp, paper and board;
cellulose mills.

(10) Building and civil engineering:
major projects for industrial estates;
major urban projects;
major tourist installations;
construction of roads, harbours, airfields;
river draining and flood relief works;
hydroelectric and irrigation dams;
impounding reservoirs;
installations for the disposal of industrial and domestic waste;
storage of scrap iron.

(11) Modifications to development project included in Annex I.

Article 6 details the form of the assessment required by reference to Annex III.

Annex III – content of the information required under Article 6

(1) The description of the proposed project and, where applicable, of the reasonable alternatives for the site and design of the project, including in particular:

the description of the physical characteristics of the main and the associated proposed projects and the land use requirements during the construction and operational phases;

6

Landfill and leachates

Sanitary landfill of domestic refuse

In some ways it may seem odd to examine this subject in a book devoted to hazardous wastes. However, the study of this topic is justified once it is recognised that the distinction between domestic refuse and hazardous waste is a nebulous one. This comment does not imply that domestic refuse is necessarily hazardous but only that experience has shown that even domestic refuse demands diligent management if problems are to be avoided. Examination of this aspect of waste management will reveal what happens as refuse degrades in a landfill site and will serve to indicate what can and can not be done when landfill of more hazardous waste is considered.

In the United Kingdom the term 'controlled tipping' is more commonly used than sanitary landfill. The Ministry of Health [1] has published suggested precautions for refuse tips which are:

(1) Every person who forms a deposit of filth, dust, ashes or rubbish of such a nature as is likely to give rise to nuisance, exceeding (i) cubic yards, must, in addition to the observance of any other requirements which are applicable, comply with the following rules:

(a) the deposit is to be made in layers;
(b) no layer is to exceed (ii) feet in depth;
(c) each layer to be covered, on all surface exposed to the air with at least 9 inches of earth or other suitable substance; provided that during the formation of any layer not more than (x) square yards may be left uncovered at any one time;
(d) no refuse to be left uncovered for more than 24 hours from the time of the deposit (iii);
(e) sufficient screens or other suitable apparatus to be provided, where

necessary, to prevent any paper or other debris from being blown by the wind away from the place of deposit.

(2) Every person who deposits any filth, dust, ashes or rubbish likely to cause a nuisance if deposited in any waters must, so far as practicable, avoid its being deposited in water.

(3) Every person who deposits any filth, dust, ashes or rubbish must take all reasonable precautions to prevent the breaking out of fires and the breeding of flies and vermin on or in such deposit.

(4) If the material deposited at any one time consists entirely or mainly of fish, animal or other organic refuse, the person making such deposit must forthwith cover it with earth or other equally suitable substance at least 2 feet in depth.

(5) Every person who deposits filth, dust, ashes or rubbish must take all practicable steps to secure that tins or other vessels or loose debris likely to give rise to nuisance are not deposited in an exposed condition on or about the place of deposit.

(6) Sufficient and competent labour must be provided in connection with the deposit to enable the necessary measures to be taken for the prevention of nuisance.

(7) So far as practicable each layer of refuse which has been laid and covered with soil must be allowed to settle before the next layer is added.

(8) Wherever practicable the person making the deposit must avoid raising the surface of the tip above the level of the adjoining ground.

(9) All refuse must be disposed of with such despatch and be so protected during transit as to avoid risk of nuisance.

In the footnotes attached to this report it was explained that:

(i) an appropriate figure is to be inserted depending on local conditions;

(ii) unless the circumstances are very exceptional the depth of the layer should not exceed 6 feet;

(iii) reduced from 72 hours mentioned in the Annual Report for 1929–30.

(iv) the appropriate figures should be inserted at (x) in section 1(c) after full consideration of the local conditions. The Ministry will be glad to advise on this point and, in any event, to be informed of the figures adopted.

The term 'sanitary landfill' originates in the USA and is defined by the American Society of Civil Engineers [2] as 'an engineered method of disposing of solid waste on land in a manner that protects the environment, by spreading the waste in thin layers, compacting it to the smallest volume and covering it with compacted soil by the end of each working day, or if necessary, more frequently'. This manual emphasises the need to 'consider leachate generation (seepage of water that might cause water pollution), gas production, odour, noise, aesthetic problems, air pollution, dust, fires, vectors (of disease) and birds' and argues that only careful attention to these factors can secure the necessary public acceptance.

The United States Environmental Protection Agency (USEPA) has published [3] general requirements for operation of a sanitary landfill which include:

(1) wastes for which the specific facility has been designed to hold are the only ones to be accepted – usually bulky wastes, digested and dewatered sludge and septic tank pumpings are restricted and/or prohibited;

(2) certain hazardous wastes should be excluded from municipal landfills – toxic chemicals (e.g. PCB's, pesticides, radioactive wastes, heavy metals, or any other which may contribute to groundwater contamination).

The method of operation of a sanitary landfill is dictated by site characteristics, waste characteristics and the rate of receipt of waste. The equipment employed on a sanitary landfill varies considerably in engineering design but its purpose is to transfer and compact materials (refuse and soil cover) with optimum efficiency. Lists of recommended and optional attachments, auxiliary equipment and supporting tools can be found in the literature [4, 5].

Two basic methods are employed in the operation of a sanitary landfill and minor modifications to these techniques will also be found. These methods are called the trench method and the area method and are shown in Fig. 1. The trench method utilises a prepared excavation into which the refuse is unloaded. This technique is particularly suitable for small operations and provides an excellent means of controlling the amount of daily cover required. The area method is applicable to the infilling of ravines, pits or other areas where much of the soil has already been removed. In both techniques, immediate compaction of the refuse is necessary and a minimum of 6 inches (150 mm) of compacted soil cover should be placed so that only a minimum part of the working face remains to be covered at the end of a day's operation. Complete coverage of all of the deposited waste by the end of the day's operations is essential. A minimum of 2 feet (0.6 m) of compacted final cover should be placed over all completed areas.

It is vital to understand the importance of placing a soil cover at the end of each day's operation over the waste that has been deposited. Its purpose is:

(1) to minimise the infiltration of water from natural sources;
(2) to reduce problems such as dust, fires, vermin and birds;
(3) to control the release of gaseous decomposition products.

The importance of soil cover is stressed repeatedly in authoritative publications and training manuals [6–10].

The degradation of refuse in a landfill

The degradation behaviour of waste in a landfill depends on several interrelated factors. These include the composition of the waste, its physical and chemical properties, the available oxygen supply, the moisture content, the temperature, the age of the fill and the presence of both aerobic and anaerobic organisms. The organic materials found in a domestic refuse

Fig. 1. Sanitary landfill by area and trench methods. (Source: Reference [2].)

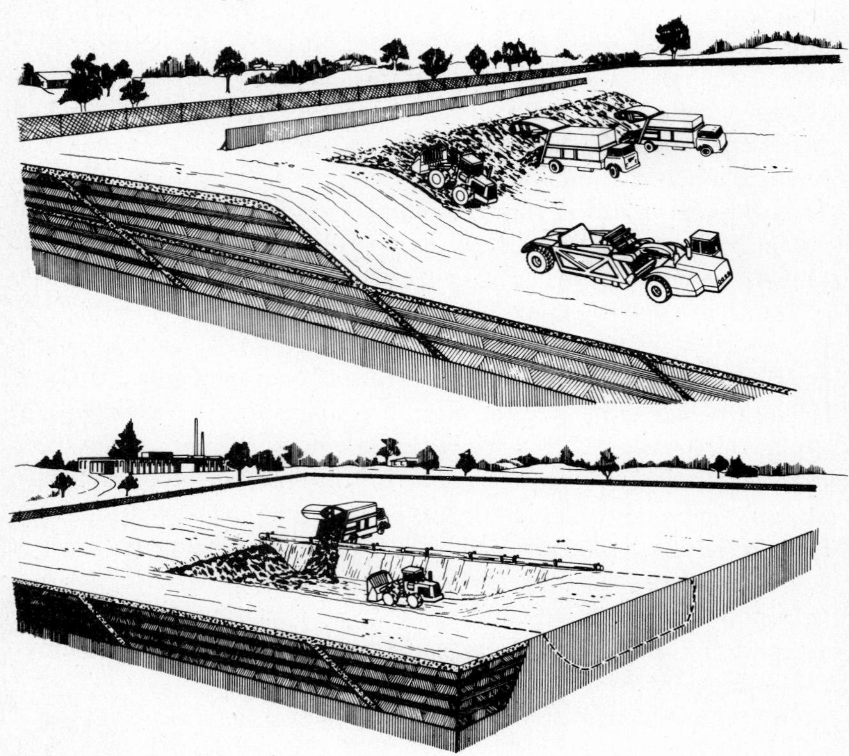

landfill decompose as a result of aerobic microbiological action and, as a result, the temperature increases. As the oxygen is quickly used up anaerobic bacteria continue the degradation process. Anaerobic degradation gives rise to soluble and gaseous compounds. Acetic, propionic and n-butyric acids are the principal soluble organic compounds formed [11].

The composition of the gases generated within a sanitary landfill varies with time. Farquhar & Rovers [12] developed the data shown in Fig. 2 in which phase 1 represents the aerobic process which occurs immediately after the waste is deposited. Phase 2 is an anaerobic process which results in the formation of acidic conditions and the liberation of carbon dioxide, phase 3 results from the action of methanogenic bacteria which decompose the fatty acids, carbon dioxide and hydrogen leading to the production of methane which continues through phase 4. Geyer [68] has published a valuable annotated bibliography on landfill gas production and migration.

Parry [13] reports that carbon monoxide, nitrogen, hydrogen sulphide, ethylene and mercaptans are also produced in trace concentrations in landfill gases. Methane is produced in the largest quantity and, though it is not directly toxic to plants, sufficient concentrations in soil can cause plant death by oxygen displacement around the roots. Hydrogen sulphide, ethylene and mercaptans which are often present in association with methane are known to be highly toxic to plants.

Fungaroli & Steiner [14] studied refuse decomposition in lysimeters constructed at Drexel University, Philadelphia. They established that once a sanitary landfill system is brought to field capacity the generation of

Fig. 2. The composition of landfill gas. (Source: Reference [12].)

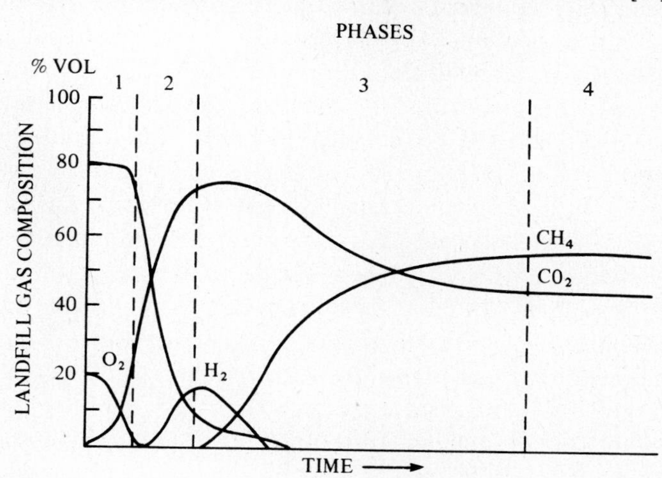

leachate bears a direct relationship to the volume of water added to the system. As water input increases leachate production also increases. Delays in initial leachate generation depend on:

(1) the initial moisture content of the various landfill components – the lower the initial moisture content, the longer the time before appearance of significant quantities of leachate;
(2) landfill bulk density – the higher the density the longer the time before the initial appearance of significant quantities of leachate;
(3) the rate of site filling – the quicker a site is filled, the longer the time before the initial appearance of significant quantities of leachate;
(4) the quantity of water infiltration – the lower the quantity of water infiltration, the longer the time before the initial appearance of significant quantities of leachate.

Leachate production can be attributed to one or all of the following:

(a) the refuse;
(b) channelling;
(c) an advanced wetting front;
(d) a main wetting front.

Before the system reaches field capacity leachate contributed by these sources is primarily due to the advanced wetting front. Once field capacity has been reached, leachate production is due to the main wetting front. It is apparent that leachate produced during the slow attainment of the system field capacity will exhibit initial pollution concentrations different from a landfill in which substantial quantities of leachate are produced immediately. Leaching solutions were generally acidic ranging from pH 5.0 to 7.0. In general, generation of large quantities of acidic leachate intensify the liquid pollution potential because the low pH reduces the exchange capabilities of renovating soil at a time when leachate quantities are high. When leachate quantities are high there are significant increases in iron concentration. The COD ranged most frequently from 20 000 to 25 000 with an early peak of 50 000 mg l^{-1}. After five years COD fell to less than 2000 mg l^{-1}. Within ten days of emplacement refuse temperatures reached 66°C (150°F). Tests established that areas of aerobic and anaerobic activity existed simultaneously side by side within the refuse mass. After five years between 75% and 90% of the water soluble components had been removed from the refuse. Upper refuse layers exhibited a higher degree of removal of inorganic leachate components than lower layers suggesting that the leaching process is progressive through the deposit. Milling of refuse increased the field capacity and the elapsed time before first appearance of leachate. An increase in the contact time between leachate and refuse increased the pollutants in the leachate. Fungaroli & Steiner concluded that the control of infiltrating water is essential in order to keep leachate contamination from migrating too rapidly through the underlying soil.

The bulk density of refuse in a landfill

The assessment of the bulk density of domestic refuse under landfill conditions is an important factor not only because it determines the quantity of waste which can be deposited in a particular void, but also because it has an impact on the rate of degradation of the refuse and the composition of gases and leachates produced.

Bevan [15] in his book *The Science and Practice of the Controlled Tipping of Refuse* reports bulk density measurements made at various locations throughout the UK. Table 1 collates his data and identifies his data sources.

Table 1. *The bulk density of refuse in a landfill*

Comment and data source	Average bulk density $kg\,m^{-3}$
Unspecified technique, P.G. Laws, Bedfordshire	387
As tipped without compaction	266
As tipped without compaction, including cover	327
Push over face and compacted with David Brown TD50 bulldozer	331
Push over face and compacted with David Brown TD50 bulldozer, including incinerator ash cover	364
Tipped into cell and compacted with Allis Chalmers HD tracked shovel	514
Tipped into cell and compacted with Allis Chalmers HD tracked shovel including cover	522
Excavation of 2-year old tip	561
Bulky refuse, push over face and compacted by bulldozer R.E. Bevan, Manchester.	243
Push over face without compaction	145
Push over face with compaction after 3 days	306
Push over face with compaction after 1 year, D.W. Jackson, Sunderland	410

Source: after Reference [15].

Table 2. *The bulk density of domestic refuse using various landfill techniques*

Refuse type	Compactor machine	Method of compaction	Type of cover	Refuse density $kg\,m^{-3}$	Refuse + cover density $kg\,m^{-3}$	Effective refuse $kg\,m^{-3}$
Crude domestic	Crawler caterpillar 955	push over face	permeable soil and brick rubble	430	510	390
				910	1010	740
				670	780	580
	Compactor caterpillar 816	onion skin method		780	950	590
				1080	1250	650
				950	1140	630
		push over face	brick cover on each layer impermeable clay cover above cell	790	860	700
				780	840	690
				780	840	690
				770	850	660
				810	880	700
				790	860	680
Bagged household	none	refuse placed into cell	none	470	—	470
Dry pulverised	Crawler caterpillar 955	minimal push over face	none	690	—	690
Crude domestic	Weatherill L61D loader	1st layer push 2nd layer thin onion skin	permeable soil	510	610	460
				820	890	730
				650	740	580
Crude domestic	Compactor caterpillar 816	push over	soil + brick + clay	650	710	520

Source: after Reference [16].

More recently Campbell & Parker [16] have conducted work on a larg landfill site and measured the compaction density of domestic refuse generated by a range of landfill equipment and methods. Their data are shown in Table 2 and they draw attention to the wide spread in the bulk density of wastes, and waste plus cover which covers a factor of about 2.5. Instead of bulk density they recommend the use of the term 'effective density' which is the weight of waste which can be deposited in a given volume occupied by refuse plus cover. Figures for effective density fall in the range 500–750 kg m^{-3}.

The 'onion skin technique' mentioned in this paper refers to a methodology described by Bratley [17] of placing waste at the base of the face and pushing and compacting it in a series of thin layers up into the face with several compactor passes. Cover stockpiled at the top of the face is pushed over the onion skin layers at the end of each working day.

Comparison of the data from Tables 1 and 2 indicates the considerable progress made in maximising the bulk density of refuse in a landfill in recent years by employing improved machinery and management techniques. However, the point should be made that even with present-day techniques the actual density achieved is likely to vary considerably within a landfill and from one site to another.

The composition of leachate from sanitary landfills

Considerable confusion exists in scientific, environmental and governmental circles about the nature and composition of leachates. For this reason it is necessary to explore this topic in some depth.

Firstly it is necessary to define leachate as the potentially polluting liquor which accumulates beneath a landfill site resulting from the infiltration of water and the natural decay of refuse. It is important to distinguish between leachate and polluted surface waters which often occur on the surface or near the perimeters of landfill sites. Such waters may well contain some leachate but frequently they have been diluted by rainfall or naturally occurring surface waters. Though such liquors might well threaten to pollute important surface streams they should not be confused with leachate.

Leachate inherits its importance since it accumulates steadily over a period of time in an unseen fashion so that few people recognise its existence or appreciate its ability to do harm. The danger is that it can seep into underground aquifers thereby affecting important sources of potable or industrial water supply. The seepage can be either slow and undetected or more rapid and dramatic, for example from the fracture of some protective barrier or change in water course.

Clearly, the greater the accumulation of leachate, the greater is the potential for environmental damage but perhaps it should be noted that no matter what steps are taken to maintain 'dry' conditions in a landfill the generation of some leachate is unavoidable. Domestic refuse itself contains water and, indeed, water is an inevitable product of the degradation process.

In the literature large differences in the composition of leachate are found which are partly due to the factors mentioned above but also to the very different ways of collecting and analysing the leachate samples. Table 3 shows data reported by Hoeks [18] which were collected from newly dumped refuse in a comparable way. Hoeks also reports the concentrations of heavy metals in domestic refuse leachate as

lead Pb 0.020–0.300 mg l^{-1};
copper Cu 0.045–0.300 mg l^{-1};

Table 3. *The composition of leachate from domestic refuse landfills*

Component		SVA (1974)	Zanoni (1973)	Mead Wilkie (1971)
COD	mg l^{-1}	63 900	—	—
BOD	mg l^{-1}	—	33 100	32 400
Cl^-	mg l^{-1}	3950	1810	2240
SO_4^{2-}	mg l^{-1}	1740	560	630
HCO_3^-	mg l^{-1}	14 430	—	—
Organic NH_3-N	mg l^{-1}	390	320	550
Inorganic NH_3-N	mg l^{-1}	1410	790	845
NO_3-N	mg l^{-1}	—	—	—
Total-PO_4^{3-}	mg l^{-1}	25.5	9.6	—
Ortho-PO_4^{3-}	mg l^{-1}	6.8	—	—
Total Fe^{2+}/Fe^{3+}	mg l^{-1}	1590	270	305
Ca^{2+}	mg l^{-1}	2625	2190	—
Mg^{2+}	mg l^{-1}	450	340	—
Na^+	mg l^{-1}	2990	1470	1805
K^+	mg l^{-1}	1800	1115	1860
pH		5.7	—	5.6
EC (25°C)	mhos cm^{-1}	32 400	—	—

Source: after References [19–21].

zinc Zn 28–30 mg l^{-1};
chromium Cr 0.120 mg l^{-1};
cadmium Cd 0.250 mg l^{-1};
arsenic AS 0.110–0.160 mg l^{-1};
nickel Ni 0.600–1.050 mg l^{-1}.

Tchobanoglous [22] reports the data in Table 4.

Robinson [28, 48] has recently reported data from Water Research Centre studies of 23 samples of leachate from 15 disposal sites in the UK which are shown in Table 5. He reports that the samples were collected, treated and analysed in the same manner but details of the techniques employed were not described. He attributes the large differences to short-term dilution effects.

Table 4. *The composition of leachate from landfills*

Constituent	Value* mg l^{-1} Range**	Typical
BOD (5-day biochemical oxygen demand)	2000–30 000	10 000
TOC (total organic carbon)	1500–20 000	6000
COD (chemical oxygen demand)	3000–45 000	18 000
Total suspended solids	200–1000	500
Organic nitrogen	10–600	200
Ammonia nitrogen	10–800	200
Nitrate	5–40	25
Total phosphorus	1–70	30
Ortho phosphorus	1–50	20
Alkalinity as $CaCO_3$	1000–10 000	3000
pH	5.3–8.5	6
Total hardness as $CaCO_3$	300–10 000	3500
Calcium	200–3000	1000
Magnesium	50–1500	250
Potassium	200–2000	300
Sodium	200–2000	500
Chloride	100–3000	500
Sulphate	100–1500	300
Total iron	50–600	60

*Except pH.
**Representative range of values. Higher maximum values have been reported in the literature for some of the constituents.
Source: after References [23–27].

More recently, Lu et al. [29] have reported data obtained as a result of USEPA surveys. Their data are reproduced in Fig. 3. Close scrutiny of the data reported above indicates some substantial differences between the British results reported by Robinson [28] on the one hand and Dutch and American data [19–29] on the other. The differences occur in pH, organic content and heavy metal content and warrant further study.

The collection and analysis of leachate samples

Since leachate from waste disposal sites is strongly reducing in nature, the collection and analysis of samples demands great care, particularly where the detection and determination of trace components is concerned. The drilling of sample boreholes into refuse sites and the strata beneath them also requires considerable expertise. Naylor et al. [30]

Table 5. *Leachates from solid domestic wastes in landfills*

All Values in mg l^{-1} except pH		
pH		6.2–7.45
COD		66–11 600
BOD		<2–8000
TOC		21–4400
Ammoniacal-N		5–730
Organic-N		ND–155
Nitrate-N		<0.5–4.9
Nitrite-N		<0.2–1.8
Ortho-P		<0.02–3.4
Chloride		70–2777
Sulphate		55–456
Sodium	Na	43–2500
Magnesium	Mg	12–480
Potassium	K	20–650
Calcium	Ca	165–1150
Chromium	Cr	<0.05–0.14
Manganese	Mn	0.32–26.5
Iron	Fe	0.09–380
Nickel	Ni	<0.05–0.16
Copper	Cu	<0.01–0.15
Zinc	Zn	<0.05–0.95
Cadmium	Cd	<0.005–0.01
Lead	Pb	<0.05–0.22

Source: after References [28, 48].

describe, in a Water Research Centre publication, the construction and use of various types of boreholes. They recommend that auger holes drilled within landfill areas should be terminated at the base of the fill material to avoid the risk of pollution of underlying aquifers. Percussion drilling and airflush and cable-tool techniques are also described. In a section relating to the recovery of samples from the landfill, these authors say that 'fluid samples from both the landfill and the groundwater zone may be collected

Fig. 3. Range of municipal landfill leachate constituent concentrations. (Source: Reference [29]. Courtesy of US Environmental Protection Agency.)

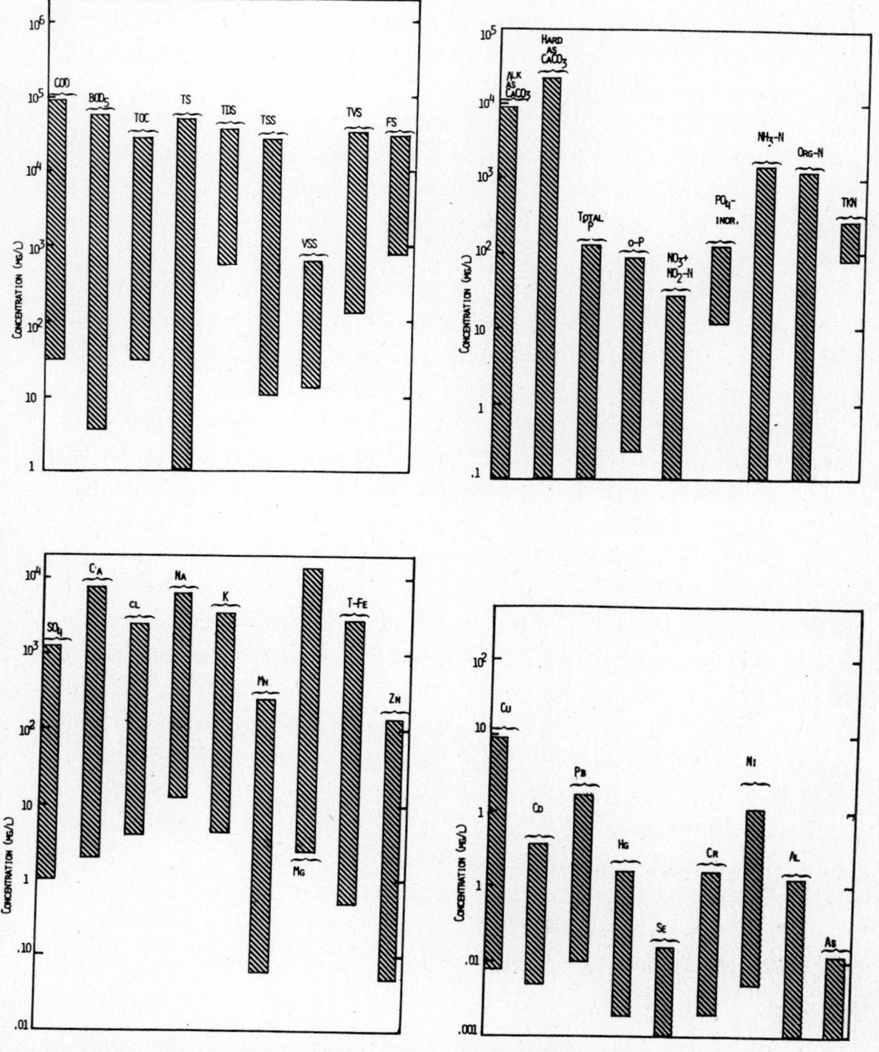

FIGURE 4 . RANGE OF MUNICIPAL LANDFILL LEACHATE CONSTITUENT CONCENTRATIONS

during the drilling by using a simple "bucket-on-a-string" type sampler. A more positive approach to the problem of representative sampling is to take samples from various levels in the borehole, using a depth sampler "triggered" to retain a water sample from that depth and recorded.' A borehole packer is described to overcome the problem of mixing in open boreholes employing a small submersible pump.

An alternative method is to use *in situ* sampling devices, sealed in a particular location within a borehole. The sampling device and a small reservoir are connected to the surface by plastic tubes and samples removed by 'suction from shallow depths or compressed gas from deep installations'. The report recommends that 'liquid samples should be preserved at low temperature in glass or plastic bottles, possibly after acidification or the addition of bactericides'. The report contains no advice on the avoidance of exposure to air.

Chian & DeWalle [31] of the University of Illinois reviewed the analytical methods reported in the literature and studied interferences in the chemical analysis due to the complex nature of leachate. Their research showed that sample collection techniques could have a profound effect on the data obtained. For example, the size of the orifice of the sample tube was found to effect suspended solid and heavy metal concentrations. The maintenance of anaerobic conditions and the exclusion of air were shown to be necessary. The materials of construction of the sampling device and sample containers needed to be specified and the detention time between sampling and analysis minimised and reported. They reported that leachate collected from a recently installed solid waste fill will have a translucent light-brown colour but it will turn dark-green or black and become turbid immediately after collection due to aerobic exposure and the subsequent oxidation of heavy metals and organics. The odour is very disagreeable and nauseating and is generally due to the presence of free volatile fatty acids such as butyric and valeric acids. Collection of leachate samples below the surface of older sites will result in the liberation of carbon dioxide gas bubbles when the sample is exposed to atmosphere. Attempts reported by these authors to preserve leachate at 4°C in a fully capped 4-litre glass bottle did not stop changes in leachate composition occurring. The oxidation of ferrous ion to ferric hydroxide contributed to the increase of colour, turbidity and suspended solids. The oxidation of organic matter and iron caused a decrease in COD, the formation of a precipitate and a decrease in conductivity. As a result, these authors recommended that several parameters should be determined directly after collection of the sample. Leachate collection should be under anaerobic conditions; a tightly stoppered glass bottle should be used for organic analysis and a polyethylene bottle for metal analysis. The sequence of parameter analyses should be oxidation-reduction potential, colour, turbidity, suspended solids, pH and conductivity. Other parameters such as COD and organic N may also change directly after sampling, but these changes may be reduced when the sample is acidified. Acidification and storage at 4°C will stop the methane fermentation responsible for free volatile fatty acid removal and slow the bacterial acid fermentation of complex organic substances. Acidification, however, enhances volatilisation of undissociated fatty acids, precipitates humic-like organics and facilitates hydrolysis of complex organics. The use of 'Standard Methods of Analysis' [32] is recommended and techniques such as the Standard Addition Method and Dilution Method are advised in order to allow for interferences.

More recently, work undertaken by Brookhaven National Laboratory in connection with the migration of radionuclides from low-level radioactive waste repositories in USA has thrown new light on the importance of sampling and analytical techniques. Low-level radioactive waste (LLW) consists of a variety of laboratory, hospital and reactor equipment, residues and trash which are contaminated with radioactive materials. For the most part, these materials include clothing, plastics, paper, ion-exchange resins, animal carcasses, decontamination agents and other materials in small quantities. In essence their nature is similar to industrial wastes generated by a wide range of manufacturing plants except that the inclusion of traces of radionuclides permits the detection and quantification of migration characteristics more readily.

Weiss & Colombo [33] surveyed four disposal sites in the USA in which LLW had been disposed of in shallow trenches backfilled with earth, compacted and clay capped. They showed that leachate accumulating in the trenches (trench water) was complex and anoxic and they demonstrated that special care was required in collecting meaningful samples of leachate. Initially leachate samples were collected by lowering plastic tubing into the trench liquors and pumping out. Samples were subjected to field measurement of specific conductance, temperature and pH within one hour of collection, filtered and then stored in glass bottles for subsequent organic analysis and plastic bottles for subsequent metal analysis. It was established that samples collected in this way were valueless due to the oxidation of ferrous iron by air contamination. This resulted in the precipitation of ferric hydroxide and radionuclides (metals) and other solutes that were originally present in the leachate may have been co-precipitated with or adsorbed onto the precipitated hydroxide.

Fig. 4. Protocol for the preparation of samples of leachate for analysis.

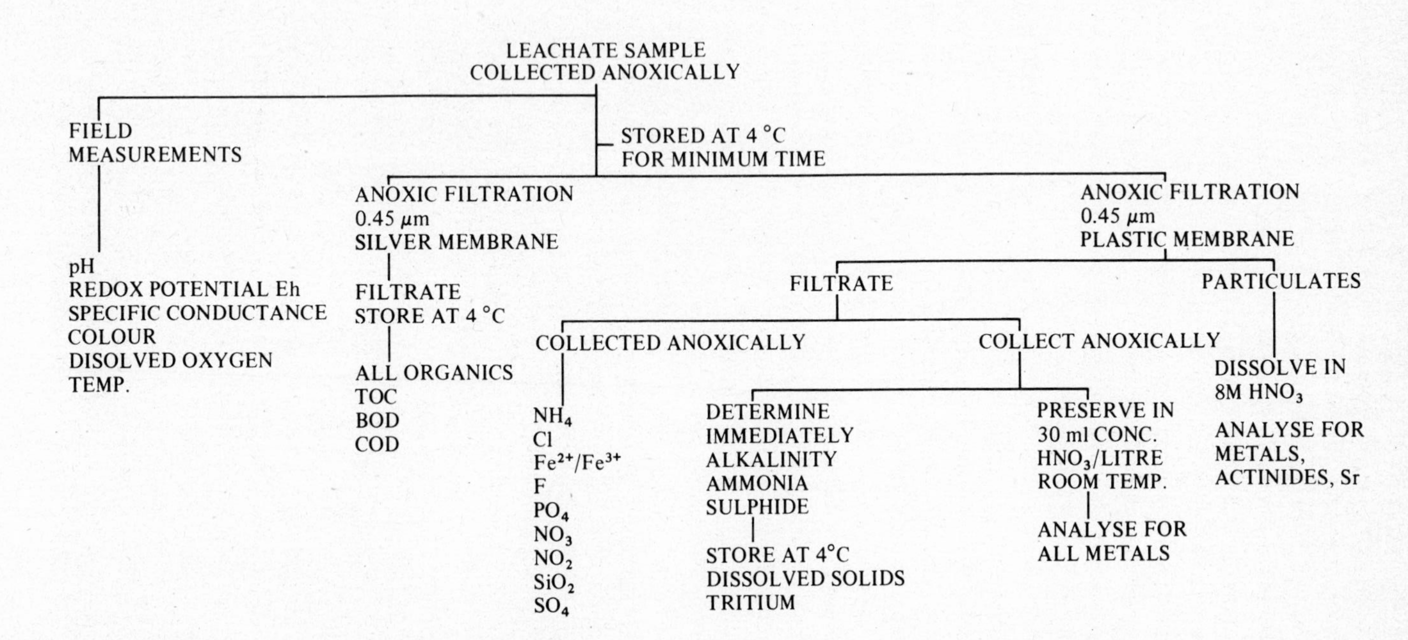

Consequently it was necessary to develop improved sampling techniques using nitrogen or argon gas which eliminated air contamination. The schematic flow diagram for anoxic sampling is shown in Fig. 4. Air barrier coils are waterfilled traps which are included to prevent air re-entering the sampling system through the waste containers. The entire collection system is purged with nitrogen or argon gas and the first several litres of recovered leachate discarded to compensate for the possible air contamination at the borehole leachate surface. It will be noted that pH, Eh, conductivity, dissolved oxygen and temperature are measured *in situ* during the sampling process. Fig. 5 shows the field equipment used.

Procedures were developed to maintain the anoxic nature of the leachates. A special bottle (the Brookhaven Bottle), Fig. 6, consisting of a 4-litre borosilicate glass vessel having two high vacuum-type stop cock valves, was developed for the anoxic collection and subsequent removal of aliquots of leachate, see Fig. 7. Leachate samples collected and stored in this way showed no visible signs of the formation of reddish-brown ferric hydroxide even after several months of storage. From the analytical regime used for partitioning the leachate samples, shown in Fig. 4, it will be noted that even the filtering procedure needs to be carried out under anoxic conditions. Inorganic components were determined using ion-specific electrodes,

Fig. 5. Brookhaven anoxic leachate sampling field kit.

colorimetric methods and atomic absorption methods. Organic compounds were determined by gas chromatographic and mass spectrometric analyses. Radiochemical techniques were used to study the radioactive components.

These researchers found that the leachates removed from the disposal trenches were all anoxic and they showed that strontium-90, caesium-137 and plutonium-238, -239 and -240 were all found as dissolved species in the trench waters. They recognised the existence of numerous organic compounds in the trench waters and identified the components listed in Table 6.

These compounds reflect the nature of the buried waste and the products of the natural biodegradation process. The low molecular weight straight

Fig. 6. The Brookhaven Bottle.

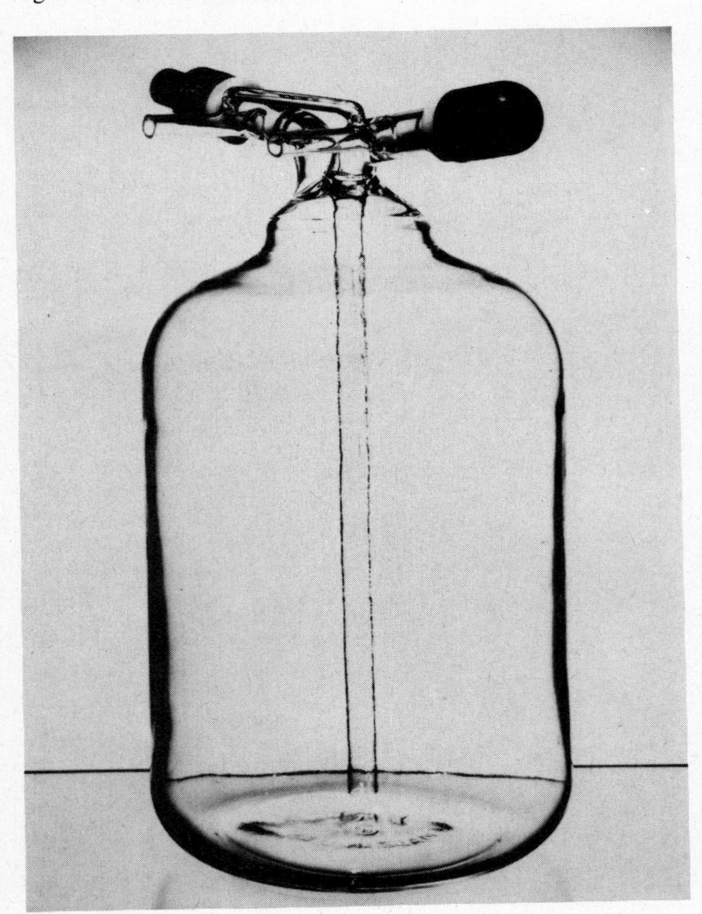

and branched chain aliphatic acids are the microbial degradation products of complex natural and synthetic materials in the buried waste. These findings would not have been possible if the researchers had used less rigorous oxic sampling conditions.

Recent studies by Cleveland & Rees [34] of the US Geological Survey confirm the inferences of Brookhaven work. Using anoxic sampling conditions, they have shown that plutonium has formed soluble organic ligands with the components of leachates at a landfill disposal site and that complexes of plutonium are not adsorbed under anoxic conditions, onto precipitated hydroxides or sediments. The authors conclude that the results indicate the importance of isolating radioactive waste from organic matter in a disposal site.

Enquiries made by this author have confirmed that leachate samples analysed by the Water Research Centre and reported in Table 5 (and references [28] and [48]) were obtained by either a 'bucket-on-a-string' technique where leachate flow existed or by collecting surface seepage. Glass sample bottles were used and no precautions were taken to exclude air. The samples were filtered before analysis. Under such conditions dilution with surface water, release of volatile organic acids, an increase in pH, the precipitation of iron and manganese hydroxides and the removal of toxic metals and some organic components by adsorption on to the precipitate is likely. Consequently, the samples are not representative of leachate existing at the base of domestic refuse landfill sites.

Differences in the methods employed for the collection of leachate samples have a profound effect on the analytical results obtained and could account for the different interpretations and conclusions developed by

Fig. 7. Schematic diagram depicting the apparatus for anoxic filtration of water. (Source: Reference [4].)

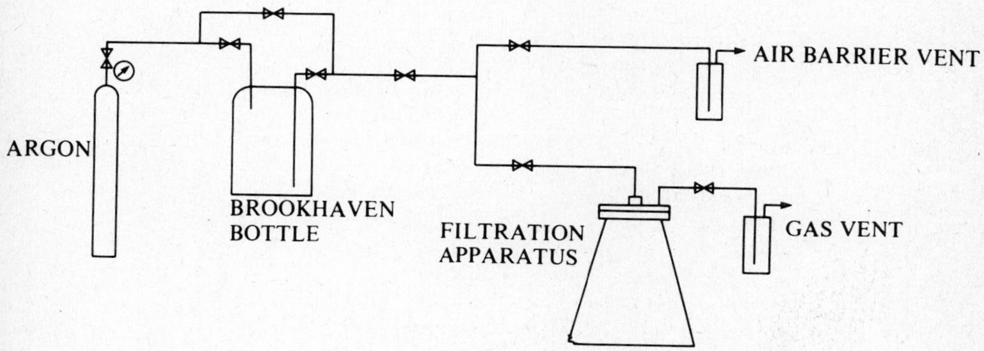

British and American authorities. Already, research work reported in OECD studies [35] is showing wider recognition of the importance of anoxic conditions and complex formation in radioactive waste studies and the time has now come to recognise that such considerations play a major role in leachate migration and attenuation in all waste disposal landfill sites.

A proposed analytical protocol for the collection and analysis of the leachate occurring beneath landfill sites is described in the Appendix of this book.

Leachate control and impoundment

Few authorities would disagree that the prevention of landfill leachate is a more certain and more economical philosophy than its treatment. Good landfill practice demands the identification, diversion or containment of all surface water flows. In the case of hazardous wastes, precautions are also necessary to seal the deposits in a confined location. In Britain, this was recognised by the Key Committee Report [36] which referred to the need for an umbrella-like surface cover to minimise infiltration. This explains the stipulation that deposits of waste in a landfill should be covered at the end of each working day by an impervious layer but, furthermore, in recent years attention has been given to the idea of

Table 6. *Organic components of the leachates found in waste burial trenches*

2 methylbutanoic acid	2 ethyl 1 hexanol
3 methylbutanoic acid	fenchone
C_6 branched acid	camphor
phenol	naphthalene
hexanoic acid	isobutyric acid
2 methyl hexanoic acid	2 methylbutyric acid
C_8 acids	3 methylbutyric acid
Cresol and isomers	pentanoic acid
2 ethyl hexanoic acid	2 methylpentanoic acid
benzoic acid	3 methylpentanoic acid
octanoic acid	phenyl propionic acid
phenyl acetic acid	xylene
phenyl hexanoic acid	cyclohexanol
p-dioxane	2 terpineol
toluene	tributyl phosphate
Bis (2 ethoxyethyl) ether	toluic acid

Source: Reference [33].

employing man-made plastic materials as an impermeable top cover. Emrich & Beck [37] describe a remedial action programme at a Connecticut landfill in which a PVC membrane seal was placed above a re-graded landfill, see Fig. 8. The membrane was a 20-mil PVC sheet with solvent welded seams covering a 10-hectare (25-acre) site. Immediately beneath the membrane was 10 cm (4 in) of compacted fine grain sand washings. The membrane was covered with 50 cm (18 in) of local fine sand and gravel which was spread in 15-cm (6 in) layers and then revegetated. Lysimeters located in the refuse were used to study leachate plume migration.

About one year after completion, seven test pits were excavated into the cover material. This examination revealed numerous indentations up to 4 cm (1.5 in) deep in the membrane but there were no obvious signs of puncture. However, removal of sections of membrane revealed many punctures on holding the membrane up to the light. The largest perforations were about 2.5 cm (1 in) in length and all of the breaks were from the top down. Rainwater, which had moved down the soil cover till it reached the membrane and then moved along the membrane, was found to be penetrating the seal at at least one test pit location. The bulk of the water, however, passed across the seal to the diversion ditches and the lysimeter studies also confirmed improvements in quality and quantity of leachate. Results to date are taken from a one-year on-going study and make no observations on gas formation but so far look encouraging.

A new development using synthetic foam as a landfill cover material appears interesting [38]. A urea–formaldehyde resin is foamed with a surfactant and jetted on to landfill surfaces as top cover or cover for vertical walls. The makers claim that the material has a high permeability to gases

Fig. 8. The use of a synthetic membrane as a top seal.

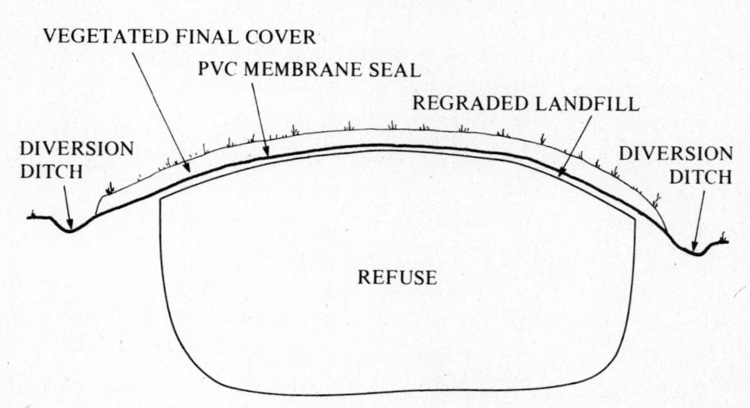

yet its hydrophobic nature and small cell size resist water penetration. Its use permits the control of dust, insects, vermin and birds and, since it collapses to a very thin layer when the next layer of waste is deposited, it releases more void volume for waste disposal.

Naturally, leachate control can also be attempted by the use of impermeable linings at the base of landfill sites and several candidate materials are available for this purpose. Fong & Haxo [39] studied the effects of exposure of 12 lining materials for 56 months to domestic refuse leachate under conditions designed to simulate those at the bottom of a domestic refuse landfill (i.e. anaerobic).

The materials studied were:

Admix materials
paving asphalt concrete;
hydraulic asphalt concrete;
soil asphalt;
soil cement.

Asphaltic membranes
bituminous seal;
emulsified asphalt on a nonwoven fabric.

Polymeric membranes
butyl rubber (63 mils);
chlorinated polyethylene (31 mils);
chlorosulphonated polyethylene (34 mils);
ethylene, propylene rubber (51 mils);
low density polyethylene (11 mils);
polyvinyl chloride (20 mils).

Exposure resulted in loss of compressive strength of the admix materials and in softening of the asphaltic materials. Most of the polymeric materials suffered from swelling and, in some cases, loss in physical properties. There was no increase in permeability in any of the liners studied.

These authors concluded that the loss of compressive strength of asphalt lined materials could lead to problems in a landfill over a long-term period (i.e. greater than 56 months). The soil cement specimens gained in strength and decreased in permeability over the test period but soil cement is brittle and large areas are subject to cracking.

The low density polyethylene sustained the least change during the exposure period but the membrane was very thin (11 mils) and would be expected to puncture and tear. Chlorinated polyethylene and chlorosulphonated polyethylene tended to swell the most. The butyl rubber and ethylene propylene rubber changed little but had the lowest initial seam strength.

Practical studies of the durability of synthetic membrane seals for landfills have also been reported by Gunkel [40]. Liners made from elasticised polyolefine, PVC, chlorinated polyethylene, reinforced chlorosulphonated polyethylene and non-woven polypropylene and nylon were carefully placed within sandwiches of varying thicknesses of crushed gravel, clayey sand, sand, gravelly sand, coarse gravel and combinations of these. Vehicular traffic in the form of a tracked bulldozer, a pneumatic tyred tractor and a cleated steel wheeled landfill compactor was driven over the test bed in straight lines. Specimens of the membranes were then examined over a light table. Punctures were detected in every case and the majority were in the upward direction, indicating that the quality of the subgrades of cover beneath the membrane played a major role. The three types of vehicle (tracked, tyred and cleated) produced similar amounts of damage but the best results were obtained when using 18 inches of sand and sandy silt subgrades beneath the membrane and 6–18 inches of these materials above the membrane. However, the authors were reluctant to specify the best membrane or its thickness from the limited data available.

Of course, clay is frequently used to impound waste products as a bottom seal, daily cover and top cover material. Its use stems from the fact that it is a plentiful, naturally occurring material and has a permeability to water as low as 10^{-7} cm per second. It is recognised that conditions at the base of landfill are quite unlike the laboratory conditions normally employed to measure permeability coefficients where distilled water or a standard permeant such as 0.01 $\underline{N}$ $CaCl_2$ or $CaSO_4$ are used. For this reason, Anderson & Brown [41] have undertaken clay permeability studies using permeants similar to those found in landfills. These authors point out that in a landfill accepting aqueous inorganic and organic wastes, organic solids and sludges, two categories of leachate are present. Firstly, there are the flowable constituents of the waste which are called 'primary leachates' and, secondly, there are flowables generated from percolating water which are called 'secondary leachates'. The nature of the primary leachate solvent phase depends on the composition of the waste and may be predominantly water or any organic fluid discarded by industry. The solvent phase of secondary leachate is water and all the components (solutes) dissolved in this water. These concepts are shown in Fig. 9. This examination shows that the components of industrial waste leachate which come into contact with clays are those shown in Fig. 10. Organic acids are any organic fluid containing an acid functional group and have the potential to be very reactive with and mobile in clay liners. Organic bases are any organic fluid capable of accepting a proton to become an ionised cation and since they are positively charged they tend to strongly adsorb onto clay surfaces causing volume changes by changing interlayer spacings and dissolving certain constituents of the clay. Neutral non-polar compounds (aliphatic and aromatic hydrocarbons) have no charge and little, if any, dipole moment and have the potential for rapid movement through clay liners, eroding the pores through which they pass and increasing permeability. Neutral polar compounds have no charge but exhibit stronger dipole moments than neutral non-polar organics. They include alcohols, aldehydes, glycols, alkyl halides and ketones. Water, of course, has a strong dipole moment and high dielectric constant and can cause a clay liner to shrink, swell, heave, crack or pipe. Water may also act to increase the hydraulic gradient that moves fluids in soils.

Anderson & Brown employed a standard permeameter to determine the permeability characteristics of compacted clays at high pressures. Special care was taken to prevent air pockets and obtain leak-proof seals in the equipment. Two smectite clay soils have been studied. Smectite minerals are composed of expandable lattices and, therefore, exhibit a large capacity for shrinking and swelling. One clay was essentially calcium saturated and

Fig. 9. Components of leachate from waste disposal sites. (Source: after Reference [41].)

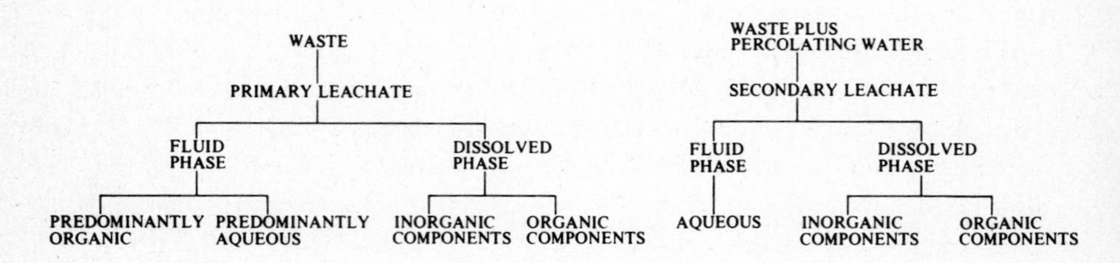

Fig. 10. Components of the solvent phase of leachate from a waste disposal site. (Source: after Reference [41].)

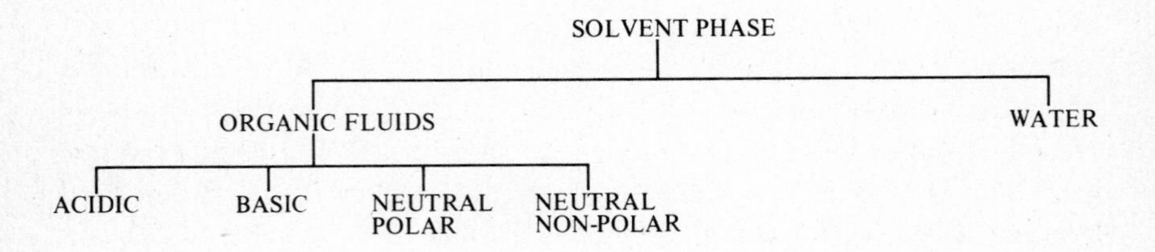

the other sodium saturated at its cation-exchange sites. In the place of actual leachates, reagent grade organic fluids were used as test permeants. These included acetic acid, aniline, acetone, ethylene glycol, xylene and heptane. Acetic acid caused a decrease in permeability, whereas aniline showed substantial increase in permeability of both clays. Acetone showed an initial decrease and then a large increase in permeability of 1000-fold and 50-fold respectively. Ethylene glycol showed 100-fold and 4-fold increases, respectively, in the permeability of the calcium and sodium saturated clays. Xylene and heptane increased permeability by greater than 1000-fold and 100-fold respectively. The authors conclude that the significant increases in permeability caused by several organic fluids points to the need to test the permeability of clay liners with the actual leachates generated by the disposed wastes.

An American company has developed a technique of employing specialised bentonite clays (which is now also available in Europe [42]) for sealing landfill sites. Various products are available consisting of chemically treated sodium bentonite. In bottom sealing applications, a 4-inch layer of the bentonite product is mechanically spread, disced into the subsoil then compacted and an 18-inch layer of native soil applied above the blanket. The manufacturers claim that the bentonite swells on contact with water and forms an impervious though flexible membrane at the base of the landfill. The membrane is reported to be unaffected by contact with landfill leachate. A similar technique is used to provide an impervious top cover to landfill sites.

This same company reports success using bentonites in slurry trenches. This technique is employed to intercept below-groundwater flows and can be used to stop ingress of water into a landfill or, alternatively, migration of leachates out of a landfill. A trench is dug around the landfill area and during this excavation a prepared slurry of bentonite is continuously pumped in. The trench must be deep enough to key into a naturally occurring impervious strata beneath the landfill. The trench is then backfilled with the excavated soil. The trench then forms an impervious cut-off and, providing it forms a satisfactory seal with the underlaying impervious strata, will act to intercept leachate seepage.

Leachate treatment

Several techniques have been used in laboratory studies, pilot scale and full scale tests for the treatment of leachate from domestic refuse landfills. These treatment possibilities will now be reviewed.

Treatment with municipal sewage

The high organic and ammonia content, the variable load, the low phosphate content and the possible presence of trace metals inhibits the biological oxidation of leachate. Palit & Qasim [43] showed that conventional activated sludge processes required the addition of phosphate-bearing nutrients and suffered problems due to sludge formation and poor liquid–solid separation. Experimental studies by Boyle & Ham [44, 45] showed that the addition of leachate with a BOD of 8800 mg l^{-1} to domestic sewage (BOD 140 mg l^{-1}) inhibited the performance of the activated sludge process. Addition of 2% leachate to the sewage had no noticeable effect but when this was increased to 5% the effluent quality was impaired and its BOD increased 50%. When the leachate content was increased beyond 5%, these authors reported substantial solids production, increased oxygen uptake rates, solid separation problems and unsatisfactory BOD levels in the effluent.

Chian & DeWalle [46, 47] also conducted laboratory scale studies on leachate–domestic sewage blends and found that leachate contents of around 2% had no effect on the BOD of the effluent but the COD of the effluent increased with increasing proportion of leachate in the influent, indicating the presence of larger quantities of refractory organic substances. When the influent contained 4% landfill leachate the oxidation process failed as shown by high effluent BOD and deteriorating sludge separation. These authors deduced that a BOD:phosphorus ratio of about 100:1 represented a limiting constraint.

These factors probably explain why only a very few British sewage works accept landfill leachate and those that do limit the input to a small fraction of the influent [48].

Aerobic biological processes

Aerobic biological processes offer an alternative potential treatment method. It is to be expected that leachates from recently deposited waste which contain mainly volatile fatty acids can be readily degraded by biological means. Boyle & Ham [44, 45] demonstrated removal of 80–93% of COD by aerating over a five-day retention period at low organic loadings but as the loading increased the process efficiency fell considerably. The high power consumption and the need for antifoam agents discouraged them and they recommended that investigation should be concentrated on anaerobic processes.

Uloth & Mavinic [49] used aeration to treat high-strength leachate and obtained 96.8% COD removal with ten-day retention times and 98.7%

COD removal with 20-day retention times. The effluent BOD's were 129 mg l^{-1} after ten days' retention and 30 mg l^{-1} after 20 days' retention. They reported that the use of antifoams and nitrogen and phosphorus-containing nutrients were necessary. Phosphate addition also proved necessary when Chian & DeWalle [46] used aerated lagoons for leachate treatment. They concluded that aerobic biological treatment of leachate was not successful at high organic loadings and low retention periods (less than seven days) due to the formation of stable refractory compounds such as fulvic acids. They argued that old stabilised landfill sites also produced fulvic-acid-containing leachates which explained why leachates from such sources were not amenable to aerobic biological treatment.

Robinson, Barber & Maris [28] have reported studies on the treatment of leachate from domestic refuse sites conducted at the Water Research Centre. The studies included aerobic biological treatment performed by aeration. It was recognised that, in Britain, winter conditions produce the greatest quantities of leachate simultaneously with the lowest temperatures and, consequently, the influence of temperature on leachate oxidation was clearly important. Aeration studies were conducted at 10°C and 20°C. Although aeration at 20°C reduced the COD by 84% and BOD by 98% over a 30-day period, aeration at 10°C was much slower with only a 20% reduction of COD in a 50-day period. The phosphorus deficiency of the leachate was recognised and the addition of Na_2HPO_4 and KH_2PO_4 to reduce the BOD:P ratio to 100:1 caused rapid reduction of COD and BOD values in the next seven days. Thus, it was established that addition of phosphorus was essential.

A series of continuous flow 20-litre aeration vessels was then established to study leachate oxidation. The influent leachate was adjusted to have a constant COD of 5000 mg l^{-1} and phosphate solutions were added. The tests were conducted at 10°C with retention periods ranging from one to 20 days. Ammonia was reduced to less than 1 mg l^{-1}. Fig. 11 shows the effect of retention time on COD reduction. The reduction in soluble BOD exceeded 99% after ten or more days' retention. The researchers reported the production of bulky filamentous sludges which interfered with reactor operation, particularly with low retention times when sludge settlement was poor, giving poorly clarified effluents. They concluded that retention periods greater than ten days are required. It was also shown that a reduction in temperature to 5°C resulted in adverse effects, particularly on the settling properties of the sludges, but substantial removal of soluble organic material could still be obtained.

In the writer's view the application of such aeration techniques to the treatment of leachates from a large scale landfill is likely to be problematical. The variations in BOD and COD of the influent leachate, the variations in flow due to uncontrolled climatic changes, the need to provide ten days' and more retention time, the poor performance in cold weather, the requirement to add phosphate nutrients and the problem of settling bulky filamentous sludges under variable flow conditions seem likely to be insuperable in all but the most sophisticated leachate treatment plants.

Anaerobic treatment

The principal advantages of anaerobic treatment of leachates when compared to aerobic processes is that they produce a much lower proportion of sludges, they generate byproduct gases and their power consumption is generally lower.

Anaerobic biological treatment degrades the complex organic compounds present in the leachates to form volatile fatty acids such as acetic, propionic and butyric acids. Methanogenic bacteria convert these by-

Fig. 11. Percentage of COD removed from solution plotted against mean period of retention in days for complete mixing aeration units, operated at a temperature of 10°C without sludge recycling.

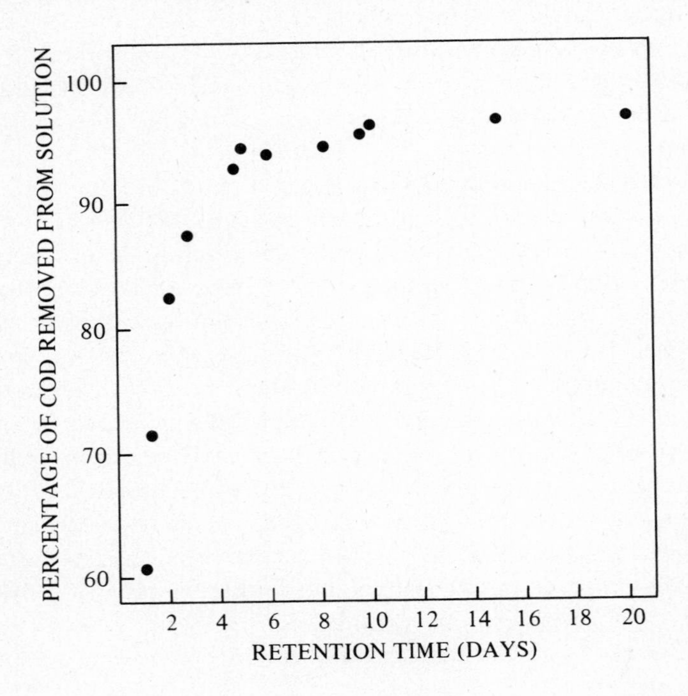

products to methane and carbon dioxide. However, methanogenic bacteria are inhibited easily by acidic pH values [50, 51] and are sensitive to the presence of metals [52, 53]. Studies at the Water Research Centre [54] have shown that the maximum acceptable metal concentrations to prevent the inhibition of anaerobic treatment are

zinc	1.1 mg l^{-1};
nickel	2.0 mg l^{-1};
lead	7.0 mg l^{-1};
cadmium	3.8 mg l^{-1};
copper	3.2 mg l^{-1};
tin	9.0 mg l^{-1}.

Attempts to overcome these disadvantages using an anaerobic filter have been made in laboratory studies. The filter consists of a bed of coarse solid which forms a base for the microorganisms. Influents are pumped upwards through the bed. Results to date [44, 45] indicate that the process is most effective at temperatures in the range 20–30°C and retention periods of ten days and more are required. The use of an upflow anerobic filter is reported to overcome problems due to toxic constituents since the metals are precipitated as sulphides, carbonates and hydroxides [55].

Full scale, on-site, anaerobic treatment of leachate has not been reported in the literature but Robinson & Maris [48] believe that insufficient data are available to permit reliable design and, in any case, some form of leachate heating would be necessary to maintain adequate working temperature in British winter conditions.

Physico–chemical treatment methods

Physico–chemical treatment methods are frequently employed to purify industrial effluent streams and many attempts have been reported to apply these techniques to the treatment of landfill leachates. The concept is that physico–chemical methods should be an adjunct to aerobic or anaerobic biological treatment methods. As a landfill stabilises with the passage of time, the biodegradable organic content of the leachate decreases and, consequently, the effectiveness of biological processes decreases and physico–chemical processes may become more appropriate.

Physico–chemical methods consist of the addition of chemicals to precipitate, flocculate and coagulate solids or to oxidise inorganic and organic components. Lime, ferric chloride and alum have been shown to have little effect on the organic content of leachates from recently emplaced wastes [55–60] but removal of metals, colour and suspended solids is achieved though large quantities of sludge are produced. Where these coagulants have been added to leachate, following biological treatment, significant reductions in colour and iron contents have been reported [44, 60] but little or no change in organic concentrations.

Leachates have been oxidised using chemical oxidants. Calcium hypochlorite proved more effective than chlorine, potassium permanganate or ozone, reduced COD by 48% and removed colour and iron when used at high dosages [44, 60]. However, the chloride concentration of the effluent was very much increased. Manufacturers of hydrogen peroxide recommend its use to remove sulphides from leachates. At pH of 8.5 or less, 1.5 parts of hydrogen peroxide per part sulphide with a contact time of 30–60 minutes has been recommended [61].

Activated carbon and ion-exchange adsorption has been used in laboratory studies to treat landfill leachates. Pohland [62] found that a mixed system of activated carbon and ion-exchange resins was necessary to remove the significant concentrations of refractory organic and inorganic residuals which persisted in landfill leachate after biological treatment. Chian & DeWalle included reverse osmosis in their studies. COD removal was effective but fouling of the membrane occurred [55, 46, 47, 63].

Although physico–chemical treatment methods offer potential for leachate purification either before or after biological processes, Cooper & Bailey [64] have concluded that they are unlikely to have widespread application for the treatment of landfill leachates because of the high capital costs, and high operating costs which must include chemicals, labour and electric power, and this view is shared by leading American researchers [65].

Recirculation and spray irrigation

The use of spray irrigation techniques to purify leachate from domestic refuse sites has been reported in the literature. Robinson, Barber & Maris [28] report on pilot scale test cells filled with pulverised domestic refuse, located in southern England and used to study leachate recirculation. The emplaced bulk density was 560 kg m^{-3} and each cell was covered with 0.25 m of soil seeded with perennial rye grass. The cells were irrigated with water to simulate natural rainfall. The leachate was collected, phosphate nutrient added, aerated and then returned to the top of the cell. The grass cover over the cells died during the experiment as a result of spraying with leachate and ponding of leachate occurred. It was established that although the volume of leachate could be reduced by evaporation at times, the overall amount of leachate available for recirculation increased with time and that this excess leachate would inevitably require a disposal route other than recirculation at some stage.

Newton [66] describes work on sanitary landfills in Gloucestershire. Liquid from a domestic refuse site was collected in a drainage ditch and sprayed onto adjacent grasslands as an experimental test. Although Newton reported a reduction in BOD and ammonia, the influent to his system was not a landfill leachate since it only had a BOD around 100 mg l^{-1}. In any event the experiment caused discoloration of the grass, ponding and had to be discontinued in wet weather. In considering the application of the spray irrigation technique, at a larger landfill in the same county, Newton found it necessary to insert a primitive activated sludge aeration plant upstream of the spray irrigator. Preliminary results were not successful and indicated the need for an expensive aeration return system and this, together with extensive ferric hydroxide precipitation and ponding on the grasslands, led to the discontinuation of the work. At this landfill even the most successful operation in summer failed to reduce BOD below several hundred milligrams per litre and the volume of liquor increased rather than decreased. The operators explain the failure of the activated sludge plant by drawing attention to the difference between domestic sewage and landfill leachate. The leachate contains high ammonia, very low phosphate, traces of heavy metals and carboxylic acids which are not found in sewage and the concentration of these ingredients varies in an uncontrolled fashion, thereby denying the operator adequate control of the process.

Rowe [67] has employed spray irrigation and similar techniques at four sites in Cornwall to leachate from domestic refuse landfills. Rowe points out that the organic content of leachate is much higher than that of sewage and argues that much of the organic content of leachate (unlike sewage) is in true solution and can not be settled out without initial treatment. Leachate discharge to surface streams gave rise to odorous and unsightly 'sewage fungus', a grey or brown fluffy growth which accumulated as long streamers on stream beds. In winter such growths extended several miles downstream of the leachate discharge point. Rowe reports that an increase in zinc, chromium and iron content and discoloration also occurred in surface streams accepting landfill leachate. He recommends that all practical steps, such as culverting, land drainage, diversion of springs and surface waters and contouring, should be taken to minimise water access to a landfill. For leachate collection and diversion Rowe advocates the use of butyl rubber or heavy duty plastic liners, and low grade grouts with bitumen coatings. Attempts to pretreat the leachate with aluminium sulphate, iron sulphate and lime with the aid of synthetic polyelectrolytes proved to be unsuccessful. The addition of sodium hypochlorite was found to be effective in oxidising iron, though it was discontinued due to the high cost and the risk of discharging 'free chlorine' to surface waters. The use of gaseous oxygen was also ruled out on cost considerations. Rowe maintains that traditional sewage treatment designs have not been developed to cope with such high BOD's or widely fluctuating flows.

For these reasons he was attracted to land treatment methods which would permit oxidation of the leachate to take place in the upper layers of the soil. Vegetation and ground cover play an important part both in the oxidation process and in the prevention of soil erosion. The aim is to allow slow movement of the leachate in the land towards a natural drainage system. Rapid run-off leads to pollution and, consequently, gradients of one tenth to one twentieth are preferred. He claims that grassland, woodland, heath and even wet marshy ground can be suitable but points out that flooding and ponding kills off even the most resistant pasture grasses. Wet peaty soils with sedge grass and dwarf gorse have allowed the application of 45 m^3 per hectare per day (4000 gallons per acre per day) without causing pollution even during winter conditions. The land should not be used for continuous irrigation but divided into at least two plots and used alternately. The measurement and control of leachate flow is necessary and the pumping system must be arranged to operate on short (5–15 minute) pumping cycles. The inclusion of large balancing tanks was found necessary. The spray system requires frequent cleaning to remove iron sludge deposits and frost damage is likely in winter. Experience at one site where liquor from the landfill was re-sprayed back on to the landfill surface lead Rowe to comment that 'completed tips may not make good irrigation plots . . . apart from the risk of uneven settlement with ponding and flooding of depressions, the downward movement of additional liquid may increase the leachate produced'. He emphasises that his work has been confined solely to domestic refuse sites and it therefore seems prudent to refrain from the application of this experience to sites containing toxic wastes.

Summary and conclusions

The study of domestic refuse landfills and their leachates is an essential first step in understanding the problems which are encountered in the parallel field of hazardous waste management. In this chapter the essentials of good landfill practice have been reviewed and the natural processes of degradation in a landfill have been examined. Many factors which effect the production of leachate have been identified. The bulk density of refuse in a landfill has been seen to vary over very wide ranges depending on both controllable and uncontrollable factors. Bulk density effects the ability of refuse to adsorb water and the time interval before leachate appears.

Confusion exists about the use of the term leachate. Leachate is the polluting liquor which is generated beneath a landfill as a result of the degradation of refuse and the infiltration of water. However, the term is frequently wrongly applied to surface waters contaminated by the seepage of liquors from a landfill site. This misunderstanding explains in part the wide variations in chemical composition of leachate which are reported in the literature. Recognition that the composition of landfill leachate naturally changes with time offers another explanation. But another critical factor is the discovery that exposure of leachate samples to air brings about changes in the pH of the liquor together with the precipitation of key toxicants. Consequently, the maintenance of anoxic conditions during the collection, preparation and analysis of samples is vital. This explains why techniques employing conventional analytical chemistry have failed to reveal the severely polluting and toxic nature of landfill leachates.

In reviewing leachate control and impoundment the importance of daily cover over the deposit has been emphasised. In addition to all the aesthetic benefits it provides, it serves to keep out extraneous water. The maxim that 'the only good landfill is a dry landfill' is readily justified. A review of synthetic covering materials and sealants has highlighted their advantages and disadvantages.

The methods available for the treatment of domestic refuse leachate have been examined. The presence of ammonia and metal toxicants, the absence of natural nutrients and the uncontrolled variations in composition explain why conventional sewage treatment processes are incompatible with refuse leachate. The incorporation of only 2% leachate (expressed as the flow of domestic sewage) is sufficient to impair the performance of a treatment works. Artificial aerobic oxidation of leachate has been found to be ineffective under winter temperatures when leachate generation is normally greatest. Anaerobic treatment generates large volumes of slurry for disposal and is susceptible to poisoning from trace contamination of the heavy metals which are present. Recirculation and spray irrigation of leachates necessarily creates an odour problem and, since British sites accumulate water from natural precipitation, tend to increase (rather than decrease) the volume of liquor involved. The benefits claimed by some workers in this field result from dilution rather than degradation of the pollutants present.

Objective assessment of all the scientific factors involved lead one to the obvious conclusion. The best cure for leachate is to prevent it.

References

[1] Ministry of Health, Annual Report, 1931–2, HMSO, London.
[2] *Sanitary Landfill*, ASCE Manuals and Reports on Engineering Practice, No. 39, New York, NY 10017, 1976.
[3] US Federal Regulations, Title 40, Part 240, 1975.
[4] *Solid Waste Management Handbook. Sanitary Landfill Operations.* Division of Solid Waste Disposal, Kentucky State Department of Health, Frankfurt, Kentucky, 1976.
[5] 'Design and Operating Guidelines for Sanitary Landfills in Ohio', Ohio Department of Health, Columbus, Ohio, 1971.
[6] USEPA, *Sanitary Landfill Design and Operations*, PB 227–565, 1972.
[7] 'Sanitary Landfill Facts', Public Health Service Publication No. 1792, Washington, 1970.
[8] 'Sanitary Landfill: One part earth to four parts refuse'. A public information cine film produced by USEPA (1972).
[9] USEPA, *Training Sanitary Landfill Employees: An Instructors Manual.* Course SW 43 cl (1973). National Audiovisual Center, Washington, DC 20409.
[10] Weiss, S., *Sanitary Landfill Technology*, Noyes Data Corporation, Park Ridge, NJ, 1974.
[11] Newton, J.R., 'Pilot scale studies of the leaching of industrial wastes in simulated landfills'. *Inst. of Water Pollution Control* (NE Branch), pp. 468–80, February 1977.
[12] Farquhar, G.J. & F.A. Rovers, 'Gas Production during Refuse Decomposition', *Water, Air and Soil Pollution*, **2**, 483–95, 1973.
[13] Parry, G.D.R., 'After use of Landfill Sites', Paper 8, Landfill Gas Symposium, Harwell Laboratories, May 1981.
[14] Fungaroli, A.A. & R.L. Steiner, 'Investigation of Sanitary Landfill Behaviour', Vol 1, USEPA–600–2–79 053*a*, July 1979, Vol 2, USEPA–600–2–79 053*b*, July 1979.
[15] Bevan, R.E., *The Science and Practice of The Controlled Tipping of Refuse*, Inst. of Public Cleansing, London, 1967.
[16] Campbell, D.J.V. & A. Parker, 'Density of Refuse after Deposition Using Various Landfill Techniques', *Solid Wastes*, Vol. LXX, No. 8, pp. 435–40, August 1980.
[17] Bratley, K.J., 'A Description of Comparative Performance Tests of Mobile Plant on a Major Landfill Site', *Solid Wastes*, **67**, 56–80, 1977.
[18] Hoeks, J., 'Pollution of Soil and Groundwater from Land Disposal of Solid Wastes', Inst. for Land and Water Management Research, Wageningen, Netherlands, Technical Bulletin 96, 1976.
[19] SVA., 'Groundwater pollution in the vicinity of the waste disposal site Ambt Delden', Report V, SVA/938, 55 pp. Amersfoort, Holland, 1974.
[20] Zanoni, A.E., 'Potential for ground water pollution from the land disposal of solid wastes', *Critical Reviews in Environmental Control*, pp. 226–60, 1973.
[21] Mead, J.S. & M. Wilkie, 'Leachate Prevention and Control from Sanitary Landfills', *Am. Inst. Chem. Engrs*, 43 pp. Texas, March 1971.

[22] Tchobanoglous, G., H. Thiesen & R. Eliassen, *Solid Wastes Engineering Principles and Management Issues*, McGraw–Hill, 1977.
[23] Brunner, D.R. & D.J. Keller, 'Sanitary Landfill Design and Operation', USEPA Publication SW–65 ts, Washington DC, 1972.
[24] County of Los Angeles, 'Development of Construction and Use Criteria for Sanitary Landfills: An Interim Report', US Dept. of Health, Education and Welfare, Public Health Service, Bureau of Solid Waste Management, Cincinnati, 1969.
[25] Cummins, R.L., 'Effects of Land Disposal of Solid Wastes on Water Quality', US Dept. of Health, Education and Welfare, Public Health Service Publication SW–2 ts, Cincinnati, 1968.
[26] Foree, E.G. & E.N. Cook, 'Aerobic Biological Stabilization of Sanitary Landfill Leachate', Dept. of Civil Engineering, University of Kentucky, Publication UKYTR58–72–CE21, Lexington, USA, 1972.
[27] California State Water Pollution Control Board, 'Report on the Investigation of Leaching of a Sanitary Landfill', Publication 10, Sacramento, California, 1954.
[28] Robinson, H.D., C. Barber & P. J. Maris, 'Generation and Treatment of Leachate from Domestic Wastes in Landfills', Inst. of Water Pollution Control, East Midlands Branch Meeting, Kegworth, Nottinghamshire, 41 pp, January 1981 (draft copy).
[29] Lu, J.C.S., R.D. Morrison & R.J. Stearns, 'Leachate Production and Management from Municipal Landfills: Summary and Assessment', USEPA, Seventh Annual Research Symposium, pp. 1–17, EPA–600/9–81–002*a*, March 1981.
[30] Naylor, J.A., C.D. Rowland, C.P. Young & C. Barber, 'The Investigation of Landfill Sites', Water Research Centre, Stevenage, Technical Report TRG1, 67 pp., October 1978.
[31] Chian, E.S.K. & F.B. DeWalle, 'Analytical Methodologies for Leachate and Gas Analysis in Gas and Leachate from Landfills', Rutgers University Conference, March 1976, pp. 44–53. PB 251–161.
[32] USEPA, *Methods for Chemical Analysis of Water and Wastes*, USEPA, Office of Technology Transfer, Washington DC, 1974.
[33] Weiss, A.J. & P. Colombo, 'Evaluation of Isotope Migration – Land Burial', NUREG, CR 1289, July 1980, Brookhaven National Labs., New York.
[34] Cleveland, J.M. & T.F. Rees, 'Characterization of Pu in Maxey Flats Radioactive Trench Leachates', *Science*, **212**, 1506–9, June 1981.
[35] for examples see Issue No. 5: July 1981, 'Newsletter: Radionuclides Migration in The Geosphere', OECD Nuclear Energy Agency, 30 Boulevard Suchet, 75016, Paris.
[36] Key Committee Report, *Disposal of Solid Toxic Waste*, HMSO, London, 1970.
[37] Emrich, G.H. & W.W. Beck, 'Top Sealing to Minimize Leachate Generation: Status Report', USEPA, Proc. Seventh Annual Research Symposium, Philadelphia, EPA–600/9–81–002*b*, pp. 291–319, March 1981.
[38] available from Borden (UK) Ltd, North Baddesley, Southampton, Beamech Ltd, Salford and Thomas Graveson Ltd, Carnforth.
[39] Fong, M.A. & H.E. Haxo, 'Assessment of Liner Materials for Municipal Solid Waste Landfills', USEPA Proc. Seventh Annual Research Symposium, pp. 138–62, EPA–600/9–81–002*a*, March 1981.
[40] Gunkel, R.C., 'Membrane Liner Systems for Hazardous Waste Landfills', USEPA Proc. Seventh Annual Research Symposium, pp. 291–319, EPA–600/9–81–002*b*, March 1981.

[41] Anderson, D. & K.W. Brown, 'Organic Leachate Effects on The Permeability of Clay Liners', USEPA Proc. Seventh Annual Research Symposium, pp. 119–30, EPA–600/9–81–002*b*, March 1981.

[42] American Colloid Co., 5100 Suffield Court, Skobic, IL 60077, American Colloid Co., Britannia House, 38 Hoghton Street, Southport, PRG OPQ, UK.

[43] Palit, T. & S.R. Qasim, 'Biological Treatment Kinetics of Landfill Leachate', *J. of Environmental Engineering Division ASCE*, **103**, 353–66, 1977.

[44] Boyle, W.C. & R.K. Ham, 'Treatability of Leachate from Sanitary Landfills', Proc. 27th Industrial Waste Conference, Purdue University Engineering Extension Series 141, Part 2, pp. 687–704, 1972.

[45] Boyle, W.C. & R.K. Ham, 'Biological Treatability of Landfill Leachate', *J. Water Pollution Control Federation*, **46**, 860–72, 1974.

[46] Chian, E.S.K. & F.B. DeWalle, 'Evaluation of Leachate Treatment, Vol. 2, Biological and Physical–Chemical Processes', USEPA–600/2–77–1866, November 1977.

[47] DeWalle, F.B. & E.S.K. Chian, 'Leachate Treatment by Biological and Physical–Chemical Methods – Summary of Laboratory Experiments', Proc. 3rd Annual Research Symposium, USEPA–600/9–77–026, St Louis, pp. 177–86, September 1977.

[48] Robinson, H.D. & P.J. Maris, 'Leachate from Domestic Waste: Generation, Composition and Treatment, A Review', Water Research Centre, Stevenage, Technical Report TR108, March 1979.

[49] Uloth, V.C. & D.S. Mavinic, 'Aerobic Biotreatment of a High Strength Leachate', *J. of Environmental Engineering Division ASCE*, **103**, 647–61, 1974.

[50] Clark, R.H. & R.E. Spence, 'The pH tolerance of Anaerobic Digestion', *Advances in Water Pollution Research* by S.K. Jenkins, Pergamon Press, Oxford, 1971.

[51] McCarty, P.L. & R.E. McKinney, 'Volatile Acid Toxicity in Anaerobic Digestion', *J. Water Pollution Control Federation*, **33**, 223–32, 1961.

[52] Mosey, F.E., J.D. Swanwick & D.A. Hughes, 'Factors Affecting the Availability of Heavy Metals to Inhibit Anaerobic Digestion', *Water Pollution Control*, **70** (6), 668–80, 1971.

[53] Mosey, F.E. & D.A. Hughes, 'The Toxicity of Heavy Metal Ions to Anaerobic Digestion', *Water Pollution Control*, **74**, 18–39, 1975.

[54] Blake, W., 'Determination of Acceptable Level of Heavy Metals in Effluents Discharged to a Foul Water Sewer', Proc. International Conference on Heavy Metals in the Environment, London, September 1979.

[55] Chian, E.S.K. & F.B. DeWalle, 'Sanitary Landfill Leachates and their Treatment', *J. of Environmental Engineering Division ASCE*, **102**, 411–31, 1976.

[56] Cook, E.N. & E.G. Folee, 'Aerobic Biostabilisation of Sanitary Landfill Leachate', *J. Water Pollution Control Federation*, **46**, 380–92, 1974.

[57] Bjorkman, V.B. & D.S. Mavinic, 'Physical-Chemical Treatment of High Strength Leachate', Proc. 32nd Annual Industrial Waste Conference, Purdue University, pp. 189–95, 1977.

[58] Thornton, R.J. & F.C. Blanc, 'Leachate Treatment by Coagulation and Precipitation', *J. Environmental Engineering Division ASCE*, 99, 535–44, 1973.

[59] Spencer, G.S. & G.J. Farquhar, 'Biological and Physical Chemical Treatment of Leachate from Sanitary Landfills', Univ. of Waterloo, Dept. of Civil Engineering, Technical Paper 75–2, 31 pp. February 1975.

[60] Ho, S., W.C. Boyle & R.K. Ham, 'Chemical Treatment of Leachates from Sanitary Landfills', *J. Water Pollution Control Federation*, **46**, 1776–91, 1974.

[61] Interox Chemicals Ltd, 'Sulphide Control with Hydrogen Peroxide; its Use in Practice', Technical Information Bulletin, 1978.
[62] Pohland, F.G., 'Landfill Management with Leachate Recycle and Treatment: An Overview', Proc. Research Symposium Rutgers University, USEPA–600/9–76–004, pp. 159–67, 1976.
[63] DeWalle, F.B. & E.S.K. Chian, 'Removal of Organic Matter by Activated Carbon Columns', *J. Environmental Engineering Division ASCE*, **100**, 1089–1104, 1974.
[64] Cooper, P.F. & D.A. Bailey, 'Physico–chemical Treatment: Prospects and Problems in the UK', *Effluent and Water Treatment Journal*, **13**, 753–61, 1973.
[65] Steiner, R.L., J.E. Keenan & A.A. Fungaroli, 'Demonstration of a Leachate Treatment Plant', US National Technical Information Service, Springfield, Va., PB/269/502, 67 pp., 1977.
[66] Newton, J.W., 'The Treatment of Leachate Arising from Domestic Refuse Sanitary Landfill Sites', *Solid Wastes*, Vol. LXIX, No. 6, pp. 254–65, June 1979.
[67] Rowe, A., 'Tip Leachate Treatment by Land Irrigation', *Solid Wastes*, Vol. LXIX, No. 12, pp. 603–23, December 1979.
[68] Geyer, J.A., 'Landfill Decomposition Gases. An Annotated Bibliography'. PB–213 487, National Environmental Research Center, Cincinnati, June 1972.

7

Leachate management from landfill and codisposal of hazardous wastes

The nature of hazardous wastes

Hazardous wastes are generated by a wide range of manufacturing industries and at various stages of the manufacturing process. They are produced in the form of liquids, solids and sludges and frequently consist of mixtures of varying and uncontrolled composition.

Although it is convenient to list categories of hazardous waste (see Chapter 1) it must be recognised that these products are often heterogeneously mixed with each other and with discarded industrial products of a non-hazardous nature. The practice of segregating hazardous wastes at their source is to be encouraged since it may permit the selection of a better environmental option or lower cost disposal method. At the present time complete segregation is rarely practised and most hazardous waste loads consist of chemical mixtures.

The composition of hazardous waste leachate

The quality of leachate produced from hazardous waste is influenced by the chemical and physical nature of the hazardous waste, the physical management of the disposal site, the access and egress of water into and out of the site, the age of the deposit and by microbiological activity. Consequently, wide variations and unpredictable changes in the quality of such leachate are to be expected. The scientific literature contains very little data on actual hazardous waste leachate and even where data are reported information on sample collection techniques is rarely itemised. Most available data pertain to sanitary landfills (which exclude hazardous waste) or to contaminated surface waters where leachate dilution and oxidation has probably occurred. Leachate composition is highly variable from site to site, at different sampling locations within a given site and at a given location over an extended period of time.

Recognising these difficulties, Touhill, Shuckrow and Associates Inc. [1] have listed leachate contaminants identified at 30 hazardous waste disposal sites in the USA. Most of their data reflected the contamination of surface and groundwater resources by migrating leachate and, therefore, do not represent the characteristics of concentrated leachate but serve instead to identify key pollutants.

Table 1. *Summary list of contaminants found in leachates in a survey of 30 hazardous waste disposal sites*

Contaminant	Pollutant group*	Concentration range reported**	No. of sites reported
Acetone	T	<0.1–62 000	3
m-acetonylanisol[α]		<3–1357	1
Ag	H, P	1–10	2
Al		124	1
Aldrin	A, H, P, S	<2–<10	2
Alkalinity, as $CaCo_3$		20.6–5400 $mg\,l^{-1}$	3
Aniline	S, T	<6.2–1900	2
Aroclor 1016/1242	H, P, S	110–1900	1
Aroclor 1016/1242/1254	H, P, S	66 $mg\,l^{-1}$–1.8 $mg\,l^{-1}$	1
Aroclor 1242/1254/1260	H, P, S	0.56–7.7	1
Aroclor 1254	H, P, S	70	1
As	H, P	0.011–<10 000 $mg\,l^{-1}$	6
Ba	H	0.1–2000 $mg\,l^{-1}$	5
Be	H, P	7	1
Benzaldehyde		P 3-100 $mg\,l^{-1}$	2
Benzene	H, P, S, T	<1.1–7370	5
Benzene hexachloride		P	1
Benzene methanol		4600 $mg\,l^{-1}$	1
Benzoic acid	S	<3–12 311	1
Benzylamine or o-toluidine		<10–471	1
Biphenyl naphthalene		P	1
Bis (2ethylhexyl) phthalate	H, P, T	53 $mg\,l^{-1}$	1
Bis (pentafluorophenyl) phenylphosphine		<38	1
B		624	1
BOD_5	C	42–10 900 $mg\,l^{-1}$	3
Bromodichloromethane	P	ND–35	1
2-Butanol		550 $mg\,l^{-1}$	1
2-Butoxyethanol		<2168	1
(1-Butylheptyl) benzene		<36	1
(1-Butylhexyl) benzene		<36	1
(1-Butyloctyl) benzene		<36	1

Table 1. (*cont.*)

Contaminant	Pollutant group*	Concentration range reported**	No. of sites reported
o-sec-butylphenol[α]		< 3–83	1
p-sec-butylphenol[α]		< 3–48	1
p-2-oxo-n-butylphenol		< 3–1546	1
C_4 alkylcyclopentadiene		P	1
C_5 substituted cyclopentadiene		P	1
Ca		164–2500 mg l^{-1}	4
Camphene		P	1
Camphor		< 10–7571	1
Carbofuran	S	P	1
Cd	H, P	5–8200	6
Chloraniline		< 10–86	1
o-chloraniline	A, H	ND–12 000	2
Chlorobenzaldehyde		P	1
Chlorobenzene	H, P, S, T	4.6–4620	5
Chlorobenzyl alcohol		P	1
Chloroform	H, P, S, T	0.02–4550	4
1-chloro-3-nitrobenzene		< 8–340	1
4-chloro-3-nitrobenzamide		440–8700	1
p-chloronitrobenzene		460–940	1
Chloronitrotoluene		ND–460	1
2-chloro-n-phenylbenzamide		< 38	1
2-chlorophenol	H, P, T	< 3–48	1
p-chlorophenylmethyl sulfide		< 10–68	1
p-chlorophenylmethyl sulfone		< 10–40	1
p-chlorophenyl methyl sulfoxide		< 10–53	1
Cl	S	3.65–9920 mg l^{-1}	6
CN	A, H	0.5–14 000	2
Co		10–220	1
COD	C	24.6–41 400 mg l^{-1}	6
Color		50–4000 color units	1
Cyclohexane	S, T	< 0.4–22.0	1
Cr	H, P	1–208 000	7
Cu	P	1–16 000	9
DDT	H, S, T	4.28–14.26	1
Dibromochloromethane	P, T	3.9	1
Dibutyl phthalate	H, P, T	21.732 mg l^{-1}	1
2-6-dichlorobenzamide		890–30 000	1
Dichlorobenzene	H, P, S	< 10–517	2
4, 4′-Dichlorobenzophenone		< 38	1
3, 3′-dichloro [1, -1′-Diphenyl]-4, 4′-diamine		< 84–1600	1

Table 1. (*cont.*)

Contaminant	Pollutant group*	Concentration range reported**	No. of sites reported
1, 1-dichloroethane	H, P, T	< 5–14 280	2
1, 2-dichloroethane	H, P, S, T	2.1–4500	5
trans-1, 2-dichloroethane	H, P	25–8150	2
Dichloroethylene	H, P, S	10 000	1
1, 1-dichloroethylene	H, P, S, T	28–19 850	5
1, 2-dichloroethylene	H, P	0.2	1
Dichloromethane	H, P, S, T	3.1–6570	4
1, 2-Dichloropropane	H, S	< 22	1
Dichloropropene	H, P, S	P	1
Dicyclopentadiene		80–1200	1
Dieldrin	A, H, P, S	< 2–4.5	1
1, 2-Diethylbenzene		7971	1
Diisopropyl-methylphosphonate		400–3600	1
Dimethyl aniline		< 10–6940	1
Dimethyl ether		10–100 $mg\ l^{-1}$	1
1, 4-Dimethyl-2-(1-methylethyl) benzene		11 913	1
1, 2-Dimethyl naphthalene		< 1453	1
Dimethyl pentene		10–100 $mg\ l^{-1}$	1
2, 3-Dimethyl-2-pentene		< 8.6	1
Dimethylphenol	S	< 3	1
2, 4-Dinitrophenol	H, P, S, T	10–99	2
Diphenyldiazine		< 36	1
Dipropyl phthalate		< 3883	1
Endrin	A, P, S	< 2–9	1
Ethanol		56 400	1
2-Ethoxyethanol[α]		3300	1
1-Ethoxypropane		87 000	1
m-ethylaniline		< 10–7640	1
Ethyl benzene	P, S	3.0–10 115	4
(1-Ethyldecyl) benzene		< 36	1
1-ethyl-2, 4-dimethyl benzene		< 1453.0	1
2-ethyl-1, 4-dimethyl benzene		< 1453.0	1
2-ethyl-1, 3-dimethyl benzene		< 1453.0	1
1-ethyl-3, 5-dimethyl benzene		12 507.0	1
4-ethyl-1, 2-dimethyl benzene		< 1453.0	1
1-ethyl-2-isopropyl benzene		< 1453.0	1
2-ethylhexanol		ND–23 000	2
2-ethyl-4-methyl-1-pentanol		22 168.0	1

Table 1. (*cont.*)

Contaminant	Pollutant group*	Concentration range reported**	No. of sites reported
(1-Ethylnonyl) benzene		<36	1
(1-Ethyloctyl) benzene		<36	1
1-ethylpropylphenol[α]		<3.0	1
1-ethyl-2, 4, 5-trimethyl benzene		<1453.0	1
5-ethyl-1, 2, 4-trimethyl benzene		<1453.0	1
F	A, H	140–1300	1
Fe		90–678 000	6
Halogenated Organics		2–15 900	1
Hardness, as $CaCo_3$		700–4650 mg l^{-1}	2
Heavy Organics		0.01–0.59 mg l^{-1}	1
Heptachlor	A, H, P, S	573	1
3-heptanone		ND–1300	1
1-Heptyl-1, 2, 3, 4-tetra hydro-4-methyl-naphthalene		<36	1
Hexachlorobenzene	H, P, T	32–<100	1
Hexachlorobutadiene	H, P, T	<20–109	2
Hexachlorocyclohexane:	H, P, T		
alpha isomer	H, P, T	ND–600	1
beta isomer	H, P, T	ND–70	1
gamma isomer	H, P, T	ND–600	1
delta isomer	H, P, T	ND–120	1
Hexachlorocyclopentadiene	H, P, S, T	<100	1
Hexane		10–100 mg l^{-1}	1
Hg	P	0.5–7.0	7
Hydrocarbons		<36–42 760	2
p-isobutylamisol[α] or p-acetonylanisol[α]		<3–86	1
Isopropanol		<100	1
Isopropylphenol[α]		3–8	1
K		6.830–961 mg l^{-1}	3
Kepone	H, S, T	2000	1
Light Organics		1.0–1000 mg l^{-1}	1
Limonene		P	1
MBAS		240	1
Methanol	T	42 400	1
1-(2-Methoxy-1-methylethoxy)-2-propanol		<2168	1
1-Methoxy-2-propanol		66 000	1
2-Methyl-2-butanol		87 000	1

Table 1. (*cont.*)

Contaminant	Pollutant group*	Concentration range reported**	No. of sites reported
Methylcyclopentane		<0.4–11	1
2-Methylcyclopentanol		1.7–2.168 mg l^{-1}	2
(1-Methyldecyl) benzene		<36	1
Methylene chloride	P	<0.3 mg l^{-1}–184 mg l^{-1}	3
Methylethyl benzene		<1453.0	1
Methylethyl ketone	H, T	53 mg l^{-1}	1
Methyl isobutyl ketone	T	2–10 mg l^{-1}	2
1-Methyl-3-(1-methylethanyl)-cyclohexane		<1453.0	1
1-Methyl-3-(1-methylethyl)-benzene		<1453.0	1
1-Methyl-4-(1-methylethyl)-benzene		<1453.0	1
Methyl naphthalene		<10–290	1
1-Methyl naphthalene		<1453.0	1
2-Methyl naphthalene		<1453.0	1
(1-Methylnonyl) benzene		<36	1
4-Methyl-2-pentanol		140 mg l^{-1}	1
4-Methyl-2-pentanone		110 mg l^{-1}	1
2-Methylphenol		<8.0–210	1
1-Methyl-4-phenoxybenzene		<8.4–670	1
(2-Methyl-1-propenyl) benzene		<1453.0	1
(1-Methylundecyl) benzene		<36	1
Mg		25–453 mg l^{-1}	3
Mn		0.010–550 mg l^{-1}	4
Mo		100–240	3
Na		4.6–1350 mg l^{-1}	5
Naphthalene	H, P, S, T	<10 mg l^{-1}–18 698	2
Nemagon		<1–8	1
NH_3-N		<0.010–1000 mg l^{-1}	3
NH_4-N		650	1
Ni	H	20–48 000	4
nicotinic acid	A, H	P	1
o-nitroaniline		170–180 mg l^{-1}	1
p-nitroaniline	A, H	32–47 mg l^{-1}	1
nitrobenzene	H, P, S, T	ND–740	1
NO_2-N		<10–100	2
NO_3-N		10–100	3
o-nitrophenol	P, S	8600–12 000	1
n-nitrosodiphenylamine	A, H	190	1
Octachlorocyclopentene		<100	1

Table 1. (*cont.*)

Contaminant	Pollutant group*	Concentration range reported**	No. of sites reported
Oil and grease	C	90 mg l^{-1}	1
Paraffins		P	1
Pb	H, P	1–19 000	6
Pentachlorophenol	A, H, P, S	2400	1
(1-Pentylheptyl) benzene		< 36	1
Perchloroethylene	P, T	ND–8200	5
Petroleum oil		P	1
pH	C	≈ 3–7.9 (pH scale)	7
Phenanthrene or anthracene	P	< 10–670	1
Phenol	H, P, S, T	< 3–17 000	4
Phenols	H, P, S, T	0.008–54.17	1
Phthalate esters		P	1
Phthalates	P	P	1
Pinene		P	1
PO_4		< 10–2740	4
Polynuclear aromatics	P	3400	1
(1-Propylheptyl) benzene		36	1
(1-Propylnonyl) benzene		36	1
(1-Propyloctyl) benzene		36	1
Sb	H, P	2000	1
Se	H, P	3–590	4
SO_4		1.2–505 mg l^{-1}	4
SOC		4200 mg l^{-1}	1
Specific Conductance (mhos cm^{-1})		80–2000	2
SS	C	< 3–1040 mg l^{-1}	4
Styrene	S	P	1
Sulphide		< 100	1
TDS		1455–15 700 mg l^{-1}	4
Temperature		58–63 °F	1
1, 1, 2, 2-Tetrachloroethane	H, P, T	< 5–1590	1
Tetrachloroethene	H, T	< 1–89, 155	3
1, 1, 2, 2-Tetrachloroethene	H, T	0.6–560	1
Tetrachloromethene	H, P, S, T	< 1–25 000	3
1, 2, 3, 5-Tetramethylbenzene		36 479	1
1, 2, 4, 5-Tetramethylbenzene	4	< 1453	1
Thiobismethane		< 1.0–290	1
TKN	C	< 1–984 mg l^{-1}	4
TOC	C	10.9–8700 mg l^{-1}	8
Toluene	H, P, S, T	< 5–100 000	7
Total inorganic carbon		71 000	1

Table 1. (*cont.*)

Contaminant	Pollutant group*	Concentration range reported**	No. of sites reported
Total P	l	< 100–3200	2
Total solids		159–1730 mg l^{-1}	1
Tribromomethane	H, P, T	0.2	1
1, 2, 4-Trichlorobenzene	H, P	< 10–28	2
Trichloroethane	H, P, T	P-35 000	2
1, 1, 1-Trichloroethane	H, P, T	1.6 g^{-1}–590 mg l^{-1}	5
1, 1, 2-Trichloroethane	H, P, T	< 5–870	2
Trichloroethene	H, T	< 3–84 000	4
Trichloroethylene	H, P, S, T	< 3–260 000	4
Trichlorofluoromethane	P, T	< 5–18	1
Trichloromethane	P, S	< 1–10 000	1
2, 4, 5-Trichlorophenol	H, S, T	P	1
Trichlorotoluenes		3300 mg l^{-1}	1
Trimethylbenzene		P	1
1, 2, 3-Trimethylbenzene		13.702 mg l^{-1}	1
1, 2, 4-Trimethylbenzene		11.239 mg l^{-1}	1
1, 3, 5-Trimethylbenzene		37.113 mg l^{-1}	1
Vinyl chloride	H, P, T	140–32 500	1
Xylene	S, T	P–5400	2
m-xylene	S, T	19.708 mg l^{-1}	1
o-xylene	S, T	1453	
p-xylene	S, T	48.170 mg l^{-1}	1

ND – not detected.
P – present, but not quantified.
α – structure not validated by actual compound.
* – codes for pollutant groups.
C – conventional pollutants (per *Clean Water Act and Treatability Manual*, vol. 3).
P – priority pollutants
A – RCRA list – acutely hazardous.
H – RCRA list – hazardous.
T – RCRA list – toxic.
S – section 311 compound.
– (a blank indicates that the compound does not fall into one of the above groups).
** – concentrations in μ g^{-1} unless otherwise noted.
Source: Reference [1].

These contaminants are listed in Table 1 and it should be recognised that conventional characteristics such as BOD, COD, TOC, TSS, oil and grease, total phenol, total phosphorus, total nitrogen and total organic chlorine need also to be taken into account.

Geraghty & Miller [2] also characterised leachate and groundwater from 43 US landfill sites and their data are shown in Table 2. Once again these data show wide variations in the concentrations of pollutants encountered.

Leachate treatment goals

Naturally, the quality objective for the treatment of leachate must reflect the selected disposal outlet. In cases where on-site treatment of hazardous waste leachate is possible and the resultant discharges are directed to an off-site domestic sewage treatment works then the leachate quality must match the requirements of the sewage works. A BOD of less than 300 mg l^{-1} and a total heavy metal content of less than 1 mg l^{-1} (in addition to other constraints) would be typical of influents accepted at UK sewage works.

Aqueous discharges to rivers and surface waters complying with the Royal Commission Standards are required to have BOD of less than 30, COD less than 30 and TSS of less than 20 mg l^{-1}.

Table 2. *Leachate and groundwater characteristics from 43 US landfill sites*

Pollutant	Concentration range	Typical concentration	No. of sites where detected
As	0.03–5.8	0.2	5
Ba	0.01–3.8	0.25	24
Cr	0.01–4.20	0.02	10
Co	0.01–0.22	0.03	11
Cu	0.01–2.8	0.04	15
CN	0.005–14	0.008	14
Pb	0.3–19	—	3
Hg	0.000 5–0.000 8	0.000 6	5
Mo	0.15–0.24	—	2
Ni	0.02–0.67	0.15	16
Se	0.01–0.59	0.04	21
Zn	0.1–240	3.0	9
Light organics	1.0–1000	80	10
Halogenated organics	0.002–15.9	0.005	5
Heavy organics	0.01–0.59	0.1	8

Source: Reference [2].

Leachate treatment trains

The high concentration of organic pollutants, the presence of inorganic contaminants and the variable and unpredictable nature of hazardous waste leachate make treatment difficult (if not impossible). From work on domestic leachate treatment (see Chapter 6) it has been shown that biological treatment is difficult and necessitates the addition of nutrients, the use of antifoams, extended contact times and the removal of substantial quantities of sludge. Even then the effluent produced contains significant concentrations of refractory organic compounds. In the case of hazardous waste leachate, biological treatment becomes even more difficult since trace metals have an inhibiting effect on the biological processes. Table 3 shows the effect of selected metals on biological treatment but this list is by no means comprehensive.

Table 3. *The effects of selected metals on biological treatment processes*

Metal	Concentration range	Effect	Reference
Barium	1–1000 000 ppm	O_2 uptake inhibited at concentrations greater than 100 ppm	[3]
Chromium Cr^{6+}	1–100 000 ppm	O_2 uptake inhibited at concentrations >100 ppm	[3]
Cobalt	0.08–0.5 ppm	inhibited biological growth	[4], [5]
Copper	5–30 ppb	stimulated biological growth	[4]
	50–560 ppb	inhibited biological growth	
Iron Fe^{2+}	10–100 ppm	O_2 uptake inhibited at concentrations >100 ppm	[3]
Iron Fe^{3+}	0.01–100 000 ppm	O_2 uptake inhibited at >100 ppm	
Lead	10–100 ppm	O_2 uptake inhibited	[3]
Manganese	12.5–50 ppm	stimulated biological growth	[3]
	50–100 ppm	inhibited biological growth	
Mercury	0–200 ppm	O_2 uptake inhibited	[5]
Zinc	0.08–0.5 ppm	inhibited biological growth	[4]
		O_2 uptake inhibited	[3]

See also Reference [6].

Jackson & Brown [7] and Stones [8] report that the individual toxic metals concentrations in sewage passing to a biological treatment phase should not exceed 1 mg l^{-1}.

Many organic compounds, particularly pesticides, are unaffected by conventional biological processes [1] some require extensive reaction times and a few cause inhibition of biological breakdown.

From this it will be seen that complex treatment trains appear necessary. Generally it is essential to remove inorganic components first to minimise effects on subsequent processes. Since practical experience and published information on the performance of leachate treatment systems are rare it is necessary to base design data on industrial effluent (wastewater) experience and laboratory scale studies. However, it should be noted that a complex leachate matrix may not behave like other wastewater systems thus affecting design and operational criteria (e.g. chemical dosage requirements) and invalidating extrapolations from other experiences [1]. In the absence of treatability data there is no guarantee that high levels of treatment can be achieved. In particular, capital investment, operating and maintenance costs are likely to be greater per unit volume treated than for industrial effluent or municipal sewage treatment.

Such considerations mean that the design of leachate treatment plant for a site where leachate exists is difficult and construction is time consuming (time which may not be available if a significant pollution threat exists). Yet for landfill sites at the planning stage where the quantity and nature of the leachate can only be guessed at, the design of leachate treatment systems is even more risky.

Nevertheless, this design challenge has to be tackled if only to establish the magnitude of the task and the quantification of capital and operating costs.

Fig. 1(*a*) shows a schematic flow diagram of suggested hazardous waste leachate treatment systems. The unit processes involved are identified in this diagram and will now be discussed in detail.

Leachate storage is an essential feature of all leachate treatment systems. The rate of leachate generation changes with the seasons, the operational management and the age of the site, but all the unit processes employed in leachate treatment (especially coagulation and sedimentation) are susceptible to changes in flow. Leachate storage is therefore necessary to permit careful control of the leachate flow to the subsequent treatment stages. Leachate pH varies considerably and needs to be controlled. Lime slurry or sodium hydroxide addition is required to elevate the pH. A continuous flow pH measuring, recording and controlling device is

Fig. 1. Schematic flow diagram of hazardous waste leachate treatment systems.

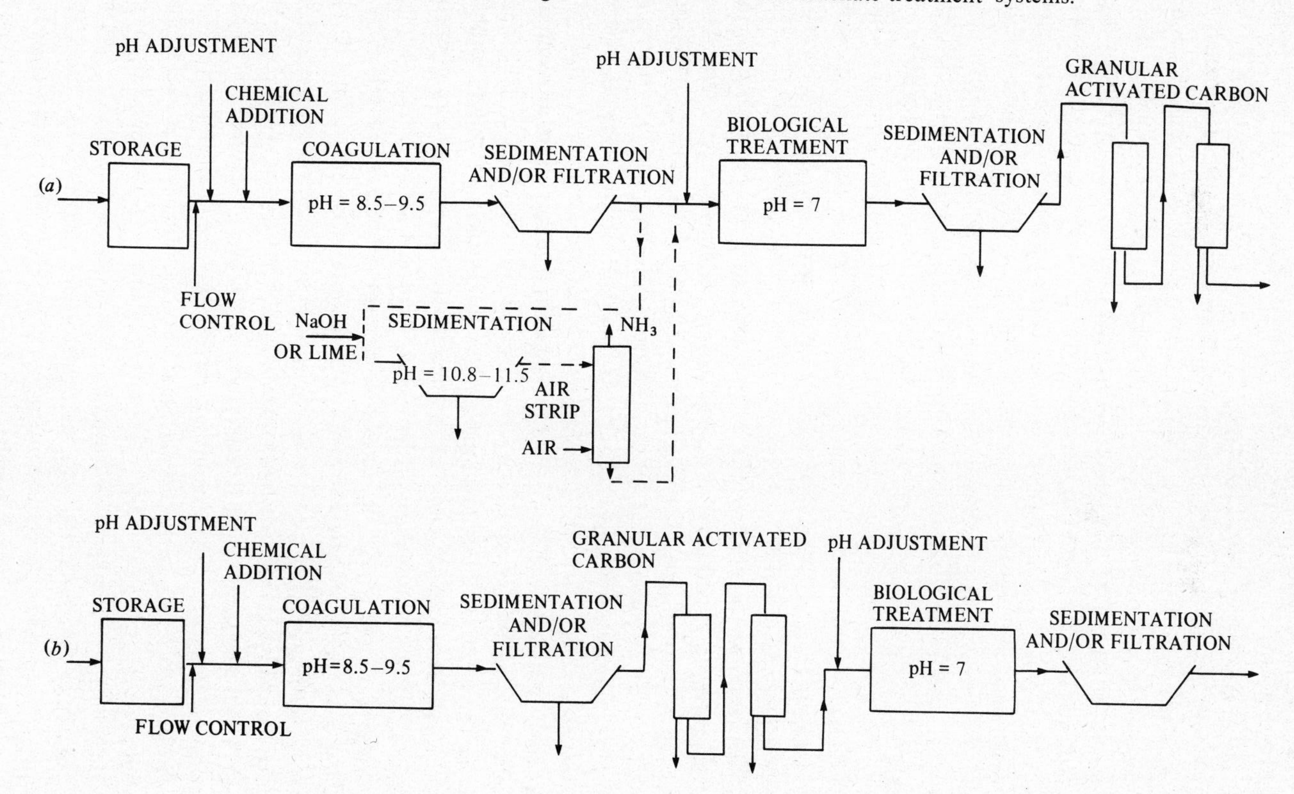

necessary and experience shows that such units require frequent cleaning and calibration.

Coagulation is included to precipitate inorganic components in an insoluble form. This is accomplished by the addition of alum, lime, iron salts (ferrous sulphate or ferric chloride) or hydrogen or sodium sulphide. Synthetic organic polyelectrolytes can be added to improve floc formation and sedimentation. It will be noted that each of these additives requires equipment to dissolve and/or control the supply rate. Adequate mixing of the additives with the leachate is required but thereafter non-turbulent flow conditions are necessary to encourage floc growth and sedimentation. When sedimentation equipment is included it needs to include provision for the regular removal of the accumulated sludge without interruption of the leachate flow. Ammonium compounds occur frequently in leachate and removal is achieved by raising the pH from 10.8 to 11.5 by lime or sodium hydroxide addition followed by sedimentation and air blowing. Consideration must be given to local air emission standards where this technique is considered.

Biological treatment needs to be preceded by a pH adjustment which requires the controlled addition of an acid. Precautions for acid handling, storage, pumping and metering need to be observed and the comments made earlier concerning pH control equipment apply again at this point. The biological treatment process could be either aerobic or anaerobic but would be subject to the process limitations described earlier in Chapter 6. As has been shown, effluent from these biological processes still contains refractory organic compounds which need to be removed to meet most discharge standards. For this purpose, granular activated carbon beds are included. The organic compounds are adsorbed most strongly by the carbon at the low concentrations to be found in this part of the process stream and, as a result, the carbon sorption cycle can be lengthened. In some cases where the leachate contains toxic organic components (i.e. toxic to the biological treatment process) it may be necessary to include the granular activated carbon beds upstream of the biological step (see Fig. 1(*b*)). In this case complete treatment by the carbon is not required and the bed is allowed to 'leak' organics. Determining the endpoint under these conditions at which carbon replacement is necessary could prove difficult. It would be expected that carbon would be removed to an off-site regenerator and replaced as necessary.

It should be recognised that the addition of pumps, valves, piping and stand-by equipment to these conceptual flow diagrams contributes significantly to the processing complexity and capital cost.

Practical and experimental data on the treatment of hazardous waste leachate is almost non-existent in the published literature. This is particularly true in Britain but in the USA, McDougal et al. have reported on leachate treatment at Love Canal [9, 10]. Consequently, this work is reviewed in detail below.

Love Canal presented an emergency situation and an immediate need for leachate management to ameliorate a public health hazard. Consequently, leachate could be subjected to treatability studies though there was insufficient time to examine all possible alternatives. Most importantly it was possible to discharge the leachate after on-site treatment to an off-site municipal sewage treatment plant which incorporated additional chemical treatment by coagulation and sedimentation. Such facilities are rare or non-existent in the UK. The treatment objective identified at Love Canal was a limit of 300 mg l^{-1} TOC which is considerably higher than would be tolerated in Britain where discharge to surface waters would be contemplated. Treatability studies showed that it was necessary to dilute the raw leachate 1:5 with water and add nutrients before biological treatment was possible and even then uncontrolled variations in the resultant effluent occurred. As a result, the investigators opted for the coagulation, filtration and carbon adsorption scheme shown in Fig. 2. Table 4 shows the analysis of the raw Love Canal leachate and this was satisfactorily treated to the treatment goals identified.

Treatment costs have been reported as \$9.80 m^{-3} (£5.05 m^{-3}) which includes replacement carbon, equipment maintenance and some (but not all) equipment amortisation. The Love Canal case study illustrates a situation where activated carbon is an effective method for removing

Fig. 2. Love Canal leachate treatment system. (Source: Reference [10].)

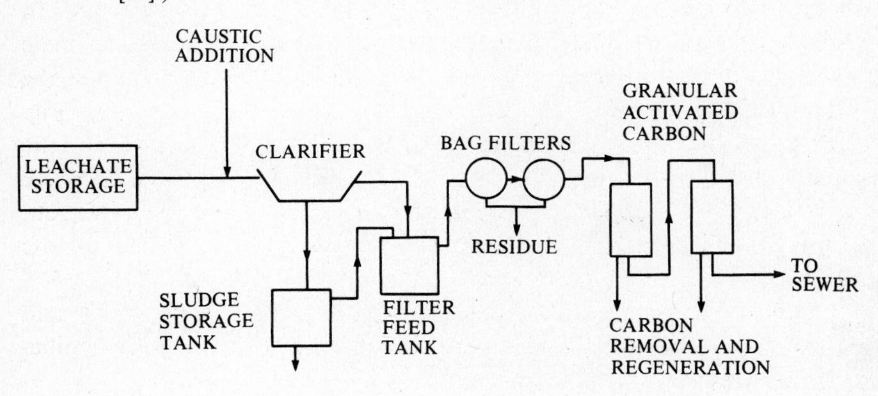

organic contaminants from hazardous waste leachate and its inclusion here is not intended to suggest that this situation is typical or that this method can be applied elsewhere.

Codisposal of hazardous wastes

Terminology

In this context codisposal is considered to be the disposal of liquid and solid hazardous waste together with domestic refuse in a landfill site. Where liquid waste is added to a landfill the term 'water balance method' has been used to describe the concept of balancing liquid input (rainfall plus liquid waste plus groundwater) with losses (due to outflows, evaporation and transpiration). The description 'dilute and disperse' is often applied to codisposal methods and, of course, is also used in connection with discharges to surface waters and emission to the atmosphere.

Practical application

The practice of codisposal is being recommended in the UK by official government bodies [11, 12] though it is discouraged or outlawed in many other countries. Fig. 3 shows a graphic representation of the

Table 4. *Love Canal raw leachate characteristics*

	In mg l^{-1} (except pH)
pH	5.6
TOC	4300
SOC	4200
COD	11 500
Oil/grease	90
Suspended solids	200
Dissolved solids	15 700
Chloride	9500
Sulphate	240
Sulphide	<0.1
Total phosphate, P	<0.1
Inorganic phosphate, P	<0.1
Sodium	1000
Calcium	2500
Iron	330
Mercury	<0.000 5
Lead	0.4

Source: Reference [9].

Fig. 3. The codisposal of hazardous waste.

DOMESTIC REFUSE IS TIPPED ONTO CLAY. THE CLAY ACTS AS A SEAL.

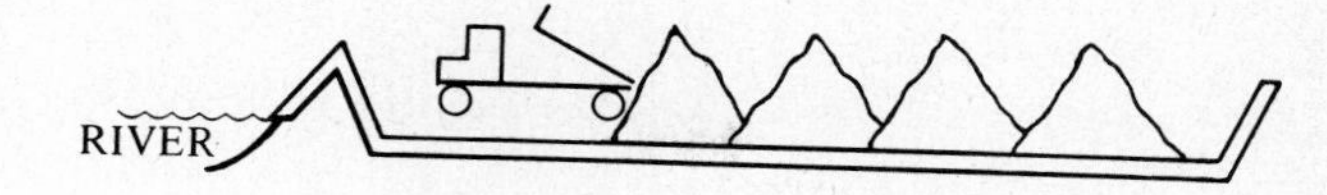

THE REFUSE IS BULLDOZED LEVEL AND COMPRESSED.

THEN THE PROCESS IS REPEATED. MORE LAYERS OF REFUSE ARE ADDED IN THE FORM OF A DOME. INDUSTRIAL LIQUID WASTE IS PUMPED INTO TRENCHES.

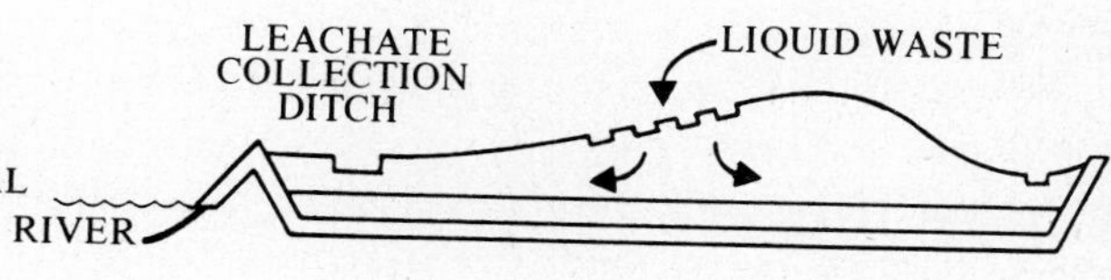

LEACHATE COLLECTS IN DITCHES AND HAS TO BE PUMPED BACK ON TO THE REFUSE. A LAGOON IS FORMED.

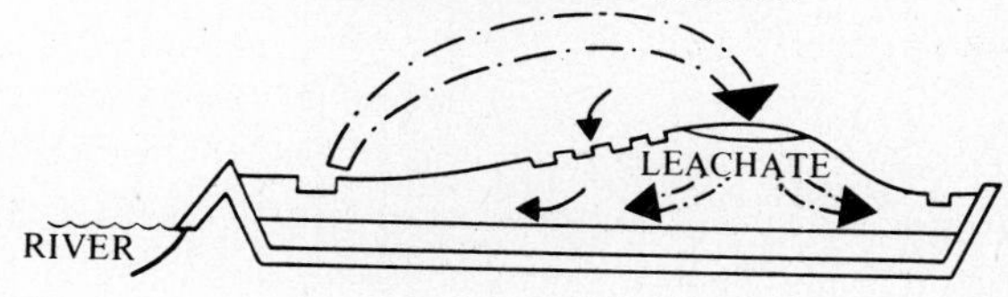

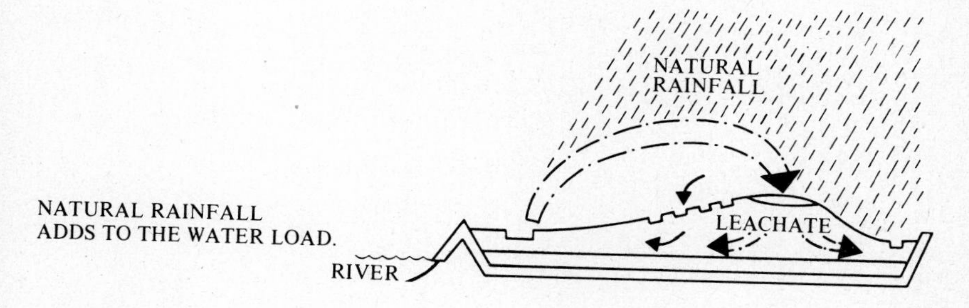

THE WATER HAS TO GO SOMEWHERE.

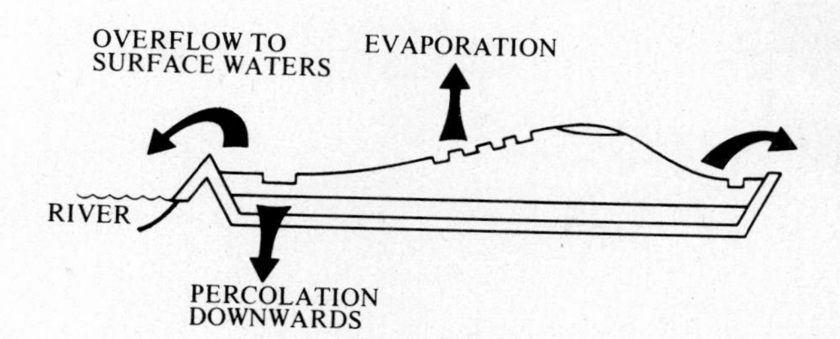

codisposal technique as used in several UK landfills.

The question of how much liquid can be safely adsorbed on to domestic refuse warrants further discussion. A good deal of research has been undertaken to determine the quantity of liquid which is adsorbed by refuse. The field capacity of refuse is defined as the volume of liquid which can be adsorbed by a given weight of refuse without the release of excess water under the forces of gravity [13]. Other definitions [14] referring to the quantity of water adsorbed by a given weight of refuse 'before significant leachate production would occur' seem to beg the question about the subjective assessment of the meaning of 'significant'.

The determination of field capacity is difficult because of the variations in the composition and bulk density of refuse in a landfill site. Table 5 shows the composition of refuse collected from dwellings in 26 local authority areas in Britain in 1974 [15]. These data compare well with American studies [16] shown in Table 6. The bulk density of domestic refuse is determined by the particle size of the components and the method of compaction.

Salvato et al. [17] report that the density of poorly compacted refuse is around 170 kg m^{-3} to 340 kg m^{-3} and that the maximum compacted density, without the use of high pressure baling or pulverisation techniques is 409 kg m^{-3}. They suggest a rule of thumb figure for landfill use of 256–290 kg m^{-3} but the use of new equipment and modern management techniques has undoubtedly increased these figures. As we have seen in Chapter 6, bulk density of refuse in a landfill ranges from 145 to 1250 kg m^{-3}. Data collected by Tchobanoglous [18] and shown in Table 7

Table 5. *The average composition of domestic refuse in Britain in 1974*

	Percentage by weight
Vegetable and putrescible	21.3
Paper	26.8
Screenings below 2 cm	19.8
Metal	8.5
Textiles	3.5
Glass	9.5
Plastics	2.9
Unclassified	6.9
Average density	160.56 kg m^{-3}

Source: Reference [15].

Table 6. *Typical composition of municipal solid wastes in USA*

	Percentage by weight Range	Typical
Food wastes	6–26	15
Paper	25–45 *	40 *
Cardboard	3–15	4
Plastics	2–8	3
Textiles	0–4	2
Rubber	0–2	0.5
Leather	0–2	0.5
Garden trimmings	0–20	12
Wood	1–4	2
Glass	4–16	8
Tin cans	2–8	6
Non ferrous metals	0–1	1
Ferrous metals	1–4	2
Dirt, ashes, brick	0–10	4

*Typically 55.3% as packaging materials.
Source: Reference [16].

Table 7. *Typical compaction factors for various solid waste components as discarded*

		Compaction factors	
	Range	Normal compaction	Well compacted
Food waste	0.2–0.5	0.35	0.33
Paper	0.1–0.4	0.2	0.15
Cardboard	0.1–0.4	0.25	0.18
Plastics	0.1–0.2	0.15	0.10
Textiles	0.1–0.4	0.18	0.15
Rubber	0.2–0.4	0.3	0.3
Leather	0.2–0.4	0.3	0.3
Garden trimmings	0.1–0.5	0.25	0.2
Wood	0.2–0.4	0.3	0.3
Glass	0.3–0.9	0.6	0.4
Tin cans	0.1–0.3	0.18	0.15
Non ferrous metals	0.1–0.3	0.18	0.15
Ferrous metals	0.2–0.6	0.35	0.3
Dirt, ashes, brick etc.	0.6–1.0	0.85	0.75

$$\text{Compaction factors} = \frac{\text{final volume after compaction}}{\text{initial volume prior to compaction}}$$

Source: Reference [18].

demonstrates that the components of refuse are compacted in a landfill at different compaction ratios.

The relationship between compaction load and bulk density is shown in Fig. 4. In using this data care must be exercised to allow for the variations in the initial moisture content of the waste and for any recoil effect.

As the particle size of the refuse decreases and/or the bulk density increases there is a considerable increase in field capacity [13].

The difficulty of establishing meaningful field capacity data is highlighted by the wide variations in British published figures. In an Interim Report [19] published in 1975, the Department of Environment quoted 0.2–$0.4\,l\,kg^{-1}$. Resulting largely from work reported by Newton [20] on fresh pulverised compacted refuse with a density of $500 \pm 70\,kg\,m^{-3}$, the Department of Environment Final Report, published in 1978 [21], modified this figure to 'at least 0.12 litres per kilogram'. Scott [22] has argued that the application of $0.817\,l\,kg^{-1}$ of water or liquid waste to refuse in a landfill is practicable. However, evidence reported by this author [23] suggests a limit of $0.04 \pm 0.04\,l\,kg^{-1}$.

The paramount importance of the determination of reliable field capacity data was recognised by the House of Lords Select Committee on Science and Technology Inquiry on Hazardous Waste Disposal in 1981. In

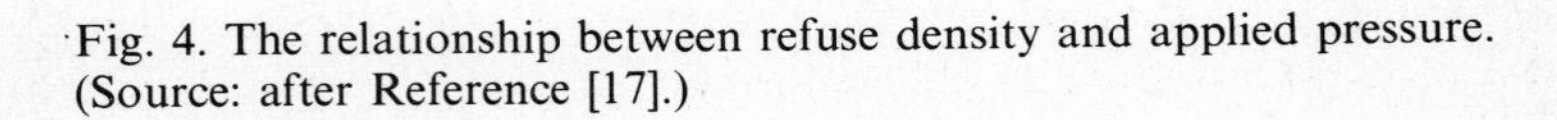

Fig. 4. The relationship between refuse density and applied pressure. (Source: after Reference [17].)

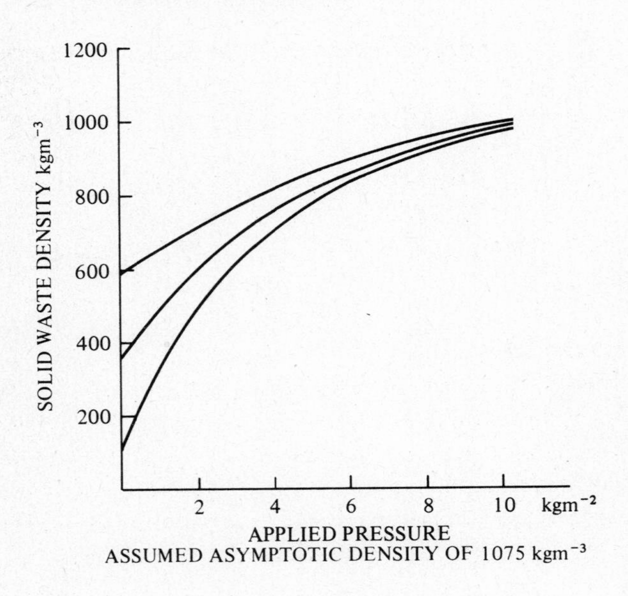

answer to questions, the government spokesman said that they could not quote a single figure that could be universally applied and thought that it was a matter for determination in particular circumstances by experienced professionals.

In the writer's experience, this task is beyond the reasonable capabilities of any landfill site operator. In order to arrive at a satisfactory answer the operator would need to make allowances for:

(1) the chemical nature and composition of waste already landfilled on the site in times past and which currently forms the base of the landfill;
(2) the age and degree of degradation of such waste;
(3) the chemical nature and composition of recently deposited refuse;
(4) the particle size of refuse components and the effect of pulverisation, shredding or milling if any;
(5) the degree of compaction of the wastes;
(6) the infiltration of natural groundwaters, surface waters and precipitation and a forecast of future meteorological variations;
(7) the natural water content of recently deposited waste;
(8) the effects of evaporation, transpiration or outflow of liquids from the site.

When it is remembered that the operator's visual evidence is restricted to observation of surface conditions and that he has no means of assessing what conditions are like below the working level of the landfill, he deserves our sympathy since the task we have given him appears impossible. As a practical matter, the site operator continues to discharge liquid waste into trenches cut into the refuse until the work area becomes saturated, thus making access to the trenches difficult. At this point, he is likely to discontinue liquid discharges and to deposit additional loads of refuse on top, thereby creating a saturated sponge within the landfill. He has now created within the body of the landfill a saturated zone which is the very condition which needs to be avoided if the attenuation capability of refuse is to be relied upon.

Research studies

Research work carried out at the Water Research Centre employing a series of test cells has been reported by Newton [20] and has been summarised by the Department of Environment [19, 21]. Fresh pulverised refuse was placed in a series of lined concrete tanks located out of doors at Stevenage. These tanks are shown in Fig. 5. For each study duplicate tanks were used; one tank permitted free access of air at its base and the second was sealed. Consequently, aerobic and anaerobic studies could be compared though the upper surfaces of the fills were left uncovered and exposed to natural rainfall. A cyanide heat treatment waste and a metal hydroxide sludge containing mainly nickel, chromium and iron were sandwiched between two layers of domestic refuse within these tanks. The waste loadings were equivalent to 0.019% CN of the total refuse for the cyanide waste and 0.27% Ni and 0.11% Cr of the total refuse for the hydroxide waste. The compacted density of the refuse fill was 500 ± 70 kg m^{-3} and its water content 26% by weight. No leachate appeared for a period of seven months but over the first year 16–23% of the natural rainfall issued as leachate. The pH of the leachate from all experiments tended to rise during the first 500 days, typically in the range 7.5–9.3 for the aerobic studies and 5.3–7.7 in the anaerobic studies. Leachate analyses were normally carried out on unfiltered samples though on a few occasions filtration was shown to reduce iron and manganese concentrations ten-fold. No precautions to exclude air during collection and analysis of leachate samples were

Fig. 5. Water Research Centre experimental test cells. (Source: Reference [20].)

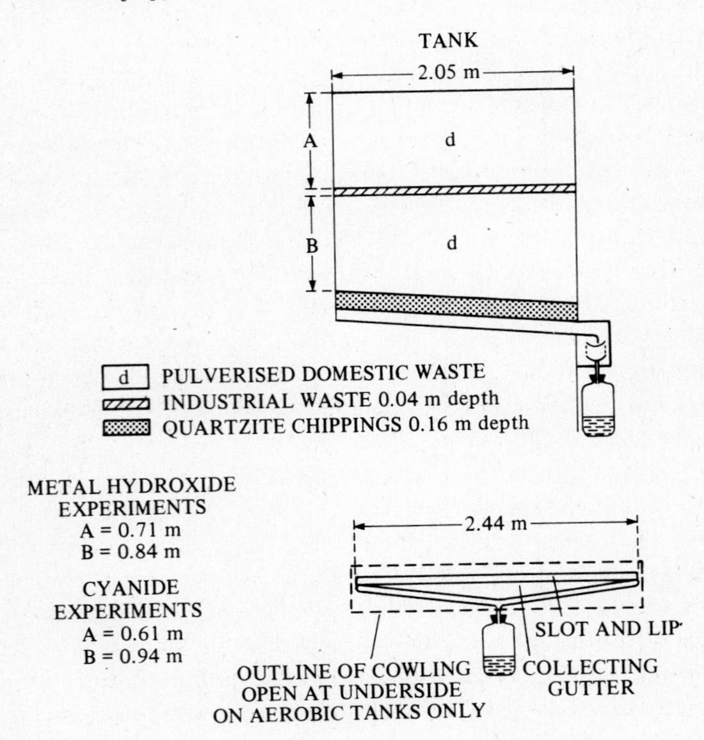

reported. Newton reports the CN content of leachate from anaerobic studies peaking at 6 mg l^{-1} during the first 30 weeks and thereafter decaying to 1 mg l^{-1} and less. Newton's data on the metal content of this leachate are more difficult to interpret. His figures are represented as mean concentrations over yearly periods which disguises the true nature of the leachate. For example, the mean concentration of nickel and chromium during the first year are quoted as 2.9 and 0.5 mg l^{-1} respectively. But elsewhere it is recorded that peak concentrations were 51 mg Ni l^{-1} and 5.7 mg Cr l^{-1} during this period. It was established that metal concentrations in leachates from anaerobic studies were generally higher than those in corresponding aerobic studies except for copper, cobalt and zinc. From parallel experiments undertaken in small scale PVC pipes it was estimated that less than 10% of all the metals present in the refuse appeared in the leachate in the first year.

In reviewing the above work this author is concerned about its relevance to large scale landfill operations. For example, fresh domestic refuse in a pulverised form might be expected to exhibit better adsorption characteristics than old non-pulverised refuse. Similarly, the bulk density of the pulverised refuse used in this study could well be higher than is customary on a landfill operation and consequently its adsorption characteristics may not be comparable. Again the use of hazardous waste loadings equivalent to 0.019% CN, 0.27% Ni and 0.11% Cr as a percentage of the total refuse present seems to represent very low concentrations which are not necessarily indicative of large codisposal landfill behaviour. The apparent absence of precautions to maintain anoxic conditions during the collection and analysis of leachate samples may also invalidate many other conclusions. The use of mean concentrations of metals in the leachate to draw conclusions omits considerations of peak emissions when the conclusions could well be quite different. Consequently, the extrapolation of data such as this to justify the application of codisposal practices on a large scale throughout Britain is, in the writer's assessment, a debatable step which with the passage of time could well prove to be most regrettable.

In this context the work of Pohland is often quoted by British researchers to substantiate claims for leachate recycle and codisposal practices. For this reason it is beneficial to examine these studies in some depth. Pohland has constructed four outdoor lysimeter columns, at the Georgia Institute of Technology, Atlanta, which are shown in Fig. 6. In his early experiments [24] the columns were filled with 2800 lb of ground refuse with a very high compacted density of 908 kg m^{-3}. The columns were covered with 0.61 metres of compacted soil and then capped with grass sod. Leachate was collected in a drain and in some experiments pumped back through a distributor installed beneath the soil cover, thus permitting recycling of leachate without exposure to air. At the start of the studies, tap water was added to the refuse to bring it to field capacity. During the first 150 days or so, the leachate had a pH in the range 4.8–5.5 and COD ranging from 4000 to 10 000 mg l^{-1}. After about one year, the COD had fallen to 1000–2000 mg l^{-1}, the pH had increased to 7.1 but this was achieved artificially by the addition of sodium hydroxide to the leachate recycle tank. Pohland reports no attempt to measure the total volume of leachate generated but indicates that rain water did penetrate the top cover. He found that the inclusion of septic tank pumpings in the refuse fill further accelerated anaerobic degradation. From this work he concluded that recirculation of leachate accelerated the degradation and settlement of the refuse and initiated anaerobic conditions earlier than test columns with no recycling. His analysis of leachate led him to conclude that 'a combination

Fig. 6. Lysimeter studies at Georgia Institute of Technology. (Source: References [24, 25, 26]. Courtesy of US Environmental Protection Agency.)

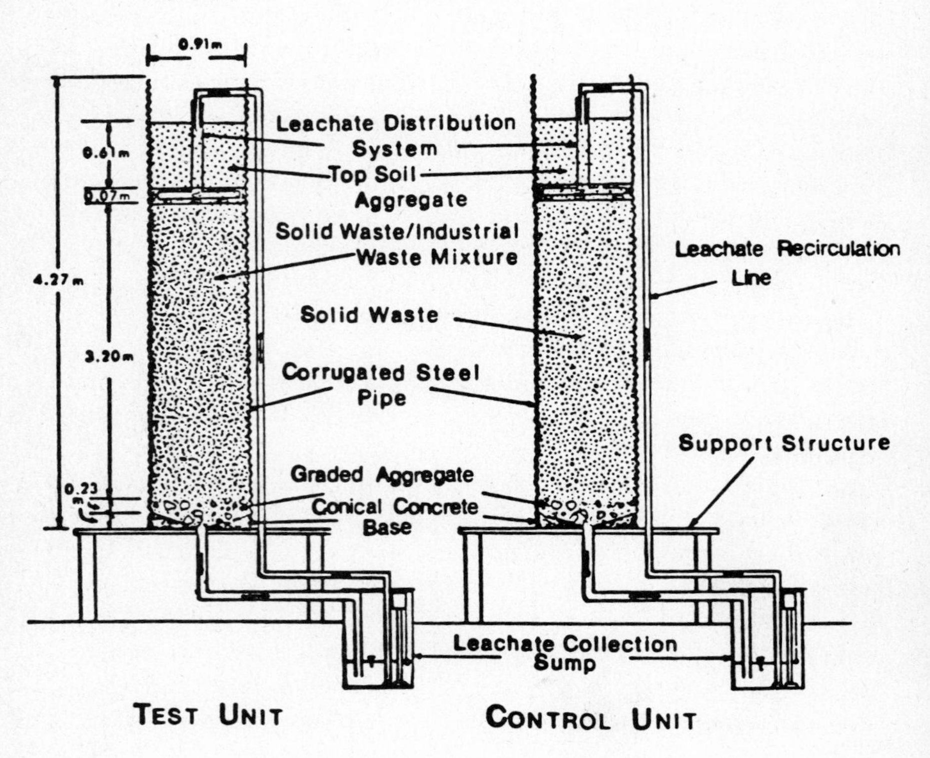

of separate biological and physico–chemical treatment methods would be necessary to reduce the pollution potential of leachate from refuse disposal sites to a condition acceptable for ultimate disposal'. He argues that separate aerobic and anaerobic biological processes would prove satisfactory for the treatment of leachate provided the residual organics and inorganics in the leachate are removed by carbon adsorption followed by a mixed bed ion exchange.

It would seem to this writer that such facilities, requiring considerable investment and supervision, hardly seem likely to be available on a landfill site. Remembering also that these inferences were made after employing artificial neutralisation, utilising ground refuse with a very high bulk density and injecting the leachate through a distributor located below the top cover material, it will be seen that this methodology is not practical for a conventional landfill site. Furthermore, on a refuse landfill, the infiltration of precipitation and/or surface waters is likely to lead to an *increase* in the total volume of leachate generated, thereby adding to leachate treatment and disposal problems.

Subsequently, Pohland [25, 26] studied codisposal aspects in these lysimeter columns by including metal plating sludges with the domestic refuse. Varying amounts of sludges containing zinc, chromium, cadmium and nickel were introduced. The columns were believed to be operating anaerobically but gas analysis is not reported to confirm this. Ingress of air to the columns may have occurred during the removal of leachate samples and no special precautions were taken to prevent air contamination of the leachate samples themselves. It is not known whether the lysimeter columns were saturated with water since this will be determined by the effectiveness of the top seal but it is clear that there has been no deliberate addition of water or other liquids to the system.

Pohland's data show that the ionic strength of the leachate falls from start to pass through a minimum around 200 days and then increases so that at around 350 days it is almost equal to that in the first 50 days of operation. Metal concentrations in the leachate also follow a similar pattern. Nickel falls from 1 $mg\,l^{-1}$ at 50 days through a minimum at 200–250 days and then rises again to 1 $mg\,l^{-1}$ at 350 days. The pH of the leachate was 5.3 after 50 days, peaked at 6.6 after 200 days and fell to 5.6 to 6.0 at 350 days.

Pohland reasons that retentive mechanisms such as sulphide precipitation can be relied upon to retain heavy metals from industrial waste within the refuse but, in the writer's view, this is not supported by Pohland's own experimental data. The release is undoubtedly slow, nevertheless measurable and sufficient to cause problems with water pollution or purification.

Leachate attenuation

When leachate emerges from the base of a landfill site, attracted downwards by the force of gravity, it encounters naturally occurring geologic materials. These materials include rocks, sands, gravel and clay-like deposits which have the potential to attenuate polluting components of the leachate. Research studies have been conducted by Harwell Laboratory staff to study this attenuation mechanism and, since their findings have been employed in the development of British government policy, a detailed review is warranted.

Black [27] describes the construction of four lysimeters on either side of an open access trench at Uffington, Oxfordshire. The site consists of Lower Greensand comprising fine sands, sandy clays and clayey sands to a thin bed of fine sand which overlays Kimmeridge Clay. In all at least 23 sedimentary layers were identified. The fine sand fraction contained quartz, feldspar, argonite/calcite, glauconite, pyrite and the clay fraction contained calcium montmorillonite (Fullers Earth). The formation is not naturally cemented and any cohesion that was found was attributed to the large amounts of dispersed clay. The site was divided into four cells each being 4.5 m × 4.5 m × 2.5 m by the construction of concrete separating walls as shown in Fig. 7 and the entire rig was enclosed within a barn. Each cell was irrigated with a synthetic solution chosen to study the attenuation of particular pollutants [28].

This review is concerned primarily with Lysimeter 4 which was irrigated with a synthetic solution having the composition shown in Table 8.

The solution was adjusted to pH 5 using ammonia. The authors report that this mixture was chosen to simulate leachate containing heavy metals obtained by the biodegradation of organic matter in refuse. The initial irrigation rate was equivalent to 1300 mm y^{-1} but experimental difficulties led to its reduction to average 705 mm y^{-1} over the first six months of operation. The aim was to create *unsaturated* conditions within the cells and to study migration by analysing samples withdrawn through suction probes located at different depths in the cells. A neutron probe and a tensiometer were used to follow the movement of the water phase and showed the existence of heterogeneous flow patterns. The researchers reported that water flow was very likely confined to specific zones of high permeability which was probably caused by burrow structure. This created difficulties in tracking the migration of pollutants by means of the suction probes. The irrigation system caused ponding on the surface of the cell and a white fungal mycelium growth developed round the ponds which may have degraded some of the fatty acids in the solution and increased the retention of metals in the upper surfaces of the cell. The researchers comment that the quality of the concrete employed in the construction of the retaining walls was not adequate to prevent seepage of liquids from the

Fig. 7. Harwell studies at Uffington, Oxfordshire. (Source: References [27–31].)

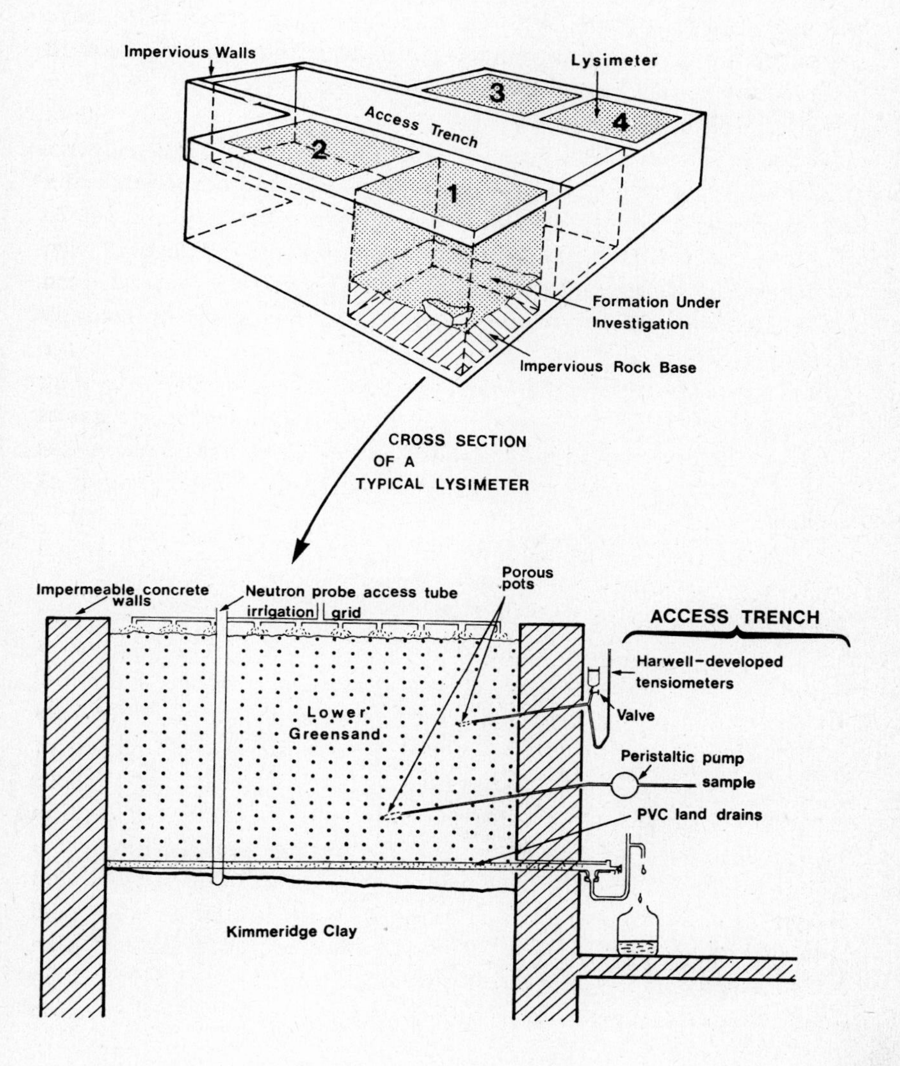

test cells into the access trench. Presumably, since it was porous, the infiltration of air into the test cells was also possible and likely (Fig. 8).

The early results of this work [28] report the composition of liquids withdrawn from the suction probes at various depths in the cell for up to 185 days from the start of the test. A probe located 40 cm below the surface showed pH ranging from 7.4 to 7.8, TOC fluctuating from 38 to 710 and metal concentrations less than 0.5 mg l^{-1}. Further down the cell at a depth of 62 cm the pH was 7.6–8.1, TOC from 10 to 170 and the concentration of heavy metals was even lower. Though the report mentions that penetration was only about 5–10 mm day^{-1}, the authors conclude that soluble heavy metals are not likely to contribute to groundwater pollution 'if the water table does not intercept the landfill base and neutral pH conditions are maintained'.

The experimental work continued and another report [29] presents data from 400 days operation. It reports that the metals had not penetrated to the uppermost suction probe (40 cm) and quotes analyses of sand/clay samples extracted from the lysimeter to demonstrate that high concentrations of metals are retained within the sands. The retention indicated that  cadmium < zinc < nickel < copper < chromium < lead is the sequence of decreasing mobility.

Further studies reported later [30] provided data from up to 606 days' duration (May 1977). After about 480 days' irrigation nickel began to appear at the first suction probe located 40 cm below the surface. At 606

Table 8. *Composition of synthetic solution used in lysimeter study*

Lysimeter No. 4			
Constituent	Ionic concentration mg l^{-1}	Constituent	Ionic concentration mg l^{-1}
Sodium	1200	Zinc	100
Potassium	900	Cadmium	100
Magnesium	100	Chromium	100
Calcium	500	Nickel	100
Iron	100	Copper	100
Lithium	20	Lead	100
Mercury	10	Acetic acid	4800
Phenol	20	Propionic acid	2700
Chloride	1700	Butyric acid	1500

Source: Reference [28].

days the nickel concentration had reached 30 mg l^{-1} but since both zinc and chromium concentrations remained very low and suction probes further down the cell showed no significant increase, the metal migration was not thought to be a problem.

About this time (early 1978), the Department of the Environment published its full report on 'Hazardous Wastes in Landfill Sites' [21] which included a review of the above work. It concluded that 'sensible landfill is

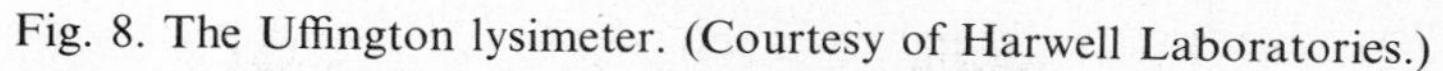

Fig. 8. The Uffington lysimeter. (Courtesy of Harwell Laboratories.)

realistic and an ultra-cautious approach to landfill of hazardous and other types of waste is unjustified'. These findings became the data base on which the DOE's policy of codisposal was based and from which gradually evolved guidelines described in many Waste Management Papers.

But the research continued and was reported in March 1979 [31]. Now it was noted that the methane content of the gases generated reached 60% and the active microbiological zone within the cell had migrated downwards corresponding with increasing concentrations of heavy metals in actual solution within the lysimeter. Nickel concentrations at the 40 cm suction probe were 26 mg l^{-1} at 600 days and increased to 55 mg l^{-1} at 850 days and remained at that level to 1000 days. Cadmium reached 10 mg l^{-1} at 950 days and zinc reached similar concentrations. The maximum chromium content was 3 mg l^{-1} but copper and lead remained below 0.1 mg l^{-1}. Probes located lower down the bed reflected these changes. The acetic, propionic and butyric acids had traversed the entire depth of the cell (2250 cm) in 800–1000 days. The order of mobility of metals was identified as

$$Ni > Cd > Zn > Cr > Cu > Pb$$

In reviewing this work in detail it is necessary to draw attention to some of the experimental conditions which have considerable impact on the results and their interpretation in terms of landfill behaviour. We have seen that the aim was to study attenuations under unsaturated conditions but such conditions are not typical of the base of a landfill where liquids (from infiltration, precipitation or codisposal of liquid wastes) accumulate. Furthermore, the conditions in these experiments were aerobic for the greater part of the time. The irrigation technique encouraged exposure to air and thus air infiltration into the body of the cells was to be expected. Indeed, gas analysis data showed aerobic conditions during the first 700 days of the test. However, soon after anaerobic conditions were established in the cell rapid migration of heavy metals and refractory organic compounds was apparent.

Field studies have been carried out on a number of hazardous waste landfill sites in the United Kingdom and reported in references [19, 21]. This work contributed to the government's view that 'an ultra-cautious approach to landfill of hazardous and other types of waste is unjustified'. Data on the migration of pollutants in these reports was obtained in one of three ways. In some cases core samples of waste and underlying strata were removed and centrifuged and the interstitial liquids analysed. New or existing boreholes were used to sample liquids and, in other cases, spring waters were sampled. In all cases it would appear that samples were exposed to air since no precautions to maintain anaerobic conditions were reported.

It would therefore be expected that increase in pH, and precipitation and adsorption of key pollutants would occur during sampling which might be interpreted to suggest that pollutant attenuation was effective.

As a result the Uffington lysimeter studies and the field measurements hardly present reassuring proof that attenuation under anaerobic conditions beneath hazardous waste and codisposal sites by naturally occurring geologic media can be relied upon to prevent migration of harmful pollutants. Could it be that British policy makers have been hasty in concluding that codisposal is a safe philosophy and that attenuation by underlying clays can be relied upon?

The release of metal ions adsorbed onto montmorillonite and kaolinite clays has been studied at the University of Aston in Birmingham. Monsef-Mirzai has demonstrated that copper ions adsorbed from solutions are desorbed readily when brought into contact with solutions containing lower concentrations of the metal [32, 33].

The importance of recognising the role that anaerobic conditions play has been emphasised in some recent American studies. Weiss & Colombo [34] used anoxically collected leachate samples and soil samples from a low-level radioactive waste disposal area to determine the sorption characteristics of certain radionuclides. It is customary to report such studies in terms of Kd where:

$$\text{Kd} = \frac{\text{soil activity/weight of soil}}{\text{liquid activity/volume of liquid}}$$

and Kd is described as a 'partition coefficient' or 'sorption coefficient'. They found that when Kd was determined under anoxic conditions the results differed by as much as an order of magnitude to those obtained under oxic conditions. They concluded that anoxic conditions in the presence of organic degradation products accelerate leaching and migration of metals and that sorption characteristics determined under oxic conditions are not reliable indications of pollutant behaviour in landfill situations.

Ames et al. [43, 44] have examined the sorption of radioactive aqueous wastes by soil and rock. They reported that

> Disposal on a specific retention basis of selected liquid waste solutions was practised at Hanford from 1944 to 1962 when it was discontinued. As used at Hanford, specific retention in practice

> represented the volume of liquid that could be discharged to a disposal pit of known dimensions without leakage to groundwater. The specific retention was expressed as a percentage of the soil column volume measured by the cross-sectional area of the disposal pit bottom and the height of the soil column between water table and pit bottom. The problems associated with specific retention as a radionuclide waste disposal technique include the fact that the forces holding the radionuclide waste solution are relatively weak and that the radionuclides which are disposed of by this technique often remain in the solution. Any water added at the top of the soil column, from natural precipitation or irrigation would lead to the migration of the retained waste solution into the groundwater. Hence specific retention as a waste disposal technique should be limited to arid areas where there is little downward migration of precipitation to the groundwater and no chance of other water addition to the top of the soil column. In addition, large soil volumes are required for specific retention disposal because very little of the soil column ion-exchange capacity is utilized.

Ames et al. studied the interaction of 19 radionuclides with soil and rock and concluded that:

> Most of the values of distribution coefficient (Kd) described in the literature were inadequate due to the failure to include comprehensive description of the system conditions. . . . Kd is an experimentally derived value that is associated with the circumstances of its determination. It can not be extended to another chemical system with the same solid exchange medium. Likewise, the Kd from one exchange medium in a given chemical system can not be used for another solid exchange medium in the same chemical system. An obstacle of Kd comparisons is the common failure to adequately characterize the solid exchange medium. Even with comparable solution compositions and derived Kd values, the solid exchange system can be completely different mineralogically, physically and chemically.

The focussing of attention on radioactive species enables greater accuracy to be achieved in laboratory measurements but, in principle, there is no difference in behaviour between a radioactive nuclide and any other isotope of the same element.

Comparison of sanitary landfill and codisposal

In the application of codisposal practices hazardous industrial wastes are deposited together with domestic refuse on a landfill site.

Consideration must now be given to the comparison of codisposal practices and sanitary landfill. Table 9 lists such a comparison. It can be clearly seen that the requirements of codisposal and sanitary landfill are diametrically opposed.

When it is remembered that the methodology of sanitary landfill was introduced in order to make disposal operations environmentally acceptable to residents living near to landfill sites, it can be expected that the decline in standards caused by permitting codisposal practices will provoke public reaction and opposition from those affected.

Table 9. *Comparison of sanitary landfill with codisposal*

Sanitary landfill	Codisposal
Provides impermeable cover at end of each working day to prevent infiltration of water	Requires permeable surface to permit infiltration of water otherwise ponds and lagoons will form
Confines waste to smallest practical area	Spreads waste over maximum area to encourage evaporation
Freedom from dust, fires, odours, vermin and birds	Large area encourages dust, fires, odours, vermin and birds
Impermeable cover and vent pipes control emission of inflammable gases	Absence of controls permits release of inflammable gases over a wide surface area
Refuse gradually reaches field capacity, i.e. 25% water by weight	Refuse becomes saturated thus creating conditions in which toxic components of hazardous waste can dissolve and move freely with leachate stream
Aerobic decomposition at first followed by anaerobic decomposition	Anaerobic decomposition rapidly sets in generating greater quantities of methane and some hydrogen sulphide
Minimises the quantity of leachate generated	Maximises the quantity of leachate generated and relies on evaporation and transpiration to control it
Operation geared to maintaining acceptable aesthetic features	Operation is less concerned with aesthetic features

Authoritative views on codisposal

The Commission of The European Communities (EEC) [35] final Council Directive on Toxic and Dangerous Wastes states in Article 6:

> Member states shall take the necessary measures to ensure that toxic and dangerous waste is collected, transported and stored separately from other matter and residues.

The American Society of Civil Engineers [36] says that:

> the addition of liquid to refuse at a sanitary landfill other than in small amounts for dust and litter control is not presently a recommended practice.

The German Federal Environmental Agency [37] states that:

> The Federal Republic of Germany tends towards separate treatment even though this method is clearly more expensive than the former (codisposal). The motive for separate identification, collection, transport, treatment and disposal of hazardous waste is undoubtedly the principle of prevention.

The USEPA [38] reports that:

> Present management of land disposal of hazardous wastes is frequently inadequate. Significant environmental impacts from such activities are not mere possibilities – actual damages to groundwater have occurred and are well documented. Although the potential for damage in general can be demonstrated, damages which would result from unrestricted landfilling at specific sites can not be predicted accurately.

and adds

> To assess the potential adverse effects of codisposal, the industrial wastes are leached with municipal landfill leachate as well as with water. Results to date indicate that when compared with water, municipal landfill leachate solubilizes greater amounts of metals from the wastes and promotes rapid migration of metals through soils.

In New South Wales, Australia, the codisposal of liquids with domestic refuse has been practised at one particular site [39]:

> While investigations were proceeding into the establishment of a neutral liquid waste treatment facility. . . . The practice is now being phased out and will be terminated in June 1981 despite the realisation that industry will be faced with additional disposal costs.

In Toronto, Canada, it is reported that [40]:

> for several years we allowed the disposal of industrial liquid waste up to 5% by weight, as well as 8% sludge, and found that we had problems of odour and leachate seepage through the sides of the landfill which finally resulted in our abandoning the practice.

John Lehman, Director of Solid Waste Programs of USEPA has reported [41]:

> Our philosophy, as the land protection group within EPA, is to minimise hazardous waste disposal to land. Consequently, we strongly support hazardous waste recycling or detoxification treatment prior to land disposal wherever possible.

In 1970 the Key Committee [42] reported to the UK government on the land disposal of solid toxic waste and said:

> Methods of preventing pollution reaching a water source . . . the most obvious method is to put a layer of some impervious material between the tip and the water at risk. To be quite safe such a layer ought to be saucer shaped so that all toxic waste is tipped within the saucer. If the cover could be made in the shape of an umbrella which extended beyond the limits of the toxic waste, then percolate would be kept from it indefinitely . . . nor would water flowing over the cover be contaminated with toxic water.

Summary and conclusions

In this chapter the heterogeneous nature of hazardous waste has been noted together with the composition of the leachate generated. Some 220 chemical compounds have been identified which have been detected in hazardous waste leachate and which are not normally associated with the leachate from domestic refuse. Additionally, the concentration of toxic heavy metals in hazardous waste leachate has been reviewed. Clearly then it is erroneous to imagine that hazardous waste leachate is identical to domestic refuse leachate.

In examining the options for the treatment of hazardous waste leachate the need for sophisticated treatment trains has been identified. These must comprise of leachate storage, flow control, pH control and adjustment, air blowing, coagulation, sedimentation (or filtration), biological treatment, more sedimentation and filtration and adsorption on activated carbon. The complexity of such a treatment system is clearly apparent. Yet experience from the Love Canal clean-up programme has indicated that even this is inadequate. The activated carbon system needs to include a regenerative facility unless it is disposed of as a solid waste. Even then it is desirable to discharge the treated effluent to a domestic sewage treatment plant which incorporates physico–chemical precipitation. Such facilities are either rare or non-existent in the UK. Consequently, the hazardous waste leachate treatment problem can be seen in its true perspective.

Codisposal methodology has been reviewed and the importance of the adsorptive capacity of domestic refuse has been highlighted. Yet this parameter has been seen to be difficult to establish and impossible to predict, depending as it does on a wide variety of factors, many of which are entirely beyond man's control.

A critical review of published research work which purports to support codisposal techniques has been made. Experimental conditions, sampling and analytical methods and interpretation of results have been criticised. Lysimeter studies on leachate attenuation and field studies of UK landfill sites have been shown to be either misleading or inconclusive. American studies have shown that Kd determinations are significantly effected by the oxic/anoxic nature of the experimental conditions. This observation throws doubt on much of the published experimental data in this field.

Codisposal methodology has been compared to sanitary landfill. The two concepts can be seen to be diametrically opposed. International attitudes towards the codisposal of hazardous liquid wastes show opposition to the technique. Yet British regulatory authorities continue to condone codisposal despite the huge body of technical evidence and world opinion which opposes this practice.

References

[1] Touhill, Shuckrow and Associates Inc., 'Management of Hazardous Waste Leachate'. USEPA, PB81–189359, Cincinnati, September 1980.

[2] Geraghty and Miller Inc., 'The Prevalence of Subsurface Migration of Hazardous Chemical Substances at Selected Industrial Waste Land Disposal Sites', USEPA, S30/SW–634, 1977.

[3] Dawson, P.S. & S.H. Jenkins, 'The Oxygen Requirements of Activated Sludge Determined by Monometric Methods, Chemical Factors Affecting Oxygen Uptake', *Sewage and Industrial Wastes*, Vol. 22, No. 4, p. 490, 1950.

[4] Loveless, J.E. & N.A. Painter, 'The Influence of Metal Ion Concentration and pH value on the Growth of a Nitro-somonas Strain Isolated from Activated Sludge', *J. General Microbiology*, Vol. 52, No. 3, pp. 1ff, 1968.

[5] Lamb, J.C., et al., 'A Technique for Evaluating the Biological Treatability of Industrial Wastes', *J. Water Pollution Control Federation*, **36**, 1263–84, 1964.

[6] Pajak, A.P., et al., 'Effect of Hazardous Material Spills on Biological Treatment Process', USEPA, 600/2–77–239, 202 pp., 1977.

[7] Jackson, S. & V.M. Brown, 'The Effect of Toxic Wastes in Treatment

Processes and Water Courses', *J. Institute Water Pollution Control*, **69**, 292, 1973.

[8] Stones, T., 'The Influence of Metallic Components on BOD of Sewage', *J. Institute Sewage Purification*, **6**, 516–20, 1961.

[9] McDougall, W.J., et al., 'Treatment of Chemical Leachate at The Love Canal Landfill Site', Proc. Twelfth Mid-Atlantic Industrial Waste Conference, Bucknell University, Lewisburg, Pennsylvania, pp. 69–75, 1980.

[10] McDougall, W.J., R.A. Fusco & R.P. O'Brien, 'Containment and Treatment of The Love Canal Landfill Leachate', *J. Water Pollution Control Federation*, Vol. 52 No. 12, pp. 2914–24, December 1980.

[11] Osmond, R.G.D., 'Landfill Research in the UK. Some Results and Implications', ISWA Congress, London, June 1980.

[12] House of Lords Select Committee on Science and Technology, 'Hazardous Waste Disposal: First Report', HMSO, London, 273–I, July 1981.

[13] Weiss, S., *Sanitary Landfill Technology*, Noyes Data Corporation, Park Ridge, NJ, 1974.

[14] Holmes, R., 'The Water Balance Method of Estimating Leachate Production from Landfill Sites', *Solid Wastes*, Vol. LXX, No. 1, pp. 20–8, January 1980.

[15] Skitt, J., *Waste Disposal Management and Practice*, Charles Knight, London, 1979.

[16] Darnay, A. & W.E. Franklin, 'The Role of Packaging in Solid Waste Management, 1966–76', US Dept. of Health, Education and Welfare, Public Health Service Publication SW–5*c*, Rockville, Maryland, 1979.

[17] Salvato, J.A., W.G. Wilkie & B.E. Mead, 'Sanitary Landfill – Leaching Prevention and Control', American Inst. of Chem. Eng., 68th National Meeting, Houston, 2084–2100, March 1971.

[18] Tchobanoglous, G., H. Theissen & R. Eliassen, *Solid Wastes*, McGraw–Hill, New York, 1977.

[19] Department of Environment, 'Programme of Research on the Behaviour of Hazardous Wastes in Landfill Sites: Interim Report', HMSO, London, July 1975.

[20] Newton, J.R., 'Pilot Scale Studies of the Leaching on Industrial Wastes in Simulated Landfills'. *J. Water Pollution Control* (NE Branch), p. 468–80, February 1977.

[21] Department of Environment, 'Cooperative Programme of Research on the Behaviour of Hazardous Wastes in Landfill, Final Report', HMSO, London, 1978.

[22] Scott, M.P., 'A Simplified Approach to Landfill Design under Conditions of Uncertainty', *Solid Wastes*, Vol. LXVII, No. 5, 1977.

[23] Cope, C.B., Evidence to House of Lords Inquiry, HMSO, London, 273–III, July 1981.

[24] Pohland, F.G., 'Sanitary Landfill Stabilization with Leachate Recycle and Residual Treatment', USEPA–600/2–75–043, 105 pp., October 1975.

[25] Pohland, F.G. & J.P. Gould, 'Stabilization of Municipal Landfills containing Industrial Wastes', USEPA, Proc. Sixth Annual Research Symposium, EPA–600/9–80–010, pp. 242–53, 1980.

[26] Pohland, F.G. & J.P. Gould, 'Containment of Heavy Metals in Landfills with Leachate Recycle', USEPA, Proc. Seventh Annual Research Symposium, Philadelphia, EPA–600/9–81–002, pp. 179–94, March 1981.

[27] Black, J.H., 'Uffington Lysimeters – Geological Aspects of Construction and Instrumentation', WLR 33, Department of the Environment, London, March 1976.

[28] Parker, A., et al., 'Uffington Lysimeters – Operation and Results, Part 1', WLR 36, Department of the Environment, London, June 1976.
[29] Campbell, D.J.V., et al., 'Uffington Lysimeters – Operation and Results, Part 2', WLR 40, Department of the Environment, London, January 1977.
[30] Campbell, D.J.V., et al., 'Uffington Lysimeters – Operation and Results, Part 3', WLR 42, Department of the Environment, London, September 1977.
[31] Rees, J.F., et al., 'Uffington Lysimeters – Operations and Results, Part 4', WLR 60/AERE–R.9425, AERE Harwell, March 1979.
[32] Anon, 'Getting the Measure of Toxic Metal Removal', *Surveyor*, pp. 10–11, 26 November 1981.
[33] Monsef-Mirzai, P., Geochemical Aspects of Waste Disposal, Ph.D. Thesis, University of Aston in Birmingham, 1981.
[34] Weiss, A.J. & P. Colombo, 'Evaluation of Isotope Migration – Land Burial', NUREG CR 1289, July 1980. Brookhaven National Labs., New York.
[35] Commission of The European Communities, 'Council Directive on Toxic and Dangerous Wastes', COM (76) 385 final, Brussels, 22 July 1976.
[36] American Society of Civil Engineers, *Manual on Engineering Practices*, No. 39, 'Sanitary Landfill', 1976.
[37] Schmitt-Tegge, Federal Environmental Agency, Germany, 'Treatment and Disposal of Industrial Wastes in the Federal Republic of Germany', ACS/CSJ Chemical Congress, April 1979, Honolulu.
[38] Fuller, W.H. (Ed.), 'Residual Management by Land Disposal', Proc. Hazardous Waste Research Symposium, USEPA, Cincinnati, PB–256 768. July 1976.
[39] Conolly, R., 'Hazardous Waste Legislation and Disposal in Australia', ACS/CSJ Chemical Congress, April 1979, Honolulu.
[40] McKerracher, I., 'Reclamation and Improvement of Land by Landfill', ISWA Conference, London, June 1980.
[41] Lehman, J.P., 'Federal Program for Hazardous Waste Management', *Waste Age*, Washington DC, September 1974.
[42] Key Committee Report, 'Disposal of Solid Toxic Waste', HMSO, London, 1970.
[43] Ames, L.L., et al., 'Radionuclide Interactions with Soil and Rock Media', Vol. 1, PB–292 460, Battelle Pacific, North West Labs., Richland WA, August 1978.
[44] Ames, L.L. et al., 'Radionuclide Interactions with Soil and Rock Media', Vol. 2, PB–292 461, Battelle Pacific, North West Labs., August 1978.

Table 1. (*cont.*)

Element	Atomic weight (g)	Content in lithosphere (ppm)	Common range for soils (ppm)	Selected average for soils: ppm	Selected average for soils: Molar conc. at 10% moisture log M
La	138.91	18	1–5000	30	−2.67
Li	6.94	65	5–200	20	−1.54
Mg	24.31	21 000	600–6000	5000	0.31
Mn	54.94	900	20–3000	600	−0.96
Mo	95.94	2.3	0.2–5	2	−3.68
N	14.01	—	200–4000	1400	0.00
Na	22.99	28 000	750–7500	6300	0.44
Ni	58.71	100	5–500	40	−2.17
O	16.00	465 000		490 000	2.49
P	30.97	1200	200–5000	600	−0.71
Pb	207.19	16	2–200	10	−3.32
Rb	85.47	280	50–500	10	−2.93
S	32.06	600	30–10 000	700	−0.66
Sc	44.96	5	5–50	7	−2.81
Se	78.96	0.09	0.1–2	0.3	−4.42
Si	28.09	276 000	230 000–350 000	320 000	2.06
Sn	118.69	40	2–200	10	−3.07
Sr	87.62	150	50–1000	200	−1.64
Ti	47.90	6000	1000–10 000	4000	−0.08
V	50.94	150	20–500	100	−1.71
Y	88.91		25–250	50	−2.25
Zn	65.37	80	10–300	50	−2.12
Zr	91.22	220	60–2000	300	−1.48

Source: Reference [91].

Data in Table 1 also report the abundance of elements in soils both as ranges and as selected averages. For example, aluminium is one of the most abundant elements constituting approximately 71 000 ppm or 7.1% by weight of the earth's crust. Silicon is the second most abundant element of the earth's crust. The soil contains 32% silicon by weight. Aluminium, silicon, and oxygen unite during weathering processes to form secondary minerals called aluminosilicates. These secondary minerals along with primary aluminosilicates represent the major part of the mineral content of soils.

Soils are mostly composed of mineral matter. Usually the organic matter is concentrated in the upper few centimeters unless it forms out of lake beds or swamps. Not all minerals change chemically by weathering. The *primary minerals* that occur in soil are there because they resist weathering as a characteristic of their chemical and structural nature. 'Sand garnets', for example, may readily be found completely unchanged in sands collecting in stream beds and ravines. The minerals in the original rock surrounding them dissolve and crack or wash away as the rock crumbles, freeing them. *Secondary minerals* form from weathering residues of primary minerals. They are often called clay minerals because of the fine particle sizes. The clays of soil have a large surface area, are colloidal in nature, and therefore represent the most chemically and physically active component of the soil. Thus soils contain both primary and secondary minerals, the proportionality of which depends on the extent of weathering as may be illustrated in Fig. 2. The majority of the minerals found in most soils fall into the technical groups listed in Table 2. Quartz, feldspars, and hornblende constitute the three most abundant primary minerals in the earth's crust.

Again, it should be emphasized that the earth's crust, originally as well as more recently, contained a wide variety of minerals in the parent rock that vary from location to location. Since minerals do not weather at the same rate, and since the materials that make up the bulk of the soil originate from these minerals, it can be expected that soils may differ somewhat according to the proportions of their original mineral components. Parent material, therefore, is a real factor in defining the soil characteristics.

Fig. 2. General conditions for the formation of the various silicate clays and oxides of iron and aluminium.

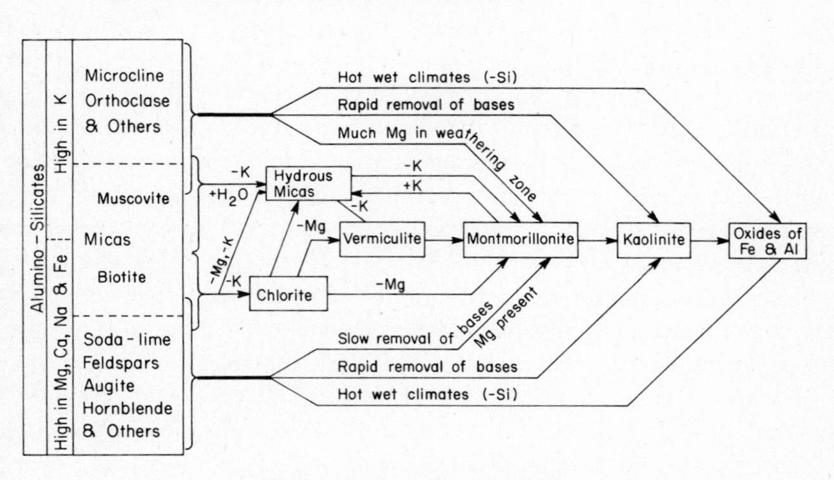

Table 2. *Representative minerals found in soils*

Oxides/Hydroxides	
Si-oxides	quartz, tridymite
Fe-oxides/hydroxides	goethite, hematite, limonite
Al-oxides/hydroxides	gibbsite, boehmite, diaspore
Silicates	
Neosilicates	olivine (Mg), garnet (Ca, Mg, Mn^{2+}, Ti, Cr), tourmaline (Na, Ca, Li, Mg, BO_3), Zircon (Zr)
Inosilicates	augite (Ca, Mg), hornblende (Na, Ca, Mg, Ti), feldspars
Phyllosilicates	talc (Mg), biotite (K, Mg, F), muscovite (K, F), clay minerals: illite (K); kaolinite, montmorillonite, vermiculite (Mg)
Tectosilicates	albite (Na), anorthite (Ca), orthoclase (K), zeolites (Ca, Na, K, Ba)
Carbonates	calcite ($CaCO_3$), dolomite ($MgCa(CO_3)_2$)
Halides	halite (NaCl), sylvite (KCl), carnallite ($KMgCl_3 \cdot 6H_2O$), ($CaCl_2nH_2O$)
Nitrates	soda-nitre ($NaNO_3$), nitre $(KNO_3)_2$, calcium nitrate ($Ca(NO_3)_2$)
Phosphates	apatite (Ca_5(F, Cl, OH) $(PO_4)_3$), vivianite ($Fe_3(PO_4)_2 \cdot 8H_2O$)
Sulfates	gypsum ($CaSO_4 \cdot 2H_2O$)

Source: Reference [14].

Aluminosilicates

The secondary aluminosilicates or clay minerals are more stable in the soil than primary aluminosilicates, since the free energy of the products of a chemical reaction must always be less than the free energy of the reactants. There are many different kinds of aluminosilicates in the soil [91, 30]. Some examples of unsubstituted and substituted aluminosilicates are compiled in Table 3.

In addition to the aluminosilicate minerals, soils contain oxides, hydrous oxides, and amorphic forms of aluminium and silicon. Hydrous oxides of iron and manganese often appear prominently along with these minerals in the clay fraction of <2 μm sizes.

Aluminosilicates play an important role in soil attenuation[1] of metal

[1] Attenuation is defined [54] by looking at the movement of a pulse of solute through a soil. As the pulse migrates, the maximum concentration decreases. Attenuation then can be defined as the decrease of the maximum concentration for some fixed time or distance traveled.

Table 3. *Selected aluminosilicates found in soil material*

Class	Name	Formula
Al-Si	sillimanite	Al_2SiO_5
	kyanite	Al_2SiO_5
	andalusite	Al_2SiO_5
	halloysite	$Al_2Si_2O_5(OH)_4$
	dickite	$Al_2Si_2O_5(OH)_4$
	kaolinite	$Al_2Si_2O_5(OH)_4$
Na Al-Si	nepheline	$NaAlSiO_4$
	albite	$NaAlSi_3O_8$
	beidellite	$Na_{0.33}\ Al_{2.33}Si_{3.67}O_{10}(OH)_2$
	hydrous mica	$K_{0.6}Mg_{0.25}Al_{2.3}Si_{3.5}O_{10}(OH)_2$
	montmorillenite	$M^{+}{}_{0.}+(Si_{3.81}Al_{1.71}Fe(III)_{0.22}Mg_{0.29})$
K Al-Si	microcline	$KAlSi_3O_8$
	muscovite	$KAl_2(AlSi_3O_{10})\ (OH)_2$
Ca Al-Si	pyroxene	$CaAlSiO_6$
	anorthite	$CaAl_2Si_2O_8$
Mg Al-Si	chlorite	$Mg_5Al_2S_3O_{10}(OH)_8$
	fluorphlogopite	$KMg_3AlSi_3O_{10}F_2$
Substituted Al-Si	vermiculite	$(Mg_{2.71}Fe(II)_{0.02}Fe(III)_{0.46}Ca_{0.06}K_{0.1})$ $Si_{2.91}{}^{-}Al_{1.14}O_{10}(OH)_2$

ions in solution because of their great characteristic surface area as clay size particles. The greater the surface area of soil particles, the greater the rates of chemical reaction. Isomorphic substitution of ions into the crystal lattices of aluminosilicates is a usual occurrence during soil formation. This can be expected to alter somewhat their free energy relationships. The clay of a given soil, therefore, not only contains several different primary and secondary mineral species, but variations of a single species, such as montmorillonite, that possess different proportions of substituted ions. The group of reference soils selected for attenuation research in Arizona, representing seven of the ten world orders, illustrates the variation one may expect in almost any soil (Table 4). Thus to base pollutant attenuation on kind of clay mineral type alone is seldom realistic.

The three most abundant secondary aluminosilicates found in the clay fraction ($<2\ \mu m$) are: kaolinite-like, illite(mica)-like, and montmorillonite-like minerals. They are concentrated more in the clay size particles than in

Table 4. *Clay ($<2\ \mu m$) and mineral composition*

Soil series	Location	Clay % of total soil	Relative amounts						
			Montmorillonite	Beidellite	Vermiculite	Chlorite	Mica	Kaolinite	Other (amount)
Davidson	N. Carolina	52	0	0	0	1	0	5	gibbsite (1)
Molokai	Hawaii	52	0	0	0	0	2	4	gibbsite (1)
Nicholson	Kentucky	49	0	0	5	0	1	1	quartz (2)
Fanno	Arizona	46	3	0	0	0	2	1	quartz (2)
Mohave (Ca)	Arizona	40	3	0	0	0	4	2	quartz (2)
Chalmers	Indiana	31	4	0	2	2	0	2	quartz (2)
Ava	Illinois	31	2	0	3	0	2	3	quartz (2)
Anthony	Arizona	15	4	0	1	1	3	2	quartz (2)
Mohave	California	11	2	0	0	0	4	3	quartz (2)
Kalkaska	Michigan	5	0	0	0	3**	1	2	quartz (2)
Wagram	N. Carolina	4	0*	0	0	3**	1	4	quartz (3)

*The amount of each mineral present is represented as: predominant 5, large 4, moderate 3, small 2, trace 1, and not detected 0.
**Chloritic intergrade = mixed layer.

silt or sand (Fig. 3). The lattice structures of kaolinite, halloysite, mica, montmorillonite, and chlorite are illustrated in Fig. 4 according to Grimm [61] and Aubet & Pinta [8]. The kaolin layer as idealized in Fig. 5 is provided in detail since the aluminium octahedral is so common as a repeating unit in other secondary clay minerals. This clay mineral is referred to as a 1:1 type in which one tetrahedral SiO_4-layer shares corners with each other and the octahedral layer of $AlO_4(OH)_2$ is located in between. This unit is about 7.2 Å thick.

In a three-layered unit of a 2:1-type clay mineral, the two layers of SiO_4 tetrahedrons again share all corners with each other and the octahedral layer of $AlO_4(OH)_2$ is located in between. Because it is three-layered, it has a thickness of about 10.1 Å.

The individual units or platelets of the clay minerals combine into multiple units to give each a distinctive chemical and physical behavior. Kaolin (1:1 type) forms tightly bonded structures as hydrogen (H) bonds between the octahedral hydroxyl-groups of one unit and the anions of the tetrahedral layer of the next unit. The number of combinations, thus, uniting to form macrounits determines the specific surface area available for chemical reactions to occur.

Montmorillonite (2:1 type) is less tightly bonded between units than kaolin clay minerals. The individual platelets, therefore, may have variable spacings depending on the abundance of different ions. Sodium, for example, is highly hydrated and, when water is available, the platelets

Fig. 3. General relationship between particle size and kinds of minerals present.

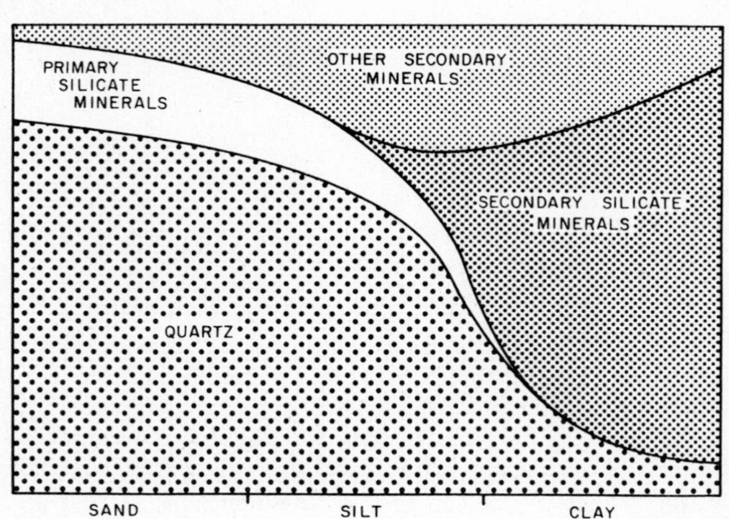

Fig. 4. Diagrammatic examples of idealized succession of layers in some lattice-type aluminosilicates (in Å units).

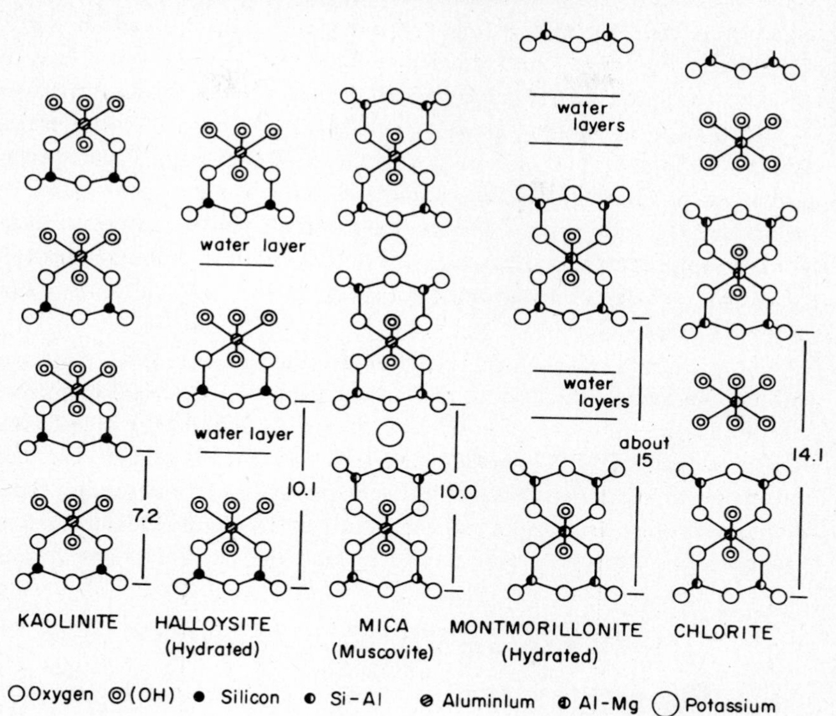

expand (disperse) to 10.1 Å. Di- and tri-valent ions allow for much less dispersion. Calcium, for example, is most often associated with the formation of larger macrounits as illustrated by flocculations.

Substitution of elements (cations of lower valence) into the clay mineral structure, and during formation, the replacement of Si by Al and Al by Mg provide wide variations in structural configuration, charge, and surface area. Such substitution results in a deficit of positive charges within the crystal lattice. This accounts for the attraction of common cations (Ca, Mg, Na, K, Al, and H) to clay surfaces. Cation exchange takes place when these ions exchange with cations in the soil solution with little or no alteration in the solid surfaces (Table 5).

Chemical characteristics

Ion exchange

Cation exchange capacity (CEC) of a soil is a measurement mostly used in the fields of plant growth and crop production. CEC has much less meaning in the field of pollutant retention or confinement since it represents exchange reactions of the major cations such as Ca, Mg, K and Na. The exchangeable cation on the particle surface is associated with negative charge sites on soil solids through largely electrostatic bonding and is therefore subject to interchange with cations in the soil solution. Ions (cations, and anions to a much lesser extent) may be bound to soil surfaces by various forces ranging from *electrostatic* to *covalent*, with related increases in bonding energy. *Specific sorption* contrasts to ion exchange and occurs when covalent bonding dominates. It represents a property of specificity for certain cations and anions. Reversible ion exchange for these ions is less than those for electrostatic bonding. CEC is a phenomenon of the hydrous oxides (Fe, Al, and Mn in particular) and of soil organic matter as well as clays.

Cation exchange capacity is found by displacing the cations from soil with various neutral salt (Na, K, NH_4) solutions of known concentrations. The most commonly used solution is 1 $\underline{N}$ NH_4OAc. The CEC is an estimate of the amount of NH_4^+ held on the soil. The exchangeable bases are considered to be those contained in the extract. Soils from arid and semi-arid lands require some modification of the standard CEC procedure in an attempt to prevent dissolution of $CaCO_3$ and $CaSO_4 \cdot 2H_2O$ and certain

Fig. 5. Idealized example of kaolin as illustrated along a axis and along b axis. The measurement of layer separation above oxygen is in Å units.

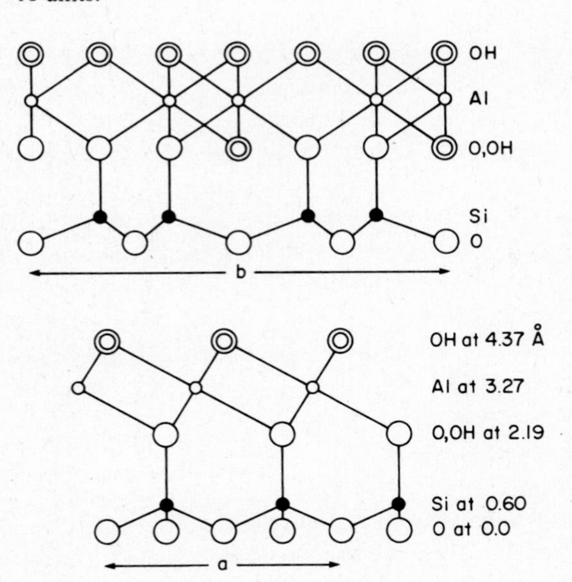

indigenous salts from contributing to the sum of exchangeable cations.

Anion exchange in soils has been given small attention compared with that of cations, primarily because the level of exchange is relatively low at the pH ranges of most natural soils. Some soils, particularly acid soils, exhibit true anion exchange, however. The anions most often involved are NO_3^-, Cl^-, SO_4^{2-}, and $PO_4^{(n)}$. Whereas CEC ranges between 0.5 and 50 meq $100\,g^{-1}$ soil, anion exchange capacity ranges between 0 and about 2.6 meq $100\,g^{-1}$ [83]. Most soils exhibit *anion exclusion* [125] because the net negative charge of soil colloid repels the anions in solution in the vicinity of the surfaces. Yet, the net negative charge of a colloidal clay system decreases with increasing acidity [36]. Kaolinite and allophane groups, for example, demonstrate positive charges at low pH values. Anion adsorption and exchange, therefore, take place at these positive charge sites. Table 5 from Schoen [113] provides data for one to make a comparison of cation and anion exchange capacities of clay minerals at very low soil pH values.

Several problems are associated with the conventional determination of exchangeable cations and cation exchange capacity as a useful tool in the prediction of soil behavior for the attenuation of hazardous pollutants from wastes. *First*, in displacement (especially by 1<u>N</u> NH_4OAc), ions in addition to those in truly exchangeable positions may contribute to the extraction solution (such as Ca^{2+} from lime and gypsum). Solution of a host of soil minerals, compounds, and complex inorganic and amorphic substances and salts may take place. *Second*, the displacement of cations with any of the monovalent cations now used does not appear to represent or describe the soil exchange capacity for heavy metals. The figures reported for exchangeable heavy metals using NH_4^+, for example, as the

Table 5. *Cation and anion exchange capacities of various clay minerals*

Mineral	Cation exchange capacity (meq $100\,g^{-1}$)	Anion exchange capacity (meq $100\,g^{-1}$)	Cation exchange to anion exchange ($c\,a^{-1}$)	Equilibrium pH
Nontronite	87	13	6.7	2.4
Bentonite	62	15	4.1	3.1
Illite	21	9	2.3	4.5
Kaolinite	27	43	0.63	4.7
Halloysite	4	7.7	0.52	4.3

Source: after reference [113].

replacing ion may be misleading [90, 119, 27, 81, 76]. Why should Zn exchange capacity be expected to be defined by NH_4^+ exchange capacity? *Third*, cations that become stored temporarily at the liquid–solid interface will again be released upon change in composition of the liquid phase either by dilution or ion concentration. Retention, thus, is not permanent as is desired for waste control. *Fourth*, the CEC is not the only factor influencing the amount of temporary ion storage. The concentration of constituents (salts, chelates, acids, bases) in the leachates, waste solutions, and solvents may actually totally offset the magnitude of the original soil exchange capacity.

On the other hand, one may expect CEC to roughly correlate with retention of some heavy metals and certain other hazardous pollutants since there often is a close relationship between particle size distribution (texture) of a soil and CEC. The surface area of a soil and/or clay content appear to be more reliable properties describing attenuation prediction than CEC, Korte et al. [84] and Fuller [41, 42].

Chemical equilibria

The great number of chemical reactions in the soil that take place at any one time proceed at various speeds ranging from those that attain equilibrium immediately to those that react so slowly they probably never do achieve equilibrium. If the reaction is reversible, then both forward and backward action is possible and there always will be a state of equilibrium existing, the position of which will depend on the composition of the system with respect to reactants and reaction products present at any one time and temperature. The ability to predict the possibility or probability that a reaction can or cannot take place can be determined through a knowledge of the chemical equilibria involved. Because the soil is constantly involved with numerous equilibrium reactions at any given time, with a host of reactants and reaction products that again become highly complicated by the myriad of waste products entering the soil system, an indepth discussion of chemical equilibria must be omitted here with reference to such excellent book reviews as Bolt & Bruggenwert [14]; Lindsay [91]; Thibodeau [124]; Novozansky et al. [103] and others readily available in well-published literature and textbooks.

Transfer of protons and/or electrons

Redox: of the many chemical characteristics of soils, oxidation and reduction reactions (redox) rank high in importance in the geochemistry of hazardous waste disposal. The transfer of electrons from donor to acceptor concerns solubility of reaction products formed when wastes and waste leachates, unnatural to the land, make contact with the natural soil body. Electron transfer is involved with every slight change in gaseous composition and movement through soil.

The redox reaction, e.g. transfer of electrons and protons from donor to acceptor, can be written:

$n_B A_{ox} + n_A B_{red} <\!\!-\!\!\!> n_B A_{red} + n_A B_{ox}$ (equivalent to the proton reaction).

The reaction can be split into the redox couples:

$$A_{ox} + n_A e^- = A_{red},\ K^\circ_{red} = \frac{(A_{red})}{(A_{ox})(e^-)^n}.$$

$$B_{ox} + n_B e^- = B_{red},\ K^\circ_{red} = \frac{(B_{red})}{(B_{ox})(e^-)^n}.$$

These half reactions are equivalent to the reactions involving the transfer of protons. Relative electron activity can also be described:

$$pe = -\log(e^-)$$

in a way like that for relative proton activity:

$$pH = -\log(H^+).$$

Strong oxidizing conditions are associated with large pe values, while strong reducing conditions are associated with small values of pe.

In the aqueous systems of soils, the dissociation of water into $H_2(g)$ or $O_2(g)$ imposes redox limits on soils. On the reducing side, Lindsay [91] provides the reactions:

$$H_2O + e^- = 1/2\, H_2(g) + OH^-$$

$$H^+ + OH^- = H_2O$$

$$\overline{H^+ + e^- = 1/2\ H_2(g)}$$

which yields the equilibrium equations:

$$K^\circ = \frac{(H_2)^{1/2}}{(H^+)(e^-)}$$

or

$$\log K^\circ = 1/2 \log H_2 - \log (H^+) - \log (e^-)$$

where the units of the gas phase are partial pressure.

The corresponding K° then equals unity ($\log K° = 0$) for standard state conditions. This sets the electron activity (e^-) at unity for the standard hydrogen half-cell reaction, and since K° is 1, then

$$pe + pH = -1/2 \log H_2$$

and when

$$H_2\ (g) = 1\ \text{atm. At Standard State}\ \rho = 1\ \text{atm.}$$
$$pe + pH = 0. \qquad T = 273K.$$

On the oxidizing side, Lindsay describes the redox limit of aqueous system by:

$$H^+ + e^- + 1/4\ O_2\ (g) = 1/2\ H_2O$$

giving the equilibrium expression

$$K° = \frac{(H_2O)^{1/2}}{(H^+)(e^-)(O_2)^{1/4}}$$

Fig. 6. Equilibrium redox relationships of aqueous systems. The inscribed area is representative of most soils. (Source: Reference [91].)

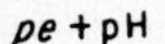

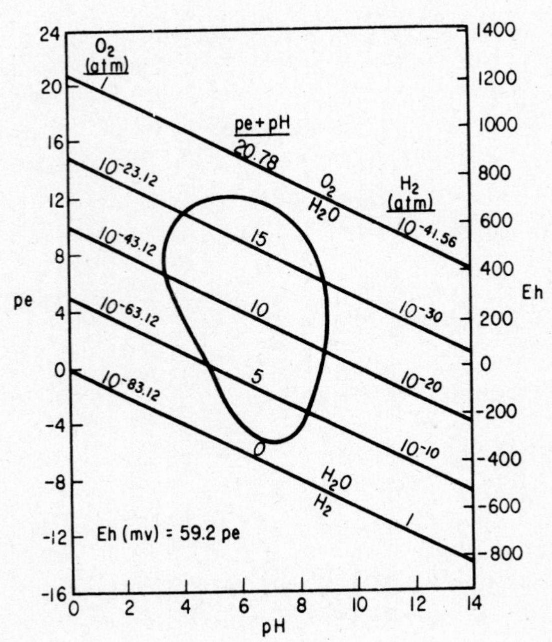

Because in dilute aqueous systems the activity of water is near unity, the log form of the equilibrium expression yields:

$$-\log (H^+) - \log (e^-) - 1/4 \log O_2 = 20.78$$

or

$$pe + pH = 20.78 + 1/4 \log O_2.$$

The equilibrium redox relationships of aqueous systems including the pe, pH and the more familiar Eh have been diagrammed by Lindsay and reproduced in Fig. 6. This figure also includes an inscribed area as representative of most soils.

pH (acidity and alkalinity)

The pH as measured by the glass electrode in soil saturated with water indicates the acidity (H-ion activity) and alkalinity (OH-ion activity) of soil. The pH is defined as either the negative logarithm of the H^+-ion concentration or as the logarithm of the reciprocal of the H^+-ion concentration. Thus $pH = \log 1/[H^+]$ where H^+-ion concentration is in mol l^{-1}.

The pH of pure water is 7.0 or 0.000 000 1 mol l^{-1}. More simply:

d ↑			i	
e	a	pH 7.0 = 0.000 000 1 mol l^{-1}	n	a
c	c	pH 6.0 = 0.000 001 mol l^{-1}	c	c
r	i	pH 5.0 = 0.000 01 mol l^{-1}	r	i
e	d	pH 4.0 = 0.000 1 mol l^{-1}	e	d
a	i	pH 3.0 = 0.001 mol l^{-1}	a	i
s	t	pH 2.0 = 0.01 mol l^{-1}	s	t
i	y	pH 1.0 = 0.1 mol l^{-1}	i	y
n			n	
g			g ↓	

Soil scientists refer to pH of soil as illustrated in Fig. 7 primarily because soils usually occupy the middle ranges of the pH scale and higher plants (crops) have adapted well to soil pH levels which usually range between 4 and 9. Many plants and economic crops are highly sensitive to soil pH and selectively are adapted to narrower ranges than 4 to 9. This also is true for microorganisms and other biological life indigenous to soil habitats. Thus pH measurements provide a valuable tool for biologists and soil scientists.

Since soil is a highly buffered medium, true acidity can be estimated if the CEC and the degree of saturation of the exchange complex with Al (reserve acidity) are known in addition to pH (active acidity). Also, the pH at the surface of soil particles (pH_s) is more acidic than the soil solution beyond the particle surface or the bulk (pH_b). Also, organic matter including the volatile acids often occurring in dilute aqueous solid waste leachates, represents constituents providing buffer capacity to soils and organic containing disposals that undergo biodegradation.

Physical characteristics

The physical characteristics of soil significantly influence the behavior of waste disposal on land and in excavated soil. The liquid and gaseous phases move and transport hazardous waste constituents in direct response to the permeability of the soil as defined by particle size distribution, bulk density, particle density, and pore size and configuration. This section briefly describes those physical processes taking place within the soil that can influence the migration rate of hazardous waste constituents. Fig. 8 by Letey [90] illustrates the interrelationship between soil–physical process which provides a point of departure for this section on physical characteristics of soils.

Particle size distribution

The soil mantle surrounding the earth is made up of particles ranging from boulders, stones, gravel, sand and silt to colloid clay. The soil scientist refers to the completely *dispersed particles* as *textural separates* (Fig. 9). Soil properties usually are characterized on the material that

Fig. 7. The pH range (acidity–alkalinity) for soils.

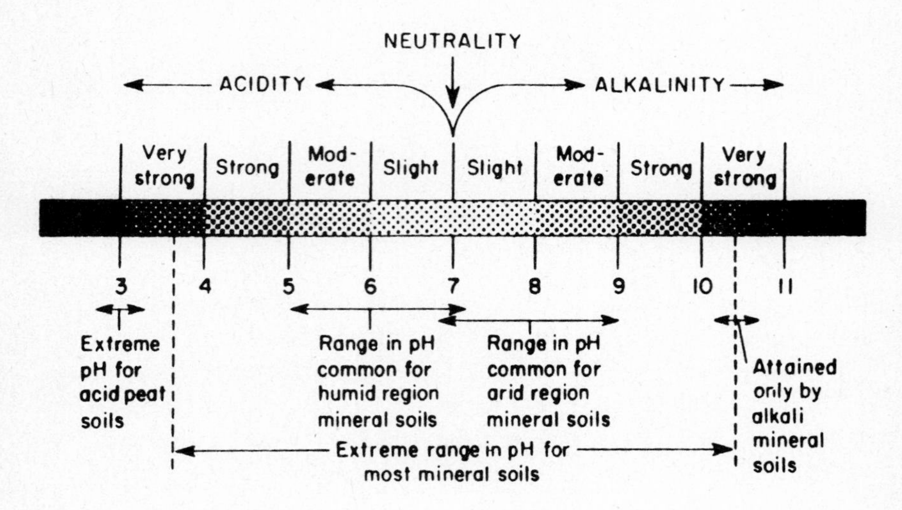

passes a 2 mm sieve when air dried and rolled to break up structural units but not the consolidated stone and gravel. The *structural units* form naturally from dispersed particles into aggregates by grouping together into units of various size and shape.

There are three systems under which soils are most likely to have been classified in the United States: The Unified Soil Classification System, the old (1938) US Department of Agriculture System, and the present US Department of Agriculture System.

The Unified Soil Classification System (USCS) serves engineering uses of soils. The criteria for soil types in the system are based on the grain (particle) size and response to physical manipulation at various water contents.

The US Department of Agriculture (USDA) System serving agricultural and other land management uses the criteria for classification in a system that is based on both chemical and physical properties of the soil. The USDA system in general use between 1938 and 1960 was based on soil genesis – how soils were formed or were thought to have been formed. The present USDA comprehensive soil classification system is based on quantitatively measurable properties of soils as they exist in the field. Although the present USDA system is incomplete and is being continually refined, it is generally accepted by US soil scientists and its nomenclature is used in most of the current literature.

Fig. 8. Some interrelationships between soil–physical processes. (Source: Reference [90].)

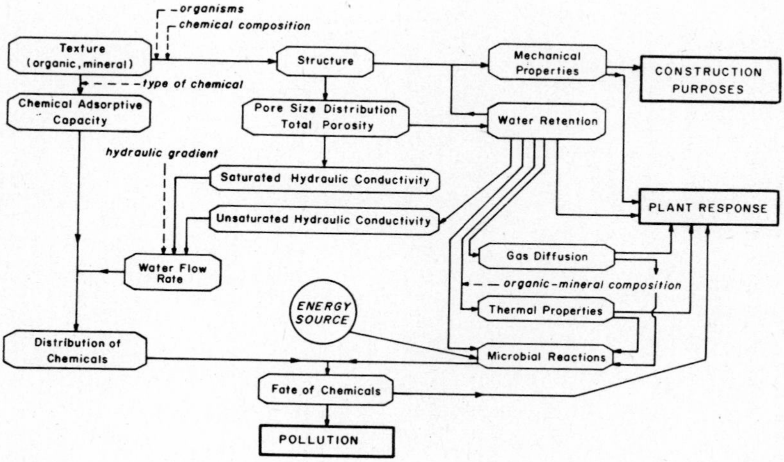

The part of the USDA classification which may be compared most directly with the soil types in the USCS system is soil texture (distribution of grain or particle size) and associated modifiers such as stoney, gravelly, mucky, diatomaceous, and micaceous. The size ranges for the USDA and the USCS particle designations (e.g. sand, gravel) are listed in Table 6. The soil texture (USDA – sandy loam, silty loam, etc.) or the soil type (USCS – GC clayey gravel, SC clayey sand, etc.) is classified on the relative amounts of different sized particles in a soil. The USDA and USCS systems for classifying soil texture are compared diagramatically in Figs. 9 and 10. A correlation of the USCS and USDA systems on the basis of texture is presented in Table 7. It should be emphasized that this correlation is not precise because texture is a high-level (major) criterion in the USCS system but is a low-level (minor) criterion in the USDA system. A soil of a given texture can be classified into only a limited number of the 15 USCS soil types while in the USDA system, soils of the same texture may be found in many of the ten orders and 43 suborders because of differences in their chemical properties or the climatic areas in which they are located.

The most active fraction is the clay (< 2 μm). Clay surfaces are the seat of most chemical, physical, and biological activity because of their large surface area per unit weight (Table 8). Particle size distribution also is an important factor in soil pore size distribution and, consequently, porosity

Table 6. *US Department of Agriculture (USDA) and Unified Soil Classification System (USCS) particle sizes*

USDA		USCS	
Particle	Size range (mm)	Particle	Size range (mm)
Cobbles	76.2–254	Cobbles	> 76.2
Gravel	2.0–76.2	Gravel	4.76–76.2
coarse gravel	12.7–76.2	coarse gravel	19.1–76.2
fine gravel	2.0–12.7	fine gravel	4.76–19.1
Sand	0.05–2.0	Sand	0.074–4.76
very coarse sand	1.0–2.0		
coarse sand	0.5–1.0	coarse sand	2.0–4.75
medium sand	0.25–0.5	medium sand	0.42–2.0
fine sand	0.1–0.25	fine sand	0.074–0.42
very fine sand	0.05–0.1		
Silt	0.002–0.05	Fines	< 0.074
Clay	< 0.002	(silt & clay)	

Source: after Reference [123].

Fig. 9. Textural chart showing soil particle sizes and related classes as given by three separate systems involved nationally and internationally.

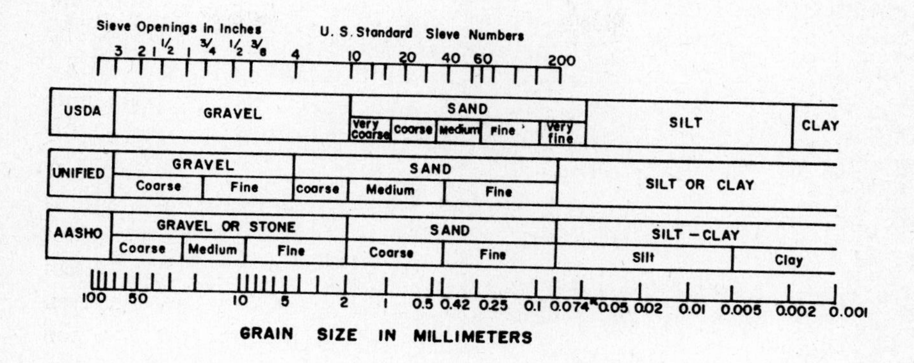

Fig. 10. Textural classification of soils comparing USDA and USCS systems based on the percentage of different sizes of particles they contain.

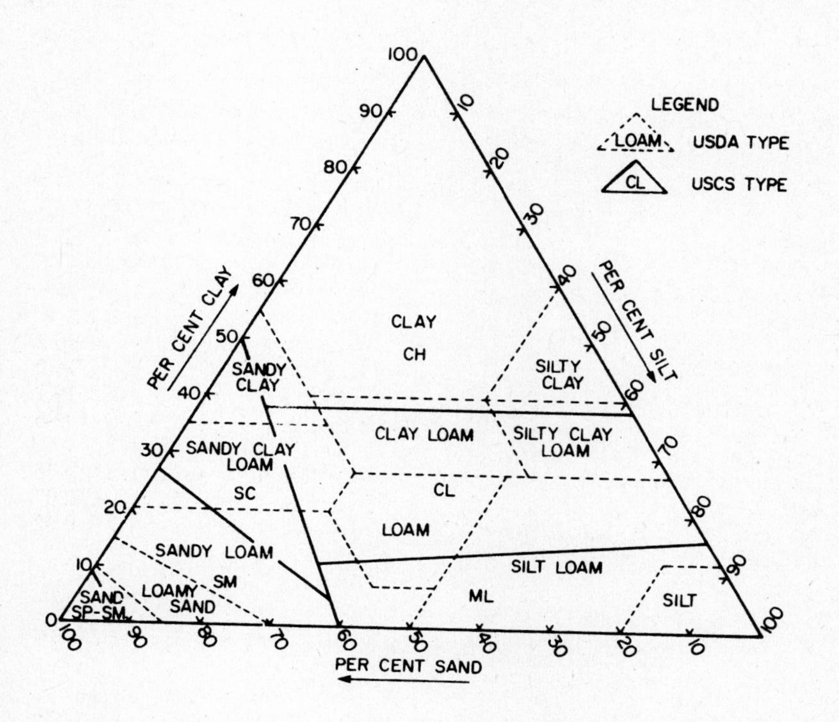

Table 7. *Corresponding USCS and USDA soil classification descriptions*

Unified Soil Classification System (USCS) soil types	Corresponding United States Department of Agriculture (USDA) soil textures
1. GW	Same as GP – gradation of gravel sizes not a criteria
2. GP	Gravel, very gravelly* sand less than 5% by weight silt and clay
3. GM	Very gravelly* sandy loam, very gravelly* loamy sand, very gravelly* silt loam, and very gravelly* loam**
4. GC	Very gravelly clay loam, very gravelly sandy clay loam, very gravelly silty clay loam, very gravelly silty clay, very gravelly clay**
5. SW	Same as GP – gradation of sand size not a criteria
6. SP	Coarse to fine sand, gravelly sand † (less than 20% very fine sand)
7. SM	Loamy sands and sandy loams (with coarse to fine sand), very fine sand, gravelly loamy sand † and gravelly sandy loam †
8. SC	Sandy clay loams and sandy clays (with coarse to fine sands), gravelly sandy clay loams and gravelly sandy clays †
9. ML	Silt, silt loam, loam, very fine sandy loam‡
10. CL	Silty clay loam, clay loam, sandy clays with <50% sand‡
11. OL	Mucky silt loam, mucky loam, mucky silty clay loam, mucky clay loam
12. MH	Highly micaceous or diatomaceous silts, silt loams – highly elastic
13. CH	Silty clay and clay‡
14. OH	Mucky silty clay
15. PT	Muck and peats

*Also includes cobbly, channery, and shaly.
**Also includes all of textures with gravelly modifiers where >1/2 of total held on No. 200 sieve is of gravel size.
†Gravelly textures included if less than 1/2 of total held on No. 200 sieve is of gravel size.
‡Also includes all of these textures with gravelly modifiers where >1/2 of the total soil passes the No. 200 sieve.

Table 8. *The relationship of surface to particle size*

Diameter of sphere	Textural name	Volume/particle ($\frac{1}{6}\pi D^3$)	Number of particles in $\frac{\pi}{6}$ cc	Total surface $\pi D^2 \times$ number of particles
1 cm	Gravel	$\frac{1}{6}\pi(1)^3$	1	3.14 cm² = 0.49 in²
0.1 cm (1 mm)	Coarse sand	$\frac{1}{6}\pi(\frac{1}{10})^3$	1×10^3	31.42cm² = 4.87 in²
0.05 cm (0.5 mm or 500 μ)	Medium sand	$\frac{1}{6}\pi(\frac{5}{100})^3$	8×10^3	62.83 cm² = 9.74 in²
0.01 cm (0.1 mm or 100 μ)	Very fine sand	$\frac{1}{6}\pi(\frac{1}{100})^3$	1×10^6	314.16 cm² = 48.67 in²
0.005 cm (0.05 mm or 50 μ)	Coarse silt	$\frac{1}{6}\pi(\frac{5}{1000})^3$	8×10^6	628.32 cm² = 97.34 in²
0.002 cm (0.02 mm or 20 μ)	Silt	$\frac{1}{6}\pi(\frac{2}{1000})^3$	125×10^6	1570.8 cm² = 1.69 ft²
0.0005 cm (0.005 mm or 5 μ)	Fine silt	$\frac{1}{6}\pi(\frac{5}{10\,000})^3$	8×10^9	6283.2 cm² = 6.76 ft²
0.0002 cm (0.002 mm or 2 μ)	Clay	$\frac{1}{6}\pi(\frac{2}{10\,000})^3$	125×10^9	15 708 cm² = 16.9 ft²
0.0001 cm (0.001 mm or 1 μ)	Clay	$\frac{1}{6}\pi(\frac{1}{10\,000})^3$	1×10^{12}	31 416 cm² = 33.8 ft²
0.000 05 cm (0.0005 mm or 500 mμ)	Clay	$\frac{1}{6}\pi(\frac{5}{100\,000})^3$	8×10^{12}	62 832 cm² = 67.6 ft²
0.000 02 cm (0.0002 mm or 200 mμ)	Colloidal clay	$\frac{1}{6}\pi(\frac{2}{100\,000})^3$	125×10^{12}	157 080 cm² = 169 ft²
0.000 01 cm (0.0001 mm or 100 mμ)	Colloidal clay	$\frac{1}{6}\pi(\frac{1}{100\,000})^3$	1×10^{15}	314 160 cm² = 338 ft²
0.000 005 cm (0.000 05 mm or 50 mμ)	Colloidal clay	$\frac{1}{6}\pi(\frac{5}{1\,000\,000})^3$	8×10^{15}	628 320 cm² = 676 ft²

and permeability of liquid and gas flow. Generally, the finer the particle sizes, the smaller the pore sizes and the larger the total porosity but slower the permeability. Sands hold less water than clays because the total pore space is not as great as for clays.

Soil structure

Soils form structural patterns from primary and secondary dispersed soil particles of all sizes (Fig. 11). Structures contain pore spaces of two major kinds: micropores located within the granules and macropores between granules. Soil structures also vary in stability and, therefore, breakdown patterns. The USDA Soil Conservation Service considers structure as referring 'to the aggregation of primary soil particles, which are separated from adjoining aggregates by surfaces of weakness'. The ultimate or individual aggregate is called a *ped*. Description of structures and their classification based on (a) shape and arrangement of peds, (b) size of peds and (c) durability as related to *type*, *class* and *grade*, respectively, appear in Table 9.

The structures in some soils may become so stable upon drying, they remain as discrete units when rewetted and permit transport of water similar to sands or fine gravel. Diffusion of liquids into and out of these

Fig. 11. Diagram of some naturally occurring and observable soil structures.

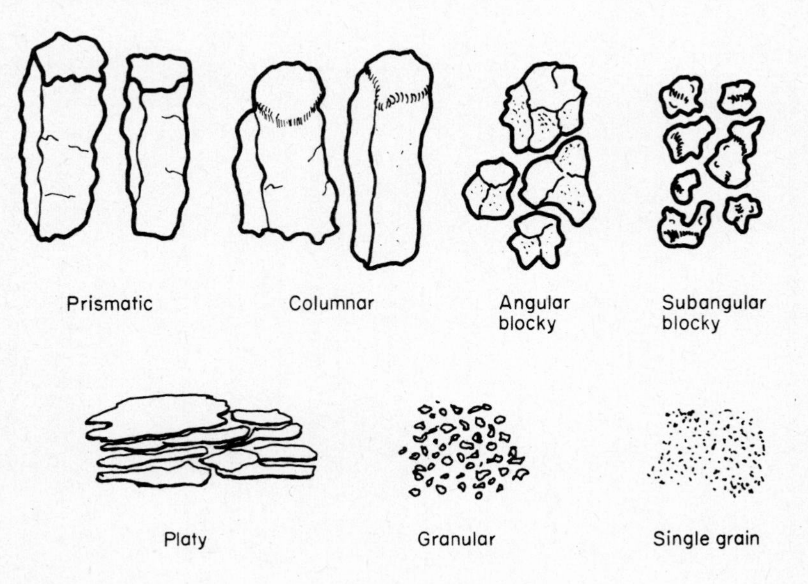

Table 9. *The classification of soil structure according to USDA soil survey*

	A-Type: Shape and arrangement of peds						
				Blocklike—Polyhedral—Spheroidal. Three approximately equal dimensions arranged around a point			
	Platelike. Horizontal axes longer than vertical. Arranged around a horizontal plane	*Prismlike*. Horizontal axes shorter than vertical. Arranged around vertical line. Vertices angular		*Blocklike—Polyhedral*. Plane or curved surfaces accommodated to faces of surrounding peds		*Spheroidal—Polyhedral*. Plane or curved surfaces not accommodated to faces of surrounding peds	
		Without rounded caps	With rounded caps	Faces flattened; vertices sharply angular	Mixed rounded, flattened faces; many rounded vertices	Relatively nonporous peds	Porous peds
B-Class: Size of peds	Platy	Prismatic	Columnar	Blocky	Subangular blocky	Granular	Crumb
(1) Very fine or very thin	< 1 mm	< 10 mm	< 10 mm	< 5 mm	< 5 mm	< 1 mm	1 mm
(2) Fine or thin	1–2 mm	10–20 mm	10–20 mm	5–10 mm	5–10 mm	1–2 mm	1–2 mm
(3) Medium	2–5 mm	20–50 mm	10–20 mm	10–20 mm	10–20 mm	2–5 mm	2–5 mm
(4) Coarse or thick	5–10 mm	50–100 mm	50–100 mm	20–50 mm	20–50 mm	5–10 mm	
(5) Very coarse or very thick	> 10 mm	> 100 mm	> 100 mm	> 50 mm	> 50 mm	> 10 mm	
C-Grade: Durability of peds	(0) Structureless	No aggregation or orderly arrangement					
	(1) Weak	Poorly formed, nondurable, indistinct peds that break into a mixture of a few entire and many broken peds and much unaggregated material					
	(2) Moderate	Well-formed, moderately durable peds, indistinct in undisturbed soil, that break into many entire and some broken peds but little unaggregated material					
	(3) Strong	Well-formed, durable, distinct peds, weakly attached to each other, that break almost completely into entire peds					

Source: Reference [123].

structures is slow compared to movement of fluids between the structures. Examples of soils that develop highly stable structures are those from tropical regions called laterites (high in clay and hydrous oxides of iron) and certain clay soils of arid and semi-arid climates. Because soils composed of mostly clay size particles offer problems of this nature as a result of wetting and drying, and freezing and thawing, they must be given special attention if used as liner material or as barriers of undisturbed soil for disposal purposes.

The chemical composition of the solution and/or solvent in soil influences the degree of soil structure (aggregate), formation and primary particle dispersal. *Flocculation* of clay may be brought about by the polyvalent cation (Ca) whereas dispersion tends to occur when monovalent cation (Na) dominates. Flocculation can take place also as a result of high electrolyte concentration, just as dispersion is associated with low electrolyte concentration.

Kind of clay mineral

The kind of clay mineral identified by X-ray analysis [75] influences the physical characteristics of the soil. The 1:1 nonexpanding lattice types, such as kaolin, do not shrink and swell upon drying and wetting as much as the 2:1 expanding lattice types, montmorillonite. Moreover, the kind of clay mineral dominating the soil determines the Atterberg limits as they relate to soil mechanics, such as plasticity, compressive strength, liquid limits, compaction, and volume change. A more comprehensive discussion of the various kinds of clay minerals appears in the section on soil minerals.

Pore size distribution and porosity

Pore size distribution affects many of the important physical processes in soil. Fluid movement most often depends more on pore size distribution than total porosity. Where the range of pore sizes is narrow as in sand (large pores) or dispersed clay (small pores), the differences are less evident than when the range of sizes is broad as in aggregated soils. Not only will the original pore size distribution influence flow processes but any alteration caused by disposal of aqueous and nonaqueous organic and inorganic fluids influences flow rates.

Mechanical plugging of pores with suspended substances from waste disposal as well as microbial (biodegradation) plugging can alter the pore size and total pore volume. On the other hand, certain organic solvents actually enhance the rate of flow through soils. Brown & Anderson [18] and Anderson & Brown [3], for example, found that permeability of two smectite clays increased substantially after the soils were permeated with certain neutral–polar, neutral–nonpolar, and basic organic fluids. The controlling mechanism was thought to be related to changes in pore size and pore size distribution that favored fluid movement.

The *total porosity* (f) of a soil requires the measurement of bulk density (P_b) and particle density (P_s) using the equation: $f = 1 - P_b/P_s$.

Bulk density may be determined by several methods, all of which use undisturbed soil cores or clods. The core method of known soil volumes includes interclod cavities whereas individual clods do not. Immersion may be made either in mercury or water. If water is used, a thin coating of paraffin wax is required before dunking.

Pore size distribution evaluations can be made by means of the *pressure–intrusion* method of Diamond [29] in which mercury is forced into pores of the dried soil. This method is most suitable for coarse grained soils. A desorption, capillary-condensation method is used for fine grained soils usually with water as the permeating fluid.

Pore size distribution may be classified into macropores and micropores, where aggregates are distinct and water stable. The macropores class represents predominantly interaggregate cavities and micropores, intraaggregate cavities. Air and water migrate primarily between aggregates or peds in the macropores and within aggregates in micropores. This separation is of course only descriptive but can be useful in attempts to explain pollutant attenuation discrepancies between soils of similar textural classes. Permeability as related to clay mineral type and amount of clay varies greatly depending on the dominance of one class over the other.

Soil particle density and bulk density

Other physical characteristics of soils serving as useful parameters in the evaluation of pollutant attenuation include *particle density* and bulk density. Particle density, P_s (g cm^{-3}), is defined as the ratio of the mass of the soil dried at 105°C to the volume of the solid particle.

Bulk density was described in the previous section under soil pore size distribution and porosity. For laboratory soil-column research, a rough estimate of bulk density is the ratio of the oven-dried mass of soil packed into the column divided by the volume of the column.

The flow of fluids in saturated soil

Laminar flow

The flow of fluids in saturated soil is mostly *laminar*. The flow prevails at relatively low velocities and in narrow channels. In laminar flow adjacent fluid layers transmit tangential stresses or drag [70]. The intermolecular attraction results in the adherence of fluid molecules to the solid wall surfaces. This adherence may be described as viscosity. To understand laminar flow, we can visualize flow through a straight, narrow cylindrical tube with a constant diameter. Fluid next to the wall does not move. The maximum rate of flow is on the axis of the tube, and constant at each concentric cylinder around the axis. Thus laminar flow derives its name from the fact that adjacent cylindrical *laminae* slide over each other while moving at different velocities and the parallel motion may be depicted as laminar (Fig. 12).

Darcy's law

The simplicity of laminar flow through straight tubes is attractive but unrealistic. Soil pores vary in size and shape and are irregular, tortuous, and complicated. Therefore, flow through soil is generally described in terms of a *macroscopic flow velocity vector* and represents more of an average of the total flow velocity over the volume of soil being considered. The detailed soil pore irregularities are not considered and the flow system is represented as a simple uniform porous body. The total flow through the entire medium, then, may be distributed uniformly over the entire cross section without differentiating between solid and pore space. It is assumed that the soil column under consideration is sufficiently large relative to the pore size and small heterogeneities to permit the averaging of velocity and potential over the cross section. If this were not the case, the soil-column technique to be described here would fail completely in its application to real situations in the field.

With this abbreviated background, let us consider Darcy's law for vertical columns, both downward and upward as in the pollutant attenuation research. Figs. 13 and 14 may be useful to this discussion.

Fig. 12. Idealized laminar flow through a narrow, straight tube.

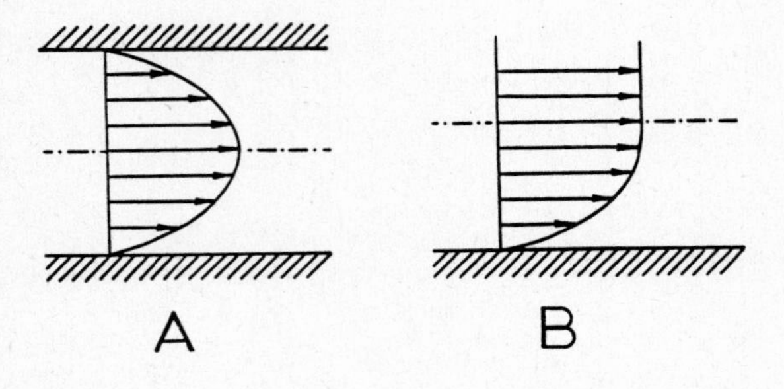

To know the flux according to Darcy's law depends on finding the hydraulic head gradient. This is the ratio of the hydraulic drop to the column length as shown in Table 10 from Hillel [70].

The Darcy equation for downward flow, therefore, becomes

$$q = K \cdot \Delta H/L = K(H_1 + L)/L, \; q = KH_1/L + K.$$

Fig. 13. Illustration of *downward* flow of water in a verticle, saturated soil column.

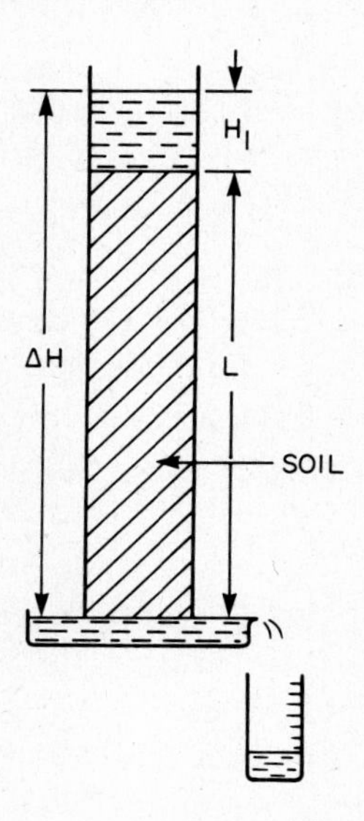

Fig. 14. Illustration of *upward* flow of water in a vertical, saturated soil column.

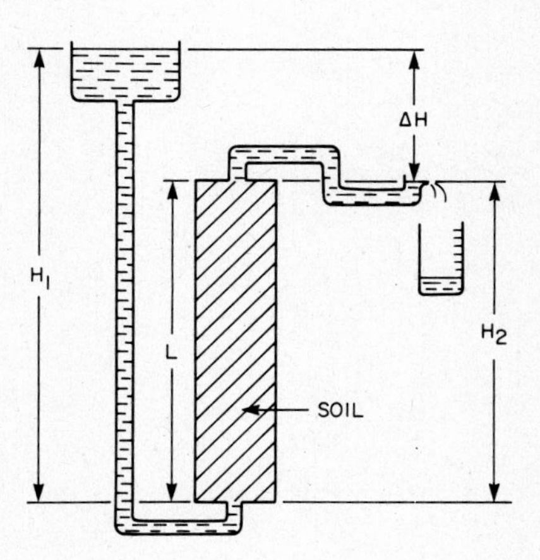

The Darcy equation for upward flow against gravity is

$$q = K(H_1 - L)/L = KH_1/L - K,\ q = K\Delta H/L$$

where

q = flux or water velocity units (L/T), in cm hr^{-1};
H_1 = submergence potential units (L), in cm;
L = length of column units (L), in cm;
K = saturated hydraulic conductivity units (L/T), in cm hr^{-1}.

Flux

Flux is used throughout this text as simplified flow velocity and is defined as the volume of water passing through a unit cross-sectional area, perpendicular to the flow direction, per unit time. Mathematically it is

$$q = V/At = L^3/L^2T = LT^{-1}$$

or length per time in cgs units, cm sec^{-1}. It is called flow velocity in some texts and may represent an average velocity of the numerous and variably shaped pores. Because of the tortuosity of the pore space and since part of the cross-sectional area may be 'dead-ended', some of the water flows a greater distance than the length of the soil column while other water enters into plugged passages prohibiting flow over the entire cross section. Tortuosity has been proposed to account for these apparent discrepancies. It is the ratio of the average length of the pore passages to the length of the soil column or specimen and is always greater than 1 but seldom greater than 2.

Table 10. *Hydraulic head at inflow and outflow boundaries and head difference for downward and upward flow of water in a vertical saturated column*

Hydraulic head		Pressure head		Gravity head
Downward flow of water in a vertical column				
At inflow boundary	$H_i =$	H_1	+	L
At outflow boundary	$H_o =$	0	+	0
Difference	$\Delta H = H_i - H_o =$	H_1	+	L
Upward flow of water in a vertical column				
At inflow boundary	$H_i =$	H_1	+	0
At outflow boundary	$H_o =$	0	+	L
Difference	$\Delta H = H_i - H_o =$	H_1	−	L

Hydraulic conductivity and permeability

Hydraulic conductivity and permeability are common terms in constant use when water movement is considered. A simple diagram (Fig. 15) [70] describes the relationships between flux (q), hydraulic gradient ($\Delta H/\Delta x$), and particle size distributions in soils. Hydraulic conductivity is the ratio of the flux to hydraulic gradient as in Fig. 15. Since flux is LT^{-1}, the dimensions of hydraulic conductivity require an assigned driving force or potential gradient. Generally, head units are used for soils. This assignment results in a saturated sand with stable structure to yield characteristic values ranging from 10^{-2} to 10^{-3} cm sec^{-1} and clay soil values ranging from about 10^{-4} to 10^{-7} cm sec^{-1}.

Since hydraulic conductivity and permeability both measure the ability of the soil to take water, they can be related to each other by the equation

$$k = K_{\eta}/\rho g$$

where

k = permeability of the soil;
ρ = fluid density or 1 g cm^{-3};
η = fluid viscosity or 0.01 cgs units;
g = gravitational acceleration constant or

$$\frac{(.01)}{(980)(1)} = 10^{-5} \text{ cm sec}^{-2};$$

K = hydraulic conductivity in cm sec^{-1}.

Technically, hydraulic conductivity has the dimensions of a velocity, and

Fig. 15. The linear dependence of flux upon hydraulic gradient, the hydraulic conductivity being the slope – i.e. the flux per unit gradient. (Source: Reference [70].)

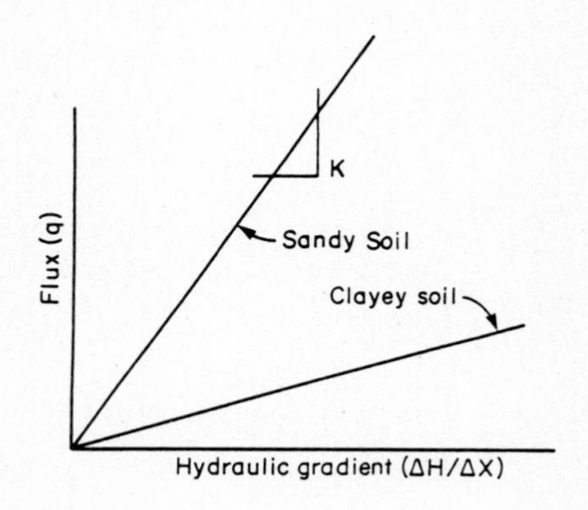

permeability has the dimensions of a length squared. Thus the square micrometer (μ^2) is the practical unit for reporting permeability which is $\mu^2 = 10^{-12}\,cm^2$.

Despite the derivation of hydraulic conductivity as a proportionality factor in Darcy's law, it is not necessarily a constant. Changes may take place in conductivity as water is passed through soil and chemical, biological, and physical processes occur. For example, hydraulic conductivity may change as a result of changes in:

(1) exchangeable ion status of a soil;
(2) concentration of solute species;
(3) electrolyte concentration of water which is not well understood;
(4) viscosity changes in the vicinity of the solid surface;
(5) physical transfer of particulate materials in water;
(6) microbial activity through biodegradation and synthesis;
(7) volume of the gaseous phase.

Examples of the wide range of hydraulic conductivities and permeabilities found in subsoils is provided by O'Neal [104] (Table 11).

The dynamic nature of soil

Soil microbial population

The soil microbial population and its activity more closely represents the dynamic aspect of the soil than any of the other systems. Yet, some dynamic functions may be ascribed to certain physical and chemical reactions in soil. The soil has life with numerous chemical reactions taking place simultaneously, and true equilibrium probably is never attained.

Table 11. *Permeability classes for saturated subsoils, and the corresponding ranges of hydraulic conductivity and permeability*

Class	Hydraulic conductivity		Permeability
	in hr^{-1}	cm hr^{-1}	cm^2
Very slow	<0.05	<0.125	$<3\times10^{-10}$
Slow	0.05–0.2	0.125–0.5	$3\times10^{-10} - 15\times10^{-10}$
Moderately slow	0.2–0.8	0.5–2.0	$15\times10^{-10} - 60\times10^{-10}$
Moderate	0.8–2.5	2.0–6.25	$60\times10^{-10} - 170\times10^{-10}$
Moderately rapid	2.5–5.0	6.25–12.5	$170\times10^{-10} - 350\times10^{-10}$
Rapid	5.0–10.0	12.5–25.0	$350\times10^{-10} - 700\times10^{-10}$
Very rapid	>10.0	>25.0	$>700\times10^{-10}$

Source: Reference [104].

Shifts occur in chemical, physical, and biological equilibria as new soil is being formed, and formed soil is being reformed and organic residues degrade and react with mineral matter. Jenny [79] focuses our attention on the dynamic nature of soil in a description of those factors associated with the earth's surface that constantly work in the formation of soil from inert mineral rock and stone. The five most prominent factors are: parent materials, topography, climate, biotic factor, and time or age. These factors and soil processes provide fundamental approaches to understanding the formation and behavior of soils.

Gross dynamic cycles of the earth

Another concept of the dynamic nature of soil is illustrated in a series of cycles of (a) the gross cycle of the outer crust of the earth as the sun contributes the original source of energy to make the cycle run (Fig. 16),

Fig. 16. The dynamic cycle of the outer thin veneer of the earth's crust that interacts with the soil dynamics and is fed by the energy of the sun.

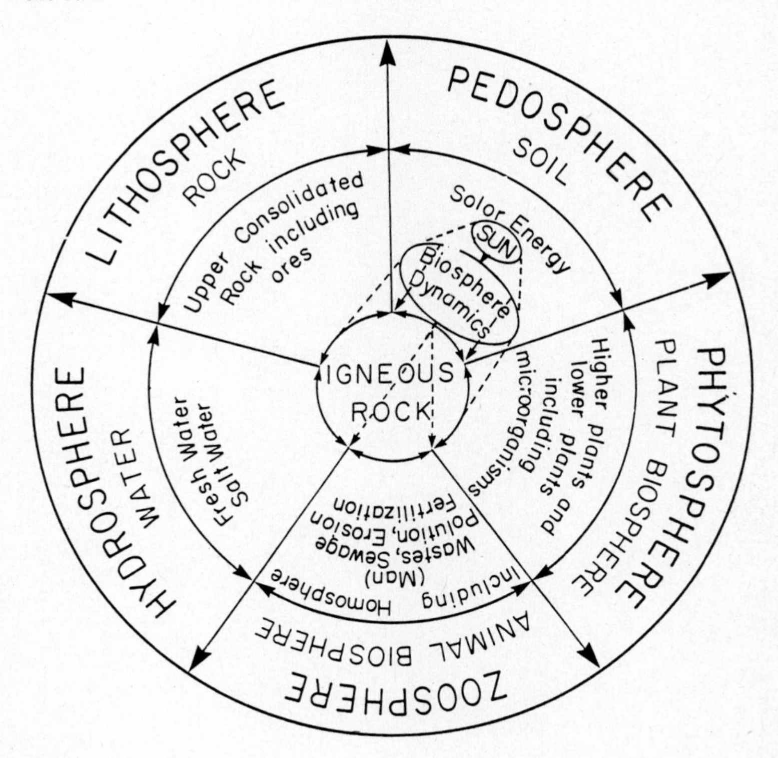

(b) the dynamic inorganic cycles in the minor lithosphere (Fig. 17), (c) the dynamic cycles involving fresh and salt waters (Fig. 18) and (d) the dynamic cycle most intimately involving the soil (Fig. 19). The gross cycle (Fig. 16) is composed of many interwoven cycles, each carrying out some specific function as a part of the earth's dynamic system. Each cycle also represents groups of cycles that interact with each other. However, each may be independent. By this is meant that each individual cycle may operate at various levels. At any time one of the components may vary a great deal while the others vary but little; they are 'independent variables' just as Jenny's soil forming factors represent 'independent variables'. Forces that vary the dynamic nature of the soil within and between cycles may be attributed largely to life processes of the biosphere which, basically, derive their energy source from the sun. The sun is the power at the hub of the gross cycle that touches all component cycles.

Fig. 17. The dynamic inorganic cycle involving wastes in the minor lithosphere.

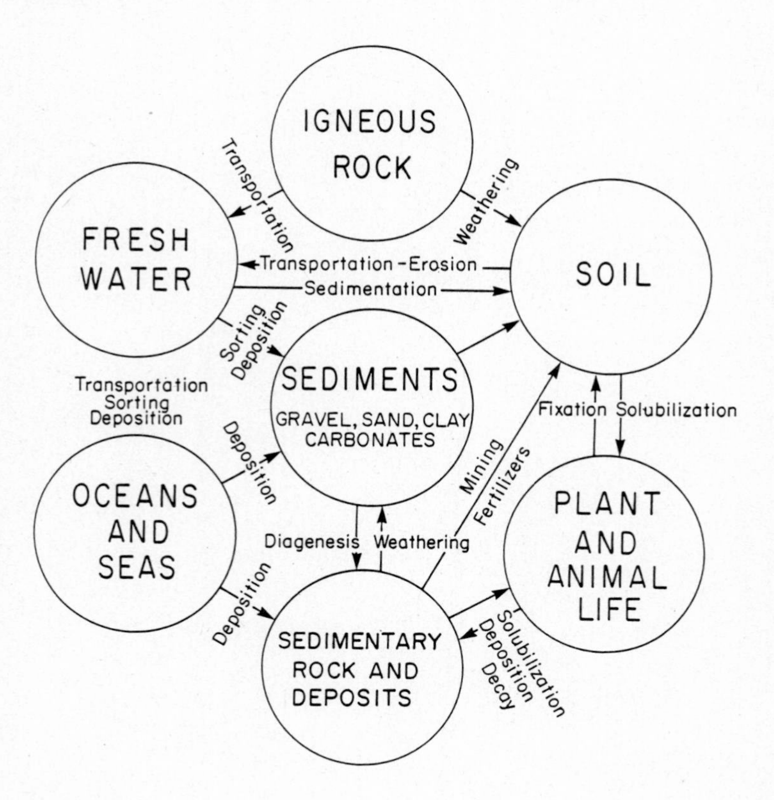

The rate-controlling processes originate largely from biological processes. Cyclical events of dissolution, growth, decay and deposition are all entered into and rate controlled by life processes, assimilation, dissimilation, and deposition.

The dynamics of inorganic cycles (Fig. 17), physical fracturing and pulverizing, swelling–shrinking, and freezing–thawing, as well as gravity and temperature, all are part of the inorganic cycle of the lithosphere that contribute to the dynamic nature of soil.

Not to be overlooked, although believed to be less closely related in the dynamic soil cycles, are the fresh and salt waters. These are transport systems and pools of accumulations (Fig. 18). Products of fresh water and sea water enter the soil. Lakes and swamps fill and become organic soils. Moreover, the bottoms of lakes, ponds, seas, and oceans are lined with a layer of soil.

In this discussion of the geochemistry of hazardous waste disposal, the

Fig. 18. Dynamic cycles involving wastes involving fresh and salt waters.

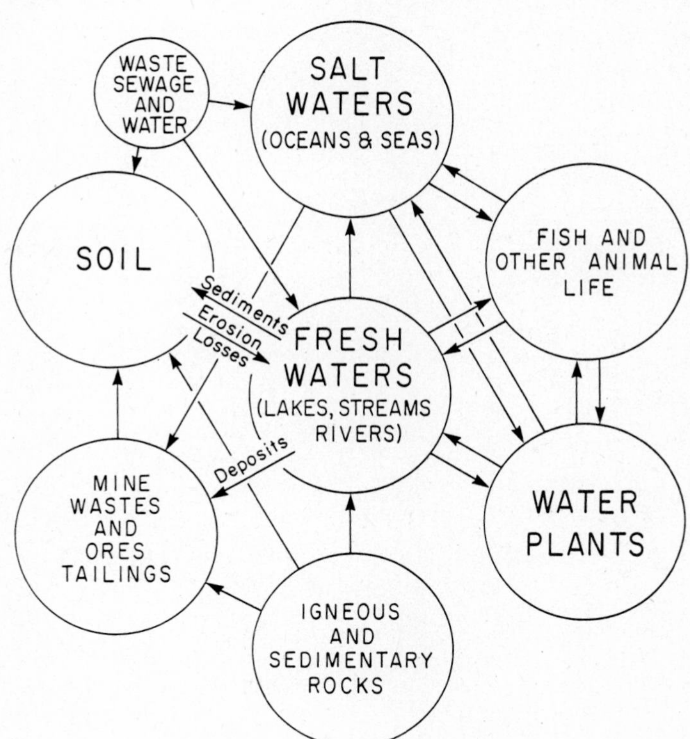

soil as a three phase system, composed of mineral, chemical, physical, and fluid flow systems that relate and react as they impinge on waste disposal, have been discussed somewhat independently. The different systems may seem like the elephant to the five blind men. We should now try to put this 'elephant' together in a dynamic system more real than perceived before.

The dynamic nature of the soil is most dramatically illustrated through the biological reactions as a function of biological activities. The soil teems with microorganisms (Table 12) that are responsible for approximately 80% of the degradation of the wastes entering the soil [63]. The kinds, number, and activity of these organisms and small animals depend on the abundance and kind of energy sources available. They depend on the physical properties, texture, structure, pore space, moisture, and temperature, and on the chemical property of pH (Table 13) and respond to shifts in redox and anion and cation concentration changes.

The microflora and microfauna system with its vast network of energy relationships is the single most important characteristic of soils giving the earth its dynamic nature. For this reason and to provide a background to illustrate the importance of microorganisms in transformations associated with hazardous waste disposal, the biochemical aspects rather than taxonomic aspects will be emphasized.

Table 12. *Approximate numbers of organisms commonly found in soils**

Organism †	Estimated numbers/g
Bacteria	3 000 000–500 000 000
Actinomycetes	1 000 000–20 000 000
Fungi	5000–900 000
Yeasts	1000–100 000
Algae	1000–500 000
Protozoa	1000–500 000
Nematodes	50–200

*The figures for bacteria, actinomycetes, fungi, and yeasts are based on plate counts and refer to viable propogules able to grow on the plating media.

†In addition to these there are large numbers of slime molds (myxomycetes), viruses, or phages of bacteria, algae, fungi, insects, and plants, arthropods, earthworms, mycoplasmas, and other organisms.

Source: Reference [63].

Table 13. *Numbers of bacteria, actinomycetes, and fungi in soil fertility plots at Riverside, CA, in relation to soil pH*

Soil pH	Bacteria and actinomycetes	Fungi
	millions g^{-1}	thousands g^{-1}
7.5	95	180
7.2	58	190
6.9	57	235
4.7	41	966
3.7	3	280
3.4	1	200

Source: Reference [63].

Fig. 19. Dynamic cycles closely involving wastes in soil.

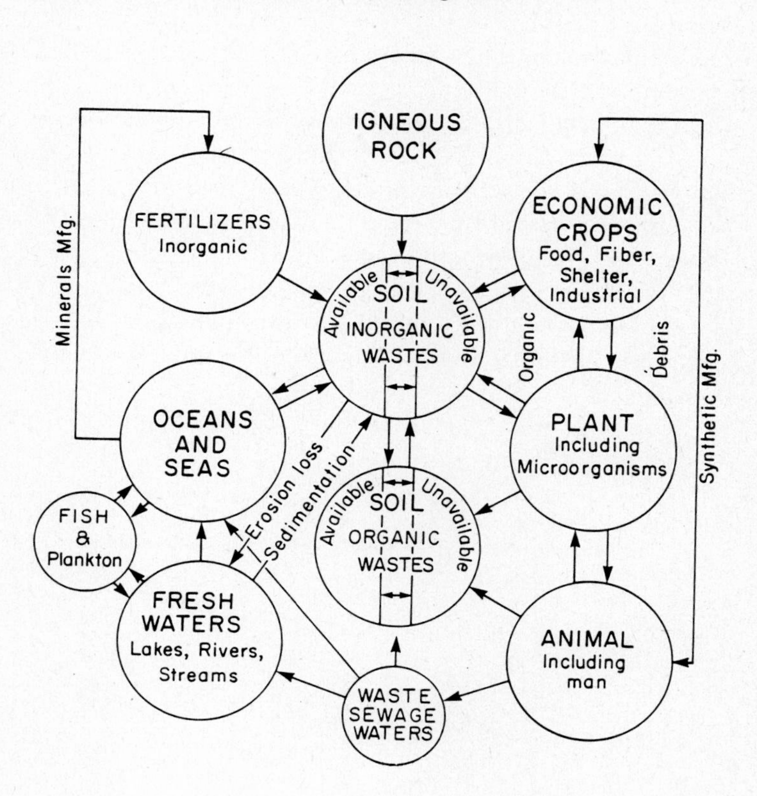

Energy relationships

One may wish to think of the soil's dynamic system as continually receiving 'reduced' compounds of carbon (derived from the energy of the sun) which are oxidized to CO_2 (a gas), water (a liquid) and mineral matter (a solid). Soil microorganisms also reduce carbon dioxide (CO_2) through photosynthesis as well as oxidize carbon compounds, although oxidation reactions dominate. Soil microorganisms are active both in anaerobic (absence of O_2) and aerobic (presence of O_2) environments. In fact, soil microorganisms are active in harsher environments than those permitting survival of higher plants.

Soil microorganisms have been roughly classified according to nutrient requirements as shown in Fig. 20.

Soil microorganisms may also be classified according to electron donor and electron acceptor reactions as illustrated in Table 14.

To provide further information on the wide versatility of soil bacteria, data in Table 15 were compiled to show inorganic compounds and ions that act readily as electron acceptors for soil microorganisms.

Some examples (Table 16) of common reactions occurring regularly in soil again may help illustrate the great versatility of the soil microflora and microfauna and their dynamic natures.

Organic residues also undergo extensive transformation by the soil microflora. Considering the great amount of major constituents such as cellulose, hemicellulose, lignin, proteins, and fats and the great diversity of other substances in lesser and trace amounts in plant residues, it is evident the soil microorganisms biodegrade almost any material, compound, or waste entering the soil under favorable environmental conditions for microbial growth and development. Even the most complex constituents like wax and lignin decompose. During the degradation process, certain

Fig. 20. Classification of soil microorganisms according to nutrient requirements.

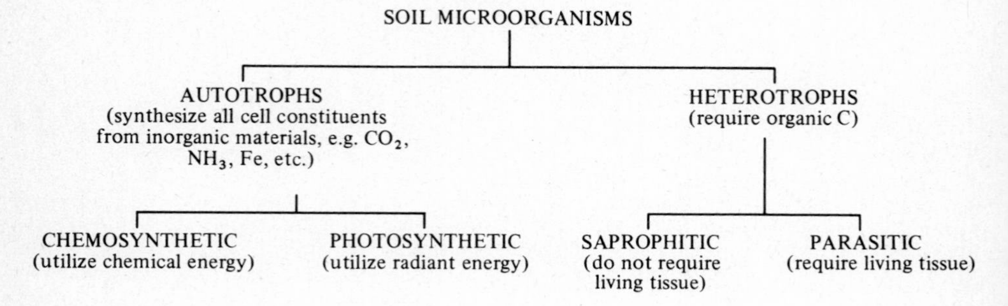

complex compounds may be released and other complexes may be synthesized in addition to microbial cells and tissues. Lignin, for example, yields a group of phenols as a result of biological attack. Some are familiar to us such as vanillin, vanillic acid, guaiacolglycerol, prolocatechuric acid, syringic acid, coniferaldehyde, p-hydroxycinnamic acid p-hydroxybenzaldehyde, to name a few. Concurrent with this dissimilation, assimilation and syntheses take place. Slimes and gums as polyuronides, for example, accumulate from certain microbial synthesis and favor soil aggregation [96, 97].

Humic and fulvic acids

Soil humus formation is a good example of microbial polymerization of a great number of carbon compounds from waste residues of plant, animal, and microorganisms (Fig. 21). Humic and fulvic acids are unique in

Table 14. *Energy considerations in classification of microbial reactions*

Class	Electron donor	Electron acceptor	Products
(1) Photoautotrophic	H_2O, H_2S, H_2R	CO_2	(HCHO) and other reduced compounds
(2) Respiration	Organic compounds	O_2	$CO_2 + H_2O$
(3) Fermentation			
(a) Homofermentative	Organic compounds	Same molecule or fragment of it	Single organic species + CO_2 in some cases
(b) Heterofermentative	Organic compounds	Same molecule or fragment of it	Mixture of organic compounds
(c) Multifermentative	Organic compounds	Different organic compound or CO_2	Organic compounds
(d) Isofermentative	Organic compounds	Another molecule of substrate	Reduced organic compound. Oxidized organic compound
(e) Metafermentative	Organic compounds	Inorganic compound	Organic compound of $CO_2 + H_2O$; more reduced inorganic compound
(4) Chemoautotrophic	Inorganic compound	O_2 or another inorganic compound	Oxidized inorganic compound

Table 15. *Inorganic compounds and ions as final electron acceptors for the soil organisms shown*

Element	Microorganism	Physiological activity[a]
As	F. ferrooxidans	As_2S_3 oxidized to AsO_3^{3-}; AsO_4^{3-}; AsO_4^{2-}(?)[b]
	Heterotrophic bacteria Achromobacter Pseudomonas Xanthomonas	AsO_3^{3-} oxidized to AsO_4^{3-}
	M. lactilyticus	AsO_4^{3-} reduced to AsO_3^{3-}
Cd	Desulfovibrio	$CdCO_3 + SO_4^{2-} + 8H^+ + 8e^- = CdS + 4H_2O + CO_3^{2-}$
Cu	T. ferrooxidans	$Cu_2S + 4H_2O = 2Cu^{2+} + 6H^+ + H_2SO_4 + 10e^-$
	F. ferrooxidans	$CuS + 4H_2O = Cu^{2+} + 6H^+ + H_2SO_4 + 8e^-$
	Desulfovibrio, C. nigrificans	Cu^{2+} and SO_4^{2-} reduced to CuS; $Cu_{10}S_9$; Cu_2S
	M. lactilyticus	$Cu(OH)_2 + H^+ + e^- = CuOH + H_2O$
Fe	T. ferrooxidans, Ferrobacillus spp., Gallionella	$Fe^{2+} = Fe^{3+} + e^-$
	Leptothrix ochracea, Sonaerotilus, Protozoa, algae	Adsorption, precipitation
	M. lactilyticus, B. circulans, B. polymyxa	$Fe^{3+} + e^- \rightleftharpoons Fe^{2+}$
	Desulfovibrio, C. nigrificans	$Fe^{3+} + SO_4^{2-} + 8H^+ + 9e^- = FeS = 4H_2O$
Ni	T. ferrooxidans	$NiS + 4H_2O = Ni^{2+} + 8H^+ + SO_4^{2-} + 8e^-$
	Desulfovibrio	$NiCO_3 + SO_4^{2-} + 8H^+ + 8e^- = NiS + 4H_2O + CO_3^{2-}$ $Ni(OH)_2 + SO_4^{2-} + 10H^+ + 3e^- = NiS + 6H_2O$
S	Thiobacteriaceae, Thiorhodaceae, Chlorobacteriaceae, Beggiatoaceae, S. natans, Achromatium, Leucothrix	$H_2S = S^0 + 2H^+ + 2e^-$ $H_2S + H_2O = SO_4^{2-} + 10H^+ + 8e^-$
	Bacteria, actinomycetes, fungi	Polysulfides reduced to thiosulfate and sulfide

Table 15 (*cont.*)

Element	Microorganism	Physiological activity[a]
	All microorganisms	$S^0 + 2e^- + 2H^+ = H_2S$
Se	M. selenicus	$H_2Se + 4H_2O = SeO_4^{2-} + 10H^+ + 8e^-$
	M. lactilyticus	$Se^0 + 2e^- + H^+ = SeH^-$
	C. pasteurianum	
	D. desulfuricans	
	Neurospora	$HSeO_3^- + 4e^- + 5H^+ = Se^0 + 3H_2O$
	C. albicans	
	Baker's yeast	
V	M. lactilyticus	
	D. desulfuricans	$H_2VO_4 + 2e^- + 2H^+ = VO(OH) + H_2O$
	C. pasteurianum	
Zn	T. ferrooxidans	$ZnS + 4H_2O = Zn^{2+} + SO_4^{2-} + 8H^+ + 8e^-$
	Desulfovibrio	$\frac{1}{5}[2ZnCO_3 3Zn(OH)_2] + SO_4^{2-} + 9\frac{1}{5}$ $H^+ + 8e^- = 5\frac{1}{5}\ H_2O + \frac{2}{5}\ CO_3^{2-} + ZnS$
Metal chelate	Heterotrophic microorganisms	Oxidation of chelating agent with precipitation of metal moiety

[a] Oxidative or reductive half-reactions listed most nearly describe the particular microbial activity cited.
[b] Proof of sulfate production lacking.
Source: Reference [67].

their universal distribution. These acids appear to be present in abundance wherever organic materials undergo decomposition, not only in soil but in sewage sludge, municipal solid wastes, and wastewaters and leachates, Chian & DeWalle [22, 23], Artiola & Fuller [in press]. The dynamic function of fulvic and humic acids has already been demonstrated and will occupy researchers' attention for a long time in the future because of their properties of attenuation and chelation.

Nutrient and metal cycles of soil

Perhaps the dynamic nature of soil can be demonstrated also through some of the many nutrient and metal cycles within the soil. The most obvious is the organic carbon cycle (Fig. 22). The decomposition of organic residues by soil microorganisms is one of the most important functions in soils. Plant nutrients and trace metals, released through decay and biodegradation, recycle to be used over and over to produce new growth. The carbon, in the form of CO_2, replenishes the atmosphere to cycle and recycle again and again.

The same is true for phosphorus (Fig. 23) and sulphur (Fig. 24). In fact, each and every element and constituent has this cyclic or turn-over principle of dynamic proportions, continually assimilating and releasing energy to the chemical equilibria in the soil solution that constantly seeks to achieve static states but seldom achieves this goal. The restless soil microorganisms find no material, substance or element beyond their capacity to react with or alter its behavior in some way. Thus numerous products and byproducts are extracted from the environment that will react with pollutants or change pollutants in some way even if one must wait a long time. Even the most toxic organic substances, manufactured to destroy biological systems, will biodegrade through microbial activity. Some examples of these appear in Table 17.

The attenuation and release mechanisms

Background

Attenuation is described [35] as any decrease in the maximum concentration or total quality of an applied chemical or biological constituent in a fixed time or distance traveled resulting from physical, chemical and/or biological reaction or transformation.

Table 16. *Common reactions occurring regularly in soil*

Organism	Reaction		kcal mole^{-1}
Nitrosomonas:	$NH_4^+ + 1\frac{1}{2} O_2$. . .	$\rightarrow NO_2^- + 2H^+ + H_2O$.	$\Delta G = -66$
Nitrobacter:	$NO_2^- + \frac{1}{2} O_2$	$\rightarrow NO_3^-$	$\Delta G = -17.5$
Thiobacillus thiooxidans:	$S^0 + 1\frac{1}{2} O_2 + H_2O$	$\rightarrow H_2SO_4$	$\Delta G = -118$
Hydrogenomonas:	$H_2 + 1\frac{1}{2} O_2$	$\rightarrow H_2O$	$\Delta G = -57$
Desulfovibrio:	$SO_4^= + 4 H_2$	$\rightarrow S^= + 4 H_2O$	$\Delta G = -83$ (-21)
Methanobacterium:	$4H_2 + CO_2$	$\rightarrow CH_4 + 2 H_2O$	$\Delta G = -31$
Ferrobacillus and Gallionella:	Fe^{2+}	$\rightarrow Fe^{3+} + e^-$	$\Delta G = -11$
Carboxyldomonas:	$CO + \frac{1}{2} O_2$	$\rightarrow CO + \frac{1}{2} O_2$	$\Delta G = -66$
Thiobacillus thioparus:	$H_2S + \frac{1}{2} O_2$	$\rightarrow H_2O + S^0$	$\Delta G = -41.5$
Actinomycetes:	$HCN + H_2O$	$\rightarrow KOH + NH_3 + CO_2$	$\Delta G =$ not calculable

For a lack of better terminology, the phrase '*factor(s) of attenuation*' will be used in identifying broad classes of reactions that appear to control pollutant migration rates. These factors may be grouped on a basis of trends in a pollutant's solubility, which is controlled by an identifiable chemical or physical condition and which has a standard means of measurement. For example, pH value limits the solubility of certain substances. The pH value of a solution or suspension can be readily monitored by a pH meter.

Fig. 21. Diagram illustrating the pathways for the formation of humus.

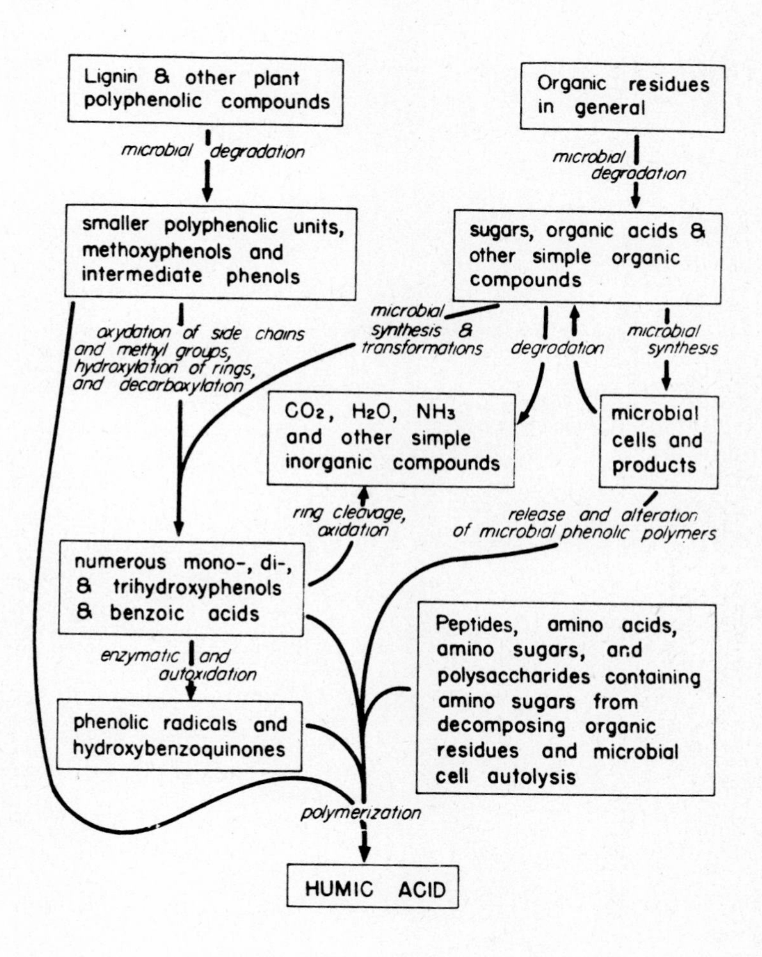

Fig. 22. The organic carbon cycle involving wastes in soil.

Fig. 23. The phosphorus cycle involving wastes in soil.

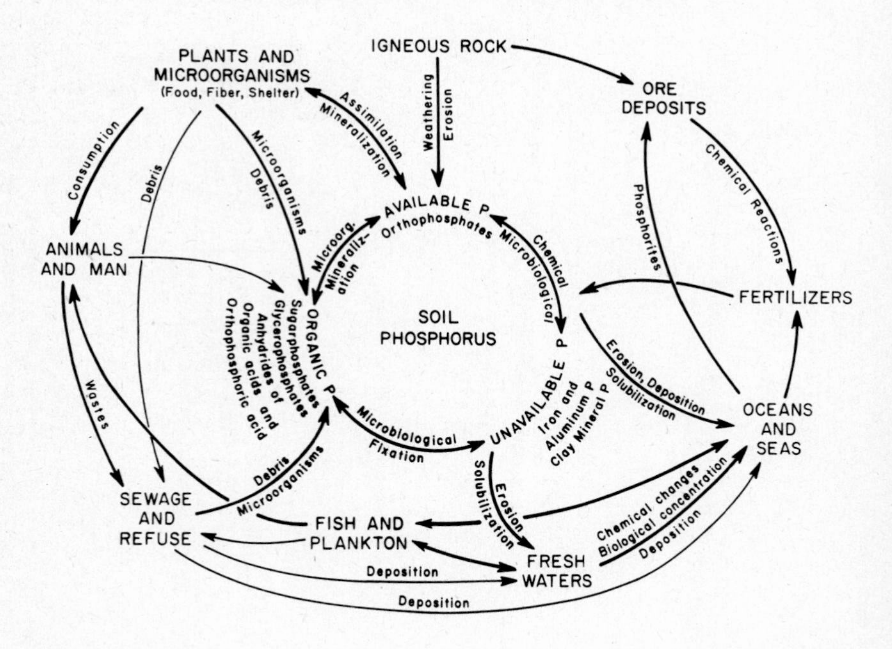

Before discussing specific reactions influencing mobility and immobility of pollutants in soils, it may be well to review some of the broad alterable factors of the soil habitat that can affect pollutant mobility. The usual waste disposal site historically has been located in subsurface soils, sand, and gravel excavations, geological materials, and shrouded in disturbed alluvial debris. The waste is left to 'digest' by anaerobiosis. Waste leachates themselves also will alter the natural soil environment. Thus, disposal management must consider the *unusual* soil condition.

Some important characteristics of the microhabitat of the soil that may be altered and, in turn, may alter pollutant movement, as it might occur in the usual aerated soil or may differ from soil to soil are:

aeration: anoxic, redox changes, waterlogged, swampy, all *reducing conditions*;

particle size distribution, texture, sand, silt, or clay content, surface area;

permeability or pore size and pore size distribution as it may influence flux of the soil solution;

pH values: either high degree of acidity or alkalinity may develop

Fig. 24. The sulphur cycle involving wastes in soil.

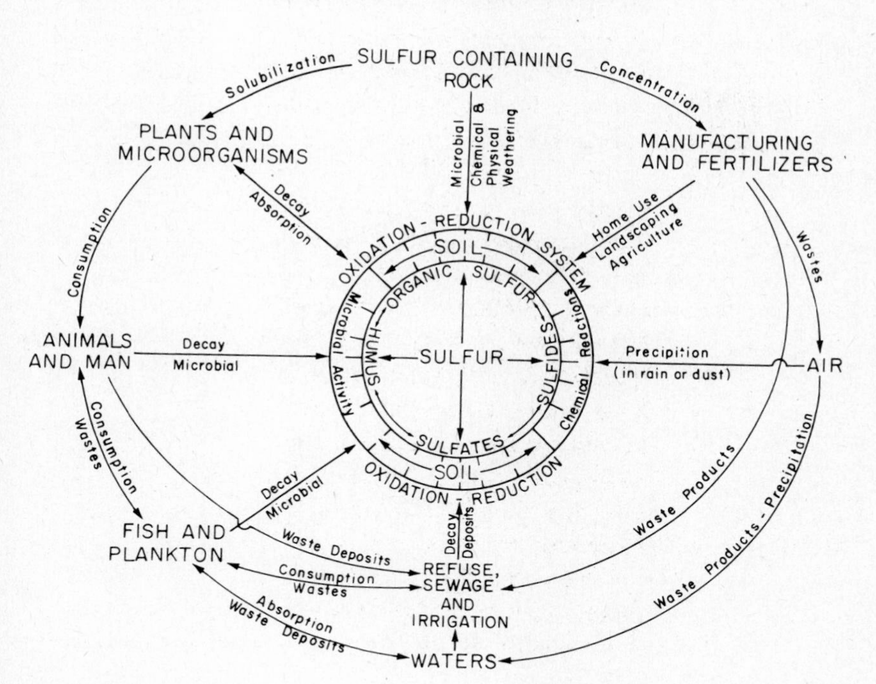

from waste disposal and acid status of leachate and waste streams may influence the solubility of complexes of potential pollutants;
lime: free soil lime, caliche, agricultural limestone, and commercial limes;
iron, aluminium, and manganese hydroxy oxides in unusually high concentration or state of reactivity or solubility;
organic carbon components in leachates and waste streams and organic soils: sequestering of heavy metals with organic complexes alters solubility and mobility as chelates and certain chemical unions can form either mobile or immobile organometallic complexes;
high specific salt concentrations, where pollutant behaviour and reactions become salt dependent.

Table 17. *Decomposition and period of persistence of several herbicides*

Name of compound	Abbreviation	Persistence in soil	Active organisms
3-(p-chlorophenyl)-1, 1-dimethylurea	Monuron	4–12 mth	Pseudomonas
4-chlorophenoxyacetic acid	4-CPA		Achromobacter Flavobacterium
2,4-dichlorophenoxyacetic acid	2,4-D	2–8 wk	Achromobacter Corynebacterium Flavobacterium
2,4,5-trichlorophenoxyacetic acid	2,4,5-T	5–11 mth	
2-methyl-4-chlorophenoxyacetic acid	MCPA	3–12 wk	Achromobacter Mycoplana
2,2-dichloropropionic acid	Dalapon	1–2 wk	Agrobacterium Pseudomonas
Dinitro-o-sec-butylphenol	DNBP	2–6 mth	Corynebacterium Pseudomonas
4,6-dinitro-o-cresol	DNOC		Corynebacterium
Isopropyl N-phenylcarbamate	IPC	2–4 wk	
Isopropyl N-(3-chlorophenyl) carbamate	CIPC	2–8 wk	
Trichloroacetic acid	TCA	2–9 wk	Pseudomonas
2,3,6-trichlorobenzoic acid	2,3,6-TBA	>2 yr	

This brief listing oversimplifies the factors in the microhabitat that influence pollutant mobility in disposal-site soils. For example, organic matter (organic carbon substances, TOC,) can attenuate heavy metals by combining with them to form very slowly biodegradable complexes or increase their mobility by forming highly soluble organo-metal-ion complexes, which differ greatly in degree of attentuation. Nevertheless, it is well to make a beginning in an empirical way, recognizing that this approach is initially necessary for practical problem solving.

Chemical reactions that may occur between the soil and the components of waste leachates or waste solvents may be grouped into (a) adsorption–desorption, (b) precipitation–solution, (c) complexation, (d) ion exchange (anion–cation) and (e) absorption–mineralization by living organisms as they relate to the dynamic chemical equilibria in soils (Fig. 25).

The soil solution

The various categories of chemical and biochemical equilibria diagrammed in Fig. 25 emphasize the key position of the *soil solution* or the soil fluid when wastes are put into the soil. For pollutants to migrate, they must be in solution or fine colloidal (i.e. $<0.45 \mu m$ and submicroscopic as viruses) and they must have some vehicle of transport. Therefore, the chemical equilibria established within each of the five groups in Fig. 25 determines the opportunity or lack of opportunity for the pollutant to migrate. The chemical equilibria, in turn, may vary according to the solution pH, Eh (redox), total organic carbon, EC (electrical conductivity) or ion concentration, and others.

To complete this line of discussion of chemical reaction within equilibria groups, let us examine the dynamic chemical equilibria occurring in soil as focused on the soil solution of Fig. 25. As hazardous metals become adsorbed to the soil surfaces (reaction 1) desorption of other elements will occur (reaction 2). Precipitation of some component (ion) in the soil solution (reaction 3) may cause dissolution of some solids (reaction 4). Nutrient uptake by organisms, complexation, and chelation (reaction 9) take place between inorganic and organic components as mineralized elements are fed (reaction 10) into the soil solution (reactions 5 and 6). Ions held on the exchange complex of the soil exchange with those in the solution (reactions 7 and 8). The rates of these reactions vary considerably. Some, such as cation exchange, though often instantaneous, may take days, months, or years and still others are so slow the rates of reaction cannot be measured with accuracy. We are most interested in those more permanent reactions for hazardous pollutant disposal which are slower than the transitory cation exchange.

Adsorption and desorption

The terms *absorption*, *adsorption*, and *sorption* frequently are confused. In this chapter absorption is confined to biological uptake of elements and constituents and sorption relates to loosely as well as strongly fixed ions or molecules by soil constituents (a host of chemical equilibria are involved). Adsorption relates to a physico–chemical process of adhesion from a fluid to solid surfaces (such as soil particles) or the immobilization processes against extraction by salt solution. *Adsorption complex* refers to a group of substances in soil capable of adsorbing other materials. Organic and inorganic colloid substances (clay, organic matter, and hydrous oxides of Fe, Al, and Mn) form the greater part of the adsorption complex; the noncolloid materials, such as silt and sand, exhibit adsorption but to a much lesser extent than the colloids. A more complete discussion of adsorption and sorption in soils is discussed by Ellis [32, 33], Gadde & Laitinen [56] and Kenney & Wildung [83].

Fig. 25. Diagrammatic representation of the dynamic chemical equilibria occurring in soils with the soil solution as the focal point.

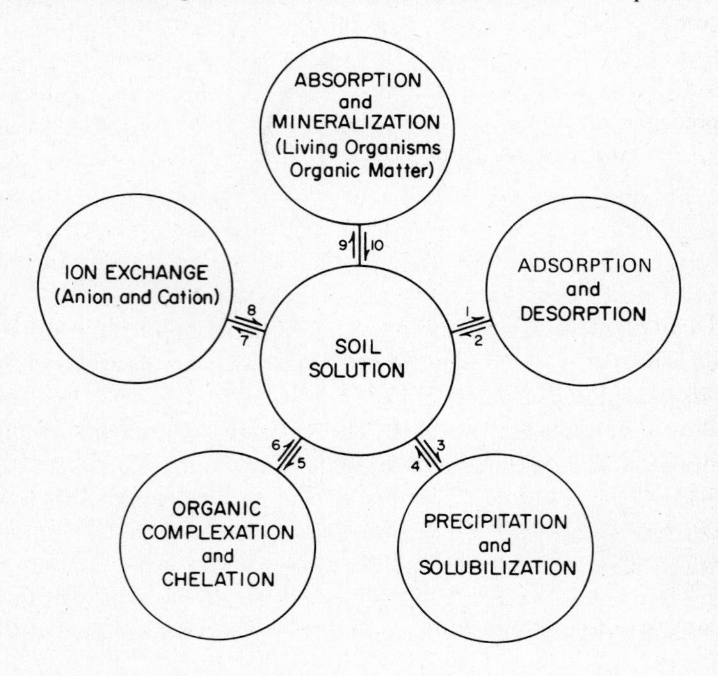

Adsorption of metals to solid surfaces of soils may be identified in quantitative or mathematical terms. The Freundlich & Langmuir isotherms, for example, have been successfully used to describe adsorption of micronutrients in water to the solid surfaces of soil [32, 129]. According to Turjoman [126], lead (Pb) strongly adsorbs on and attenuates with calcareous soils according to the Langmuir isotherm.

What are those soil conditions that favor adsorption of pollutants? Almost all of the important characteristics of the microhabitat of the soil that alter with time influence adsorption. Some of the most influential factors were listed at the beginning of this section. Perhaps one may speculate at this point to focus attention on the high degree of dependence of adsorption on the availability of the surface area of soil. The amount of clay in a soil, therefore, should be an important factor in determining the extent of attenuation or retention of pollutants by soil. Conversely, sandy and gravelly soils with low surface area per unit weight or volume should be avoided as undesirable disposal sites for wastes.

Precipitation and solubilization

Precipitation of hazardous waste pollutants is one of the most effective means of migration control. Precipitation takes place in soils naturally when wastes and waste fluids and aqueous leachates come in contact with soil. A shift in equilibrium from a soluble state to solid state (precipitation) is governed by a great number of varying properties indigenous to soils. Some of the factors controlling the relation of precipitation to adsorption are:

(1) concentration of pollutant (metal) in solution;
(2) pH, alkalinity and acidity;
(3) ion pair formation;
(4) organo-pollutant (metal) complexation;
(5) Eh or pe (redox).

Those metals most likely to be controlled by precipitation are Fe, Mn, and Pb [33]. Jenne's [78] and Lindsay's [91] reviews of the application of mineral solubility data to metal solubility concluded that, except for Fe and Mn, predictions were not particularly useful. Precipitation–solution equilibrium reactions in soils are complex and depend on the environmental habitat as it relates to the nature of three components: (a) the porous medium through which the pollutant moves, (b) the fluid transport system and (c) the specific pollutant itself [41, 42, 43].

The extent of retention of pollutants by soil as a result of inorganic complexation and chelation is a controversial subject, not because good evidence for both reactions is lacking, but because the mobility of pollutants associated with these reactions is unclear. Except for Fe, Chian & DeWalle [22] indicate that chelation of metals by refactory organics in MSW leachates is not a major mechanism in metal attenuation processes. Certain competitive effects of Fe and Ca were thought to prevent such metals as Cu from forming chelates in MSW landfill-type leachates. Cu attenuation in an aerobic system is believed to be due more to carbonate and hydroxide formation (precipitation) than complexation. Under anaerobic conditions metal sulphide and carbonate precipitation are important reactions in metal retention by soil.

Organic complexation and chelation

Significance

Only since the early sixties have chemical reactions involving organic complexation of metal ions in soils been recognized as having any significant effect on the solubility of soil solution components: Kenney & Wildung [83]; Schnitzer & Skinner [112]; Schnitzer [110]; Schnitzer & Kahn [136]; Stevenson & Ardakani [120]; Fuller et al., [54]. Difficulty in describing pollutant migration in quantitative terms is due mostly to (a) the only recent emergence of understanding of the chemistry of soil organic matter, (b) the great number and diversity of organics contained in waste fluids and (c) the unusually great variability in the levels of potentially hazardous components in MSW leachates as well as in industrial waste streams. Moreover, chemists, particularly soil chemists, give little attention to organic complexation of waste components not essential to the nutrition of higher plants. Evaluation of organic complexation as an important factor in pollutant attenuation becomes even more complicated by the growth and activity of soil microorganisms and their effects on the chemical reactions in soil. Solid wastes containing readily available carbonaceous food sources for microorganisms such as those found in MSW landfill leachates generate a great variety of organic compounds, complexes, and chelates during biodegradation processes. Some of these are slimes, gums, and tissues that tend to reduce the rate of flow of fluids through soil as well as interact with components of hazardous wastes, whereas others form chelates with the pollutants that greatly enhance the risk of migration of hazardous substances, toxic metals, etc., through soils to groundwater chains.

Part of the inability to fully evaluate the significance of the interactions of heavy metals and organic matter as a factor in pollutant attenuation also centers around the dilemma as to whether the metals or pollutants in aquatic systems (a) exist in solution as free ions, (b) are chelated by organic components, or (c) adsorbed to suspended particles. All three forms appear to exist but the proportionality is not well known. Working with MSW landfill-type leachates, Chian & DeWalle [22] concluded chelation of metals by refractory organics is not a major mechanism in metal attenuation processes although it does occur readily. Their conclusion was drawn after extensive fractionation of the different molecular weight (MW) fractions present in the leachate showed that the majority of the metals permeated a 500 MW UF membrane. They concede, however, that Fe appeared to deviate from this general pattern of metal solubility since Fe was found associated with the 100 000 MW UF retentate. Because of the large amounts of competing Ca^{2+} and Mg^{2+} in aquatic systems both Stumm & Bilinsky [121] and Siegel [115] believe that a chelator has an impact only when the stability constant with the trace metal is many times (about 10^6) higher than the competing Ca^{2+} and Mg^{2+} complexes. Other mechanisms, however, take over in brines (thermal) such as chloride effects that keep heavy metals very mobile [9].

Evidence of complexation of heavy metals with humic and organic substances is presented by Barsdate & Matson [11] working with Co, Pb and Zn; Slowey et al. [116] working with Cu; and Hobson et al. [71] and Geering et al. [59] working with Cu, Mn, and Zn. Stevenson & Ardakani [120] identified a number of organic ligands from soil with potential for metal complexation. Mercury is a classical example of the formation of organometallic complexes in soil: Hepler & Olofsson [69]; Alesii & Fuller [1]; Rogers [108]; Summers & Silver [122]. Lindsay [91] has reason to believe that natural chelates are present in most soils. The major problems in the development of a more workable knowledge of the natural chelates in soils may fall into one or all of the following categories associated with the almost insurmountable difficulty in (a) identifying them, (b) determining how long they exist and (c) obtaining meaningful equilibrium constants for reactions in which they participate.

The large volume of literature documenting numerous cases of accelerated movement through soils of heavy metals and certain other pollutants attributed to chelation and organic complexation reactions justifies a disposal practice of isolating hazardous wastes from organic matter when possible. For example, the level of total organic carbon constituent in MSW leachates is conclusively demonstrated to be one of the most important factors influencing heavy metal migration through soils. The higher the concentration, the higher the rate of migration and the lower the attenuation [44, 45, 49, 84].

Polluting and hazardous metals may be incorporated directly into tissues and cells of microorganisms both under oxic as well as anoxic conditions. Elements such as radioactive phosphorus (^{32}P) have been shown to move readily through soils as microbial cells and debris or as known organically complexed compounds [64, 65], whereas inorganic P (and ^{32}P) notoriously becomes fixed and moves very little. The same effect is reported for ^{45}Ca and ^{89}Sr incorporated directly into microbial cells and tissues [52, 53]. Movement of ^{89}Sr-^{90}Sr(^{90}Y) through a calcareous soil was shown to be increased as a result of chelation with chemical ligands. Further evidence of accelerated movement associated with organic complexation was substantiated by tracing the migration of algal- and fungal-bound radiostrontium as chelate complexes through soil [55].

Plutonium in trench leachates at radioactive waste disposal sites was found by Cleveland & Rees [26] to remain in a soluble state when chelated with EDTA and to resist sorption or attenuation by the soil material. Similar data are reported by Weiss & Colombo [135] working with certain radioisotope wastes. Both authors warn of the serious consequences in the codisposal of radioactive waste with organic materials.

Foremost in recent investigations designed to evaluate the availability to plants of hazardous and nutrient metals contained in municipal sewage and other sludges is the well-known extraction technique using aqueous solutions of ligands such as EDTA and DTPA. More metals are made soluble with these ligands than by water alone. The metals made soluble are considered to be chelated [134, 122, 68, 111]. Schnitzer & Kodama [111] also confirmed the removal and solubilization of Cu sorbed on montmorillonite by fulvic acid in solution. Fulvic acid acted as a strong ligand for chelating the Cu and other metals.

We now know that (a) waste streams often contain substantial levels of organic molecules, some of which behave as strong chelating agents, (b) the soil at the disposal site often contains natural organic matter, (c) many organic compounds complex readily with trace and heavy metals and (d) mixing of wastes and waste streams is a disposal practice to be avoided, particularly where organic matter is involved, since the overall attenuation of waste disposal pollutants is lessened rather than enhanced.

Humic substances

The identification of substantial amounts of *humic* and *fulvic acids* in biodigesting wastes renews the argument for organic complexation as a viable 'mechanism' in pollutant solubility counteracting attenuation, particularly metals (Artiola & Fuller [in press]). Certain identifiable groups, namely, phenolic and carboxyl attached to large organic molecules like humic and fulvic acids and heterocyclic compounds, appear to play an important role in complexation (Artiola & Fuller [in press] and Broadbent & Bradford [16]. Broadbent, for example, provides evidence that carboxyl groups and certain other groups complex both Cu and Ca but react selectively with Cu in preference to Ca. Iron frequently has been found closely associated with fulvic acids. Shapiro [114] and Christman [24] provide data indicating Fe complexes with all size fractions of organic matter. The possibility exists also that iron hydroxides are associated with organic molecules and they in turn have a strong affinity for heavy metals [13].

Schnitzer & Kodama [111] believe that fulvic acid in aqueous solutions interacts with di- and tri-valent metal ions forming stable complexes by two types of reactions. One reaction is a major effect, simultaneously involving both COOH and phenolic OH groups; the other reaction is a minor effect in which adjacent COOH groups participate [112, 57]. They conclude, further, that humic substances are major components of soil and water, that they contain (per unit weight) relatively large numbers of O-containing functional groups such as COOH, phenolic OH and $C=O$, and that they are natural polyelectrolytes. Schnitzer & Kodama [111] suggest further that humic compounds are capable of attacking soil minerals by complexing and dissolving metals and transporting them within soils and waters. They also believe that at low metal-to-fulvic acid or humic acid ratios, the metal–organic complexes are water soluble; but at high ratios, they are no longer soluble in water. Humic and fulvic substances also were found in our laboratories as the major component of MSW landfill leachates (Artiola & Fuller [in press]).

Nonhumic organics

Even though the nature of the association between metals and organic complexes is not well understood, the current belief is that the carboxylic groups of humic and fulvic acids are highly reactive and cations, minerals, and certain organic constituents may be bound to them through these groups. Basic functional groups of organic matter, i.e. amines and heterocyclic nitrogen may also contribute to metal retention [15]. Kenney & Wildung [83] suggest that lower molecular weight biochemicals (non-humic) of recent origin can be involved in complexation and solubilization in soil. Biodegradation processes of solid wastes are capable of producing large quantities of organic acids and amines and other organics with highly reactive groups. Some of these are:

(1) the low molecular weight volatile acids (acetic, butyric, propionic);
(2) carboxylic acids derived from simple saccharides (uronic acids, gluconic, and α-ketogluconic);
(3) acids of the citric acid cycle (citric, succinic, pyruvic, maleic);
(4) aromatic acids – many fragments from lignin decomposition (vanillic, syringic, p-hydroxybenzene, catechuic);
(5) amino acids (aspartic, glutamic);
(6) phenols and polyphenols (polychlorinated biphenyls, phenol, pentachlorophenol);
(7) phosphate related compounds (phytic acid, inositols, ADP, and ATP).

Sewages

Municipal sewage sludges provide good examples of complexations between organic heavy metals in concentrations well in excess of those found in natural soil and organic matter (Table 18). Sewage sludges are rich in both humic and fulvic acid compounds. For example, Rebhun & Manka [107] calculated about 39–45% of the COD of activated sludge effluent to be humic substances. The proportion of heavy metals in sewage sludge that resists solubilization by water, DTPA, or dilute acid is considerable. Yet little information is available distinguishing between organic and inorganic mechanisms of retention. Sewage sludge includes

Table 18. *Cadmium, copper, nickel, and zinc contents in some representative municipal sludges of the USA*

Location	Cadmium	Copper	Nickel	Zinc
		$mg\ l^{-1}$*		
Cincinnati, Ohio (Millcreek)	<40	4200	600	9000
Dayton, Ohio	830	6020	<200	8390
Monterey, California	<220	720	200	3400
Tahoe, California	40	1150	<400	1700
		$mg\ kg^{-1}$ †		
Chicago, Illinois	190	1219	612	3736
Kokomo, Indiana	1139	3166	701	22 292
Tucson, Arizona	13	318	63	1117

*Saturated suspension.
†Air-dry weight basis.
Source: Reference [21].

inorganic and organic systems. Much of the sludge is soil, refractory, or inert material that bacteria very slowly degrade, and polymerized waste products of bacteria. Humic substances in sludges are able to chelate heavy metals some of which are temporarily retained in the sludge against migration except as fine particulates. The availability of the heavy metals to extraction solutions and to plants varies from one sewage sludge to another [117, 118, 21]. The presence of pollutants as organic complexes has been postulated but further evidence is necessary to confirm and to evaluate the extent of attenuation and mobility attributable to complexation.

Animal wastes

Animal manures contain significant amounts of trace nutrient elements and heavy metals, i.e., those having densities greater than 5, Artiola-Fortuny & Fuller [4] and Meeks et al. [99]. Manganese and zinc concentrations in farm manure exceed that of native soil organic matter and food crops. The practice of enriching animal feeds with heavy metals, such as $CuSO_4$ in swine and poultry rations, accentuates the metal levels appearing in manures. Some feeds are fortified with disease-controlling, growth-promoting, potentially toxic metals such as As, Co, Mo, and B. The extent of complexation of the metals with the organic components has received only minor attention. Yet there is a general opinion that chelation and other complexations account for much of the behavior of the metals of animal manures toward plant absorption and biological mineralization rates. For some answers, we can look to leads from the vast amount of research data on municipal sewage sludge that began to accumulate in the literature during the seventies.

Soil organic matter

Some inferences as to the importance of organic complexation can be derived from the extensive research by Schnitzer [110] who describes Fe and Al organo-metal complexes of both surface and podzol B horizon associations. The B horizon of podzols enrich with metal complexes (mainly Fe) as a result of natural eluviation from the surface of constituents released by decomposing plant residues and Fe and Al released by the solubilization of Fe- and Al-containing minerals.

Clark [25] estimates that just the live weight of the soil microflora alone may vary from 0.5 to over 4 metric tonnes in the surface 15 cm of 1 hectare. Judging from this, a great part of the leachate from readily biodegradable organic wastes must be in the form of live microflora. This type of immobilization, although somewhat transient for a single organism, is a significant factor for certain pollutants that are absorbed and utilized by the microflora.

The absorption and mineralization

The microflora in soil and in biodigesting fluids of wastes absorb nutrient metals by choice as well as some nonnutrients by chance. Plants growing in waste disposals also may immobilize heavy metals in this manner of incorporating them into cells and tissues. Residues of microbial activity that constitute highly resistant organic substances are called refractory or inert material and are composed of (a) inert residues (some lignin-like) that organisms cannot further degrade readily, (b) microbial cell and tissue debris and inert material from lysed cells and (c) polymerized excretory and synthesized waste products. Schnitzer & Kahn [136] suggest that some of these very slowly degradable residues are humic in nature and possess very high molecular weights. Such substances have the capacity to chelate and complex heavy metals and often prevent them from leaving the solid sludge residue.

Mineralization results when microorganisms decompose (biodegrade) organic residues of plants and animals, sewage sludges, and synthesized organics releasing the metals to the soil or other fluids. The simple inorganic elements that become freed from organic combinations then may be recycled into other synthesized substances to support new growth, undergo recombination into complexes, or take other avenues of attenuation.

Ion exchange

Ion exchange between the solid surfaces of the soil and the soil solution involves both cations and anions. Since most soil surfaces are negative in charge, cation exchange dominates. The forces that hold pollutants onto the soil surface range from *electrostatic*, considered to be associated with conventional exchange reactions, to *covalent*. Only the latter demonstrates significant ion specificity (specific sorption) and significant levels of irreversible 'exchange ions'.

As implied earlier, transitory retention of metals is not very useful for confining waste components to the disposal site. At best, this type of reaction functions only as a delaying action for pollutants on their way to contaminating the environment.

Covalent bonding assures less opportunity for reversible ion exchange and, therefore, is of more interest to waste disposal programs. There also is

(silt and clay) textured soils. Salt effects and ion suppression also relate here to particle sizes.

Structure

The importance of soil structure to attenuation has received attention. It is defined as the arrangement of the mineral and organic components. The mineral particles are grouped into subunits called aggregates or peds. Thus soil particles are neither uniformly nor randomly distributed. Although similar structures tend to form under similar climatic and vegetative conditions, they may be continually forming and reforming. The microorganisms tend to support stable structures by producing cementing agents such as slimes, gums (polyuronides), decomposing organic matter, and readily decomposable organic wastes. Salts (electrolytes) in contrast contribute to flocculation or dispersion (transitory structure) as the composition of the soil solution is altered by addition of chemical wastes, acids, lime, etc. The polyvalent cations (e.g. Ca) tend to flocculate soils and monovalent cations (e.g. Na) tends to disperse soils.

Soils with highly solvent or water-stable structures permit high rates of fluid flow and thereby limit the capacity of the soil to retain pollutants. The internal part of the soil peds may be penetrated at such a slow rate that attenuation is minimal at best. Highly dispersed soils also may be so deflocculated as to inhibit fluid movement. Somewhere between these two extremes of highly aggregated and highly dispersed soils, a suitable unchanneled flow rate is achieved for maximum pollutant retention.

Permeability/hydraulic conductivity

If the disposal fluids do not permeate and infiltrate the liners (or compacted soils), there is no transport and attenuation is no longer a factor in pollution control. The question that is asked most often relates hydraulic conductivity or permeability to attenuation of a pollutant. Soluble pollutants migrate at a rate less than that of the intrinsic permeability. How much less depends on extent of attenuation. The greater the attenuation, the greater the difference between the velocities (time or distance traveled).

Another method for arriving at comparisons between velocities of water and pollutant migration through soils, for example, is to compare fluid flux and migration of selected metals in laboratory soil columns (Alesii et al. [2]). Natural MSW leachates with Be, Cd, Cr, Fe, Ni and Zn were used to perfuse nine soils representing seven major soil orders. Soil columns were under anaerobic conditions and flow rates of leachate were regulated to deliver fluxes ranging from 1 to 15 ml hr^{-1}. Effluent from the columns was collected at the end of each hour and analyzed for the presence of the metal ions being studied. Solution displacements continued until the concentration of the metal in the effluent equalled that of the influent.

Flux significantly influenced the attenuation of Al, Be, Cr(VI) and Fe(II) in all soils, but was of little importance for Ni and Zn (Table 21). The concentration of the trace or heavy metal (such as Cd) in the leachate influenced the absolute amount retained by the soil more than any single factor except, perhaps, clay content. Flux control appears as one of several modification techniques to minimize movement of pollutants from solid waste operations. By removing the soil, lining the excavation, and compacting it to known densities, flow rate of leachates from solid waste disposal sites can be controlled.

Associated with hydraulic conductivity and intrinsic permeability are such physical factors of stratification, bulk density (compaction), natural cementation, and accretions. All of these affect the rate of flow through the soil. There is no way to know what the physical condition of the soil is except by boring holes and logging changes as soil cores are taken at frequent intervals with depth. Changes in texture can cause real problems of infiltration (Fig. 26). Addition of waste residues produces similar effects. Some of these layers are impermeable or transmit water so slowly they act as impermeable layers. Water may follow laterally along such layers until it finds an outlet and contaminates adjacent land or reaches groundwater. Sites having adverse physical structures should be avoided unless there is assurance of containment of leachates carrying pollutants.

This discussion presumes the fluid involved is aqueous or predominantely aqueous. The hydraulic conductivity and permeability of organic solvents, strong acids, and bases presents a wholly different set of circumstances. In a review of the literature on the effects of organic chemicals on clay liner permeability, Brown & Anderson [18] state that there is a near-complete lack of knowledge about the possible impact that waste impoundment has on the permeability through clays. The general consensus is that many organic solvents not only increase soil permeability but are responsible for many serious clay liner failure mechanisms. Brown & Anderson list three failure mechanisms that cause excessive permeability of clay soils: (a) dissolution, (b) volume change and (c) soil piping.

Dissolution, volume change, and piping

Both acids and bases are claimed to be responsible for *dissolution failures* permitting excessive permeability with a consequent unchecked pollutant migration. *Volume change* appears to be accompanied by clay

Table 21. *Effect of MSW leachate flux on the retention of selected metals in MSW leachates passed through nine soils*

Soil	Flux range	Metal retained by soil				
		Cd	Cr (VI)	Ni	Zn	Be
		metal/unit of soil-unit of leachate				
	cm/day	$\mu g\ g^{-1}-cm$ ‡				
Davidson clay	3.0–6.6	2.96	3.81	3.55	4.44	29.28
	6.7–10.7	2.44	1.20	3.43	4.09	15.90
	13.7–16.1	2.33	nd †	2.73	3.57	13.81
Molokai clay	3.4–7.3	4.98	nd	5.27	5.01	48.35
	9.2–10.2	4.32	nd	5.48	3.61	43.15
	14.2–18.0	4.29	nd	5.42	3.44	37.61
Nicholson silty clay	3.0–4.8	3.64	5.42	3.11	3.99	28.32
	7.0–9.8	nd	2.14	nd	3.63	19.93
	14.2–17.1	2.34	1.11	2.80	3.51	12.36
Fanno clay	3.3–4.3	4.39	3.36	4.42	4.22	13.55
	6.8–8.3	nd	1.96	4.38	4.18	11.11
	12.7–16.1	3.35	1.14	3.69	3.44	9.10
Mohave (Ca) clay loam	3.4–6.2	2.25	1.18	nd	nd	nd
	7.2	nd	1.10	nd	nd	nd
	14.2–19.0	1.80	0.97	nd	nd	nd
Ava silty clay loam	1.2–4.6	5.11	nd	4.10	3.76	14.59
	7.3	5.00	nd	nd	3.24	nd
	13.2–15.0	4.58	nd	3.32	2.96	11.34
Anthony sandy loam	3.4–4.1	2.60	nd	3.00	2.23	13.08
	7.3–8.3	2.38	nd	2.69	2.06	nd
	11.5–16.8	2.28	nd	2.41	1.75	7.05
Kalkaska sand	2.4–4.4	3.11	3.70	2.77	2.26	9.51
	6.5–7.1	2.90	2.66	2.36	2.08	7.22
	13.8–16.5	2.83	2.20	2.32	2.01	4.10
Wagram loamy sand	3.6–5.5	1.63	nd	1.47	2.17	9.25
	7.2–7.3	1.56	nd	nd	2.09	nd
	14.3–14.5	1.42	nd	1.27	2.09	1.95

†nd means not determined.
‡$\mu g\ g^{-1}-cm$ means μg of metal retained per gram of soil and cm depth of leachate passed through the 5×10 cm soil column. To convert $\mu g\ g^{-1}-cm$ (pore volume of effluent) multiply above column by 10 cm and divide by porosity.
Source: after Reference [1].

dehydration by organic solvents. *Piping* is a 'tunneling erosion' where the subsoil loosens and flows when the soil becomes saturated with a fluid (water). It is a form of heterogeneous permeability [38]. Piping causes sinking and caving of soil in a 'tubular' or 'pipelike' depression (Fig. 27).

Piping is a common natural phenomenon in alluvial soils along rivers and streams. Alluvium, where differential particle size distribution during deposition has taken place and where subsoils are highly dispersed by certain salts, is susceptible to piping. Earth shrinkage and cracking due to removal of underground water over long periods of pumping may initiate piping formation. Wind and water can then deposit fine sandy material in the resulting cracks that differs in texture from the surrounding soil.

Because natural soil piping is due primarily to differential particle size distribution in the peculiar pipe pattern, it may be controlled by any method that will mix the soil to a uniform, homogeneous mass. Where piping is suspected, deep plowing, ripping, or tilling with blades or farm

Fig. 26. Water movement in the soil is influenced by changes in texture and development of a nonhomogeneous system. Water and air move slowly across junctions between different textures whether they are mineral or organic.

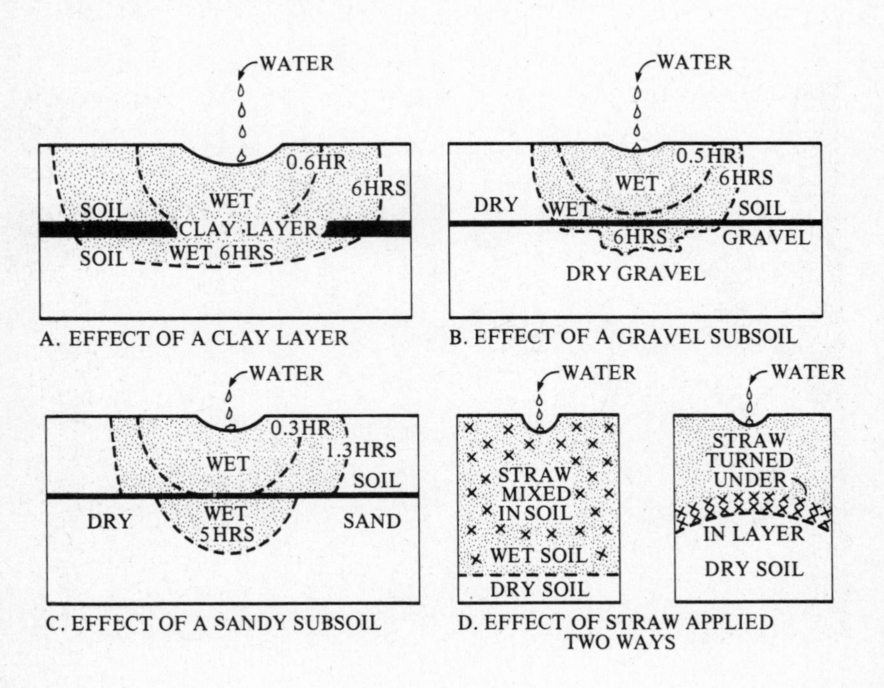

implements is suggested before building foundations are laid and landscapings begun.

Schramm et al. [in press] evaluated the saturated hydraulic conductivities and intrinsic permeabilities for eight soils and five solvents: kerosene, ethylene glycol, isopropyl alcohol, xylene, and water. Values of intrinsic permeability varied for the same soil depending on the solvent. Generally, water gave lower values. Large variations in intrinsic permeability were related more to soils rather than to solvents. The multiple regression equation provided a model for intrinsic permeability as a function of silt, clay, and bulk density area with r^2 values of 0.80.

Freezing and dehydration

Freezing–thawing and wetting–drying affect the concentration of soluble components in waste fluids and leachates ([12], Tables 22 and 23). Upon rewetting and thawing the precipitated components only partially dissolve. This is desirable from the standpoint of aiding in the prevention of pollutant migration. The more of a component that precipitates and remains insoluble, the less that is available to migrate through the soil

Fig. 27. An example of soil piping. Note the arrow pointing to the hole in the side of the ditch where water has forced the soil to move in the pipe and out the exposed end.

Table 22. *Extent of solubilization of components after rewetting the dehydrated (lyophilized) MSW landfill leachate residue*

Leachate	pH	TOC	Mg	Ca	Cd	Zn	Fe	Na	K
					% of original				
Leachate I	8.6	66.2*	71.4	11.1**	nd	2.0**	0.02**	100	95.0
Leachate II	8.4	52.2**	55.9*	16.7*	nd	4.3**	0.66**	100	86.9
Leachate III	5.4	62.7**	67.3**	97.1	nd	51.7**	0.01**	81.8**	91.9**
Enriched leachate III	5.5	56.4**	nd	nd	74.5**	81.7**	0.5**	nd	nd

*Significant at the 5% level.
**Significant at the 1% level.
nd – not determined.
Source: Reference [12].

(Fig. 28) [48]. Freezing can be important in cold regions. Water freezes initially with the remaining, more concentrated solution freezing at a lower temperature. Upon thawing only part of the components reenter the aqueous phase. The concentrations continue to decrease with each freeze–thaw cycle.

Climate

Since there is little one can do to modify area climate, design and management input must make the necessary adjustments for successful disposal. The most critical climatic characteristics that require design and management input include (a) precipitation (both total amount and distribution), (b) temperature, (c) wind and (d) evaporation.

Rainfall

Precipitation requires the first and most attention. Rain provides water as the vehicle of transport. Infiltrating water not only transports the pollutant but modifies the soil–waste interactions and establishes the position of chemical equilibria. There is no doubt that rainwater must be

Table 23. *Total organic carbon (TOC) and Cd, Fe and Zn soluble after repeated freezing and thawing of three natural MSW leachates*

	MSW leachates						
	I	II	III				
Freezes	TOC	TOC	TOC	Cd	Fe	Zn	pH
	% of original remaining						
Original	100	100	100*	100	100	100	5.3
Freeze 1	74	56	91	85	83	97	5.4
Freeze 2	63	55	90	84	79	93	5.3
Freeze 3	61	55	91	82	74	89	5.3
Freeze 4	31	36	86	81	64	88	5.3
Freeze 5	30	38	84	80	52	...	5.3

*Mean of 4 reps. Least significant difference method at 1% level of significance.
Source: after Reference [12].

controlled as a factor (a) of surface erosion, transport, and surface water pollution, and as a factor (b) within the soil profile and geological material all the way to the groundwater, lateral seepage, and upward capillary rise to the surface again.

Rainfall distribution patterns influence the *rate* of soil microbial activity and the *route*. Seasonal wetness can turn land disposal from an aerobic to an anaerobic process. The biochemistry of these processes differ greatly and in turn influence soil chemical reactions. Anaerobiosis usually is associated with low redox values, reducing conditions, and usually acid accumulation, all of which limit effective attenuation of metals and many organic compounds. Migration of most pollutants is maximized. Seasonally dry weather also affects microbial activity. Microorganisms require moisture to be active in biodegradation. For example, pentachlorophenol (PCP), a wood preservative, may remain relatively unchanged in the soil surface as long as it is dry. Yet PCP decomposes to

Fig. 28. The effect of successive freeze–thaw action on the solubility of Cd, Ni, Zn and Fe, and total organic carbon and salt (EC) in MSW landfill-type leachate.

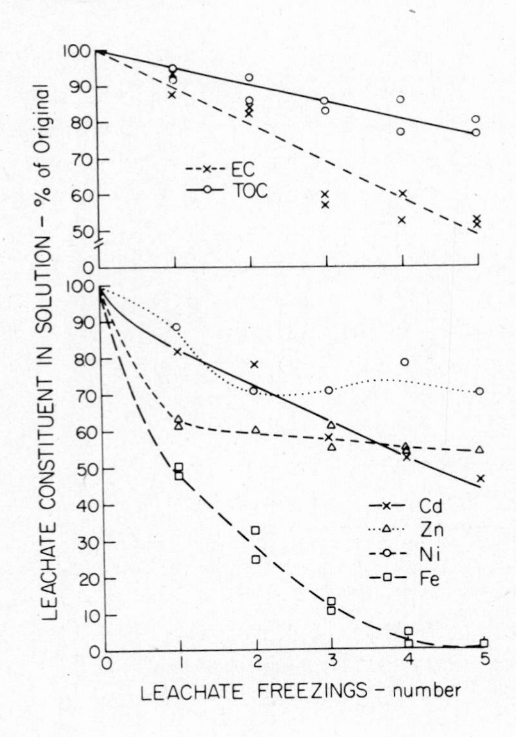

harmless constituents in a few weeks when the soil is wet (Rao [106]). Numerous other effects of rainfall and rainfall distribution patterns can be cited but time and space will not permit such liberty at this time.

Temperature

Temperature also significantly influences the design and management of a disposal facility. Microbial activity discontinues when soil and biodigesting waste disposals remain below a threshold of about 5°C. Increasing the temperature from this level through the thermophilic range about doubles biological and chemical activity every 10°C. Total biodigestion or biodegradation again falls off rapidly as temperatures rise further. Such activity is responsible for the second cycles often observed in disposal sites digestion rates, CO_2, H_2 and CH_4 production. Climate, therefore, dictates a different kind of disposal design in arid hot lands than in humid temperate or humid cold lands. Depth to water tables, native deep water reservoirs, run-off, and recharge all are part of the temperature–moisture climatic regimes affecting leachate accumulations. More care in control design and operations must accompany lands of shallow water tables than sources many meters deep.

Wind

The hazards to the public should be minimized by selection of sites with favorable wind directions away from population centers. Volatilizations from disposals may or may not be toxic but most odours are unpleasant. Atmospheric transport of pollutants by (a) volatilizations of the actual waste components, (b) dispersion of aerosols from spray operations and (c) distribution of dust, debris, and general dirt from traffic in and out of the site, and gases from heat generation, small combustions, and small fires (as well as mental anguish over odourless CO_2, H_2 and CH_4) can not be overlooked. The effects of drying from wind action on freshly tilled soil near highways can cause dangerous dust problems.

Chemical Factors

The introduction of such unnatural materials, fluids, slurries, and solid wastes in such high concentrations and in such known and unknown variety to soils confronts us with a highly complicated system to understand and predict. Geochemistry of natural soil and soil material is just beginning to be understood after about 100 years of concentrated effort. The most characteristic feature of the soil geochemistry in the early 1980s is the overwhelming lack of knowledge about specific reactions taking place in soils subjected to waste disposal. Empirical methodologies, useful for predictive purposes to explain what may occur when a certain waste is land laid, unhappily emerge as a result of this dilemma. Some of these problems are not unlike those associated with soil fertility, fertilizers, and plant nutrition.

To understand the geochemical aspects of waste disposals in soils, three main points of departure are employed:

(1) survey of established waste disposal sites;
(2) laboratory soil-column and batch techniques;
(3) identification of specific equilibria reactions.

The geochemical aspect of pollutant attenuation in soils is predominantly related to:

(1) pH (acidity and alkalinity)
(2) redox (anaerobic and aerobic)
(3) total dissolved solids (inorganic salts)
(4) amount of reactivity of surfaces of clay minerals and hydrous oxides of Fe, Al, and Mn
(5) organic reactions, complexation, and nutrient cycling.

Soil pH

The pH of the soil can influence attenuation in those instances where the waste fluids appreciably alter the natural pH. The buffering capacity of the soil is substantial, yet it is alterable and can become dominated by large volumes of acidic and basic waste solutions. The natural pH range of soils is much narrower than that of most waste streams, leachates, and solvents. To attempt the maintenance of the original soil pH at a site may not always be feasible and/or economical. When pH control is feasible and it is necessary to control pollutant migration, the method most often recommended is to add lime to the system to counteract excessive acidity and sulphur or sulphuric acid to counteract basicity. Sulphur will not be very effective, however, if the environment is not favorable for aerobic microbiological activity. The reaction may occur as previously described.

$$2S + 3O_2O + 2H_2O \xrightarrow[\text{thiooxidans}]{\text{Thiobacillus}} 2\ H_2SO_4,\ \Delta G = -83\ \text{k cal}$$

or

$$5S + 6KNO_3 + 2CaCO_3 \xrightarrow[\text{thiooxidans}]{\text{Thiobacillus}} 3K_2SO_4 + 2CaSO_4$$

$$+ 2CO_2\uparrow + 3N_2\uparrow,\ \Delta G = -660\ \text{k cal}$$

or

$$FeS_2 + 3\tfrac{1}{2}\ O_2 + H_2O \xrightarrow{\text{F. ferrooxidans}} FeSO_4 + H_2SO_4.$$

$$H_2S + 2O_2 \xrightarrow{\text{oxidation}} H_2SO_4 \text{ (under conditions of limited } H_2S).$$

Soil chemical reactions respond readily to changes in pH. Many rates of reactions are pH dependant. Attenuation, therefore, is affected by pH. In general, the solubility of heavy metals and, indeed, many organic pollutants, is higher at low pH (below 4) than near or slightly above neutral (pH 7). Certain elements (e.g. Be, Cu, Fe, Mn, and Zn) become quite insoluble in dilute aqueous solutions near and above pH 8.4. The final pH at the interface of soil and waste is a highly significant factor in retention of pollutants. The waste stream pH level is more important in chemistry of the waste disposal site than soil pH. Since soil pH changes so little within a given area, the opportunity for selection of a disposal site based on pH levels of the soil is not often available.

Griffin et al. [62] considered soil pH to significantly affect removal of heavy metals from pure solutions and MSW leachates by clay minerals. Adsorption of the cationic metals Pb, Cd, Zn and Cr(III) increased as the pH increased. Adsorption of the anionic heavy metals, Cr(VI), As and Se decreased as the pH increased. Also when MSW leachate (pH neutral) was substituted for water, the amounts of cationic heavy metals removed from solution by the clay was reduced by as much as 85% of the original concentrations. The MSW leachate had a relatively small effect on the amount of anionic heavy metals removed by clays. Since this was primarily a cation exchange system, one may have expected the latter results. On the other hand they remind us that generalizations must be made with caution and predictions require a full knowledge of the specific disposal system.

Reduction/oxidation (redox)

A reaction in which an electron transfer takes place is called an oxidation-reduction process. An element or substance accepting electrons is reduced. For example, $Fe^{3+} + e \rightleftarrows Fe^{2+}$ or $Ox + ne \rightleftarrows Red$.

The intensity of the oxidizing or reducing action of a system is determined by a platinum-reference electrode pair in the solution. The potential difference between the electrode pair corrected to the standard hydrogen electrode is called the redox potential, Eh. Redox potentials can also be expressed as pe_1 or log of electron activity, just as pH is based on $mol\,l^{-1}$ of H^+. Lindsay [91] feels this expression would make the use of redox much more palatable to many scientists including soil scientists. I agree with Lindsay, since electrons may be considered as other reactants and products. Both chemical and electrochemical equilibria can be expressed by the sum of pe + pH.

Under reducing conditions, chemical and biological reactions consume oxygen at a greater rate than it is being transported into the system. Microorganisms find substitute electron acceptors for O_2 in metabolic processes. These substitutes may contain combined oxygen as nitrates (NO_3^-) or sulphates (SO_4^{2-}) or may involve electron transfer without the involvement of oxygen as with ferric (Fe^{3+}) or manganic (Mn^{4+}) substances.

Oxidizing conditions favor attenuation as opposed to reducing conditions. The transfer of electrons need not include oxygen. Cyanide is an exception. When it is oxidized to NO_3^-, it is highly mobile. The reduced form, NH_3, is less mobile. Precipitates develop in anoxic landfill leachates upon exposure to air (oxic conditions). Many of the trace contaminants become a part of this insoluble residue (Korte et al. [84]).

Reducing (anoxic) conditions favor accelerated migration of heavy metals as compared with oxidative (oxic) conditions. For example, the trace contaminants As, Be, Cr, Cu, Fe, Ni, Se, V, Zn are much more mobile under anaerobic than aerobic soil conditions, all other factors being the same. Water saturation of the soil such as occurs below the leachate level of sanitary landfills favors accelerated mobility of most trace constituents. An exception should be noted, however, since H_2S production can greatly reduce the migration of many heavy metals with which H_2S combines readily to form highly insoluble precipitates. The effect of aerobic (oxidizing) and anaerobic (reducing) conditions on mobility is discussed in greater detail later in this section.

When wastes are disposed of in ponds, lagoons, or deep fills, management of redox status is not always practicable. However, in land spreading or spray irrigation operations, oxidizing (aerobic) conditions can be promoted by allowing the soil–waste system to dry between waste applications. The determination of the tendency for soils towards oxidation or reduction conditions should play a part in establishing management practices for waste and wastewater disposal sites.

Oxygen-stressed soil and geological material will receive greater research attention as soil scientists shift effort from the historical emphasis on food and fibre production to controlled waste disposal. Soils and geological materials surrounding and below sanitary landfills, and ponds, and lagoons developed for animal waste disposal are highly anoxic. Deep-well injections also create anoxic waste disposal conditions. Excessive sprinkling or irrigation with sewage effluents and aqueous wastewaters may develop intermittent oxygen-stressed soil as a result of water saturation and/or partial soil saturation. Even in unsaturated soil conditions a great demand for oxygen by biodegradation processes and limited access to air by filled as well as partially filled pore spaces may create areas of anaerobic microspace and reducing conditions. Thus, in waste and leachate disposal, oxygen-stressed soil and geological material is a common and expected situation for which special attention is required.

Mobility of heavy metals and trace elements, in general, will be accelerated in oxygen-stressed compared with oxygen-rich soil. Some of the factors which influence mobility will be briefly reviewed to illustrate how anaerobic conditions may promote mobility of some contaminants while not affecting the mobility of others.

Some characteristics of reducing (anaerobic) systems are:

(a) Gas production, CO_2, CH_4, and H_2S predominates; other gases, such as H_2 and N_2, form in low concentrations and occur mostly under special circumstances, or are not significantly involved in contaminant mobility.

(i) Methane (CH_4) – methane is one of the most characteristic gases produced during microbial decomposition of organic compounds under anaerobic conditions. The methane bacteria are most active between pH values of 6.4 to 7.2. Below pH 6 and above pH 8 the growth rate (and gas production) falls off rapidly. Methane does not measurably affect trace contaminant mobility.

(ii) Hydrogen sulphide (H_2S) – under reducing conditions in the absence of 'free' O_2, hydrogen sulphide is produced by reduction (to H_2S) in soils. Unlike CH_4, H_2S is highly reactive with certain trace contaminants. The solubility products of some of the sulphides are:

$FeS = 3 \times 10^{-17} - 1 \times 10^{-16}$; $ZnS = 1 \times 10^{-24}$;
$CdS = 5 \times 10^{-27}$; $PbS = 7 \times 10^{-28}$;
$CuS = 4 \times 10^{-36}$; $Hg_2S = 1 \times 10^{-45}$;

and

$HgS = 3 \times 10^{-52}$,

in a saturated solution of H_2S at $[H^+] = 1$. This gives a relative indication of attenuation expected by sulphide formation of the various selected elements in anaerobic soils where H_2S is produced. Exceptions can occur under special circumstances. For example, sodium sulphide may make mercuric sulphide more soluble if present in any appreciable quantity.

(iii) Carbon dioxide (CO_2) – is produced in abundance in all landfills and in landfill leachates as well as other wastes having a favorable microbiological habitat and a decomposable carbon source. This gas unites with water to form carbonic acid. Carbonic acid production reduces the pH (often to as low as 5.5) and in this way contributes to accelerated migration. The abundance of bicarbonate ions (HCO_3^-) aids in keeping certain trace contaminants more soluble. Selenium and probably arsenic are not influenced as much by bicarbonate as are other trace constituents. Neither asbestos nor cyanide mobility is affected by bicarbonate or acid development. Maintenance of relatively high CO_2 concentrations in the leachate from sanitary landfills is necessary to keep trace and heavy metals from precipitating during sampling and experimentation. Carbon dioxide production thus tends to enhance mobility [84].

(iv) Other gases – volatile compounds of Hg and As may be produced under anaerobic conditions. Amounts of mono-mercury and methylmercury produced in an anaerobic system are dependent on temperature, pH, organic loading, mercury compounds and microbial species present [80]. Trimethylarsine also is a volatile reduction product of microorganisms. The reaction is:

$$As(OH)_3 \xrightarrow{RCH_3} H_3C - \overset{\overset{O}{\|}}{\underset{\underset{O}{|}}{As}} - OH \xrightarrow{RCH_3} H_3C - \overset{\overset{O}{\|}}{\underset{\underset{CH_3}{|}}{AS}} - OH \xrightarrow{RCH_3} \underset{\text{Trimethylarsine}}{As(CH_3)_3}$$

(v) Hydrogen (H_2) – small amounts of hydrogen gas evolve from anaerobic leachates. It does not appear to be an important factor in attenuation of the trace contaminants as such.

(b) Reducing conditions (redox effect) – reducing conditions in soil promote mobility of most of the trace contaminants. Cd, Pb, and Hg mobility are less affected by lack of oxygen than As, Be, Cr, Cu, CN, Fe and Zn. Cyanide, a readily mobile constituent, will denitrify and evolve as N_2 gas. The relative mobility of As, Be, Cr, Cu, Fe, Zn, and cyanide, for 'usual' aerated soil conditions, probably will not change much with time if conditions become reducing. Although the actual mobility of Cd, Pb and Hg will not change, their position relative to the other elements will be changed. Cyanide will be at the top of the list as originally, its mobility also being little altered by the presence or absence of O_2 in the soil.

(c) Organic acid production (pH effect) – organic acids will be produced when organic materials decompose in a limited oxygen environment. This is the case in a municipal sanitary landfill or in an industrial waste landfill so long as the concentration of salts or toxic materials does not limit microbiological activity. Organic acids enhance the rate of mobility of most of the trace metals through the soil. Even a slight lowering of the pH value of the soil solution in the region from pH 8 to pH 6 will markedly influence attenuation of most heavy elements. Asbestos, As, Se, and probably CN mobility should not be appreciably influenced within this range. Lowering the soil pH to 3 or below (somewhat unlikely in most soils except sands and gravels), however, will greatly increase solubility of heavy metals. Organic acids (fulvic, humic, uronic and others) contain carboxy (−COOH) and acid hydroxy (−OH) radicals that react with heavy and trace metals. The downward movement of the trace contaminant metals through soil and geological material as soluble metal–organic matter complexes can be of considerable importance. Fraser [37], for example, describes Cu metal–organic-complex mobility as a factor in contaminant accumulation in underground seepage water. The organic acids produced under anaerobic conditions form chelates with many heavy and trace metals. These metals are then protected and available for accelerated movement through soils. Our knowledge in this area is still fragmentary, leaving little opportunity for quantitative evaluation.

(d) Retardation of biodegradation – anaerobic degradation of organic matter is slower than aerobic degradation. In addition it often stops at some intermediate stage of oxidation leaving an accumulation of incompletely-oxidized organic products. Slime and gum accumulation, however, tend to be less severe under strictly anoxic as compared with oxic conditions. Clogging and filling of the soil pore spaces may be expected to be less severe under anaerobic conditions as compared to aerobic or partly aerobic conditions. The nature of anaerobic flora is less conducive to clogging of pore spaces than when fungi, algae, and slime bacteria accumulate in the presence of oxygen. All of the anaerobic conditions tend to permit water to pass through the soil more rapidly. This effect may be counterbalanced by accumulation of unoxidized sludges and original organic debris which would tend to have an opposing influence.

(e) Water movement, not retarded by slime or gum formation, is expected to be impeded less under anaerobic than aerobic conditions. If the waste is strictly inorganic, water movement should not be affected by the presence or absence of oxygen.

As indicated previously, it is only practicable to manage the aeration status of the soil–waste environment when the disposal operation is conducted at the soil surface (land spreading or spray irrigation of wastes). However, the effects of anaerobic conditions should be taken into consideration in any disposal operation, particularly for wastes with a significant content of heavy metals or organics. For example, heavy metal-bearing wastewater treatment sludges formed by alkaline precipitation should not be disposed of in a municipal refuse or other organic waste environment; the organic acids formed during anaerobic decomposition would interact with the sludge, reversing retention and freeing the heavy metals for migration.

Total dissolved solids (salts)

The wide variety of reactions trace contaminants might undergo makes relative concentrations of ions very important with respect to mobility. Low concentrations of salts favor more complete attenuation by simple attachment to soil ion adsorption positions. Also, many of the trace contaminants form more insoluble precipitates at low concentration. Examples are the arsenates and sulphides of Pb^{3+} and Cd^{2+}.

Where concentrations of some salts are high, the effect of competing reactions can be especially important. In a leachate containing hazardous pollutants in small amounts, in the presence of a large amount of Ca^{2+}, the Ca^{2+} might effectively dominate the exchange reaction to the exclusion of the trace contaminants. On the other hand, if sulphides were also present, even in the presence of high salt concentrations, the trace elements would be immobilized by precipitation.

High concentrations of certain ions can also dramatically reduce solubilities due to the 'common ion' effect. A leachate high in sulphate or chloride would shift the equilibrium of those elements precipitated by chloride or sulphate far to the left or to the less soluble state, in what is known as 'salting out'.

Concentration of ions or salts may either increase or decrease attenuation depending on (a) the kinds and concentrations of ions present in the soil solution, (b) the concentration of the trace contaminant in the soil solution and the leachate from the waste and (c) hydrogen ion activity or pH. Each case has to be evaluated separately.

One generalization concerning salt concentration appears to hold for aqueous effluents and leachates such as the MSW landfill-type: the greater the common inorganic ion concentration, the greater is the tendency for certain pollutants such as heavy metals to migrate through the soil.

Mixing wastes to promote formation of insoluble compounds is a possibility to reduce pollutant migration. Attenuation observed for a given soil and concentration of contaminant in a specific waste may not be a constant; attenuation is likely to be different if another waste with the same contaminant concentration but differing composition of other substances is disposed of on that soil. Attenuation by mixing two different wastes, although attractive, is a two-way street: some pollutants may be made less mobile whereas others may be made more mobile. Reacting constituents must be known before recommending mixing.

Adsorption (physico–chemical)

Chemical adsorption received attention earlier under chemical reactions but the more physico–chemical oriented reactions were not emphasized. Discussions of adsorption appearing in the literature often do not include a clear delineation between chemical and physical adsorption processes. Perhaps it is too much to expect, because the mechanisms of adsorption are not well established. To illustrate, Murrmann & Koutz [102] state that: 'Adsorption is the most important process by which chemicals are removed from wastewater applied to soil . . . they are not well understood.' The adsorption forces which often dominate behavior of heavy metals in soils, particularly when the metals are present in the soil solution in small amounts, may be of various types. Only a small proportion of the reactions of heavy metals usually quantified as adsorptive can be accounted for by chemical bonding. Solid-state diffusion of specific ions (Cu and Zn) into octahedral positions in layer silicates may be possible if open crystal structure prevails [10], but Jenne [78] believes this to be a minor mechanism of immobilization for most heavy metals.

Heavy metal adsorption most often has been related to sorption by the layer silicates which include (a) surface complex ion formation, (b) surface reactions, not ion exchange, (c) lattice penetration or inclusion and (d) ion exchange. Other hypotheses concerning heavy metal adsorption include reactions with organic matter, lime, and hydrous oxides. There is enough information available now to broadly classify adsorption into four main groups:

(1) layer silicates
(2) hydrous oxides
(3) organic matter
(4) lime (carbonates).

Layer silicates

Solid-state diffusion known as 'lattice penetration' is not a likely mechanism of consequence for heavy metals. Heavy metals appear to diffuse to a distance of a few atomic planes only at the crystal edges and along fractures. This is more of a fixation mechanism (i.e., retention against removal by salt solutions) than a sorption mechanism.

Layer silicates may bind soluble chemical species through two principal mechanisms: (a) ion exchange as the result of electrostatic attractions and

(b) ion adsorption through covalent bonding. Ion exchange is relatively nonspecific and readily reversible while ion adsorption may be very specific and at times irreversible. Ellis & Knezek [32] have discussed mechanisms of Zn and Cu reactions with clay surfaces. Involved in these latter reactions are ion exchange (both with Zn^{2+} and Cu^{2+} ions as well as with the $ZnOH^+$ and $CuOH^+$ ions), precipitation of insoluble Zn and Cu salts, and some very specific covalent bonding between Zn, Cu, and the layer silicates.

Hydrous oxides of Fe, Al, and Mn

Adsorption reactions with hydrous oxides of Fe^{2+}, Al^{3+} and Mn^{2+}, in general, are considered by some investigators Jenne [78], Fuller [41, 42] and Leeper [89] to furnish a major mechanism for the attenuation of heavy metals in soils. The abundance of Fe, Al, and Mn in soils and their chemistry which is sensitive to slight changes in redox make them prime components for removal of trace contaminants from circulation. Some of the trace contaminants at first may be adsorbed on the surface of the hydrous oxide and later buried by the continued layer formation of hydrous oxide–metal combinations. Leeper [88], for example, writes an equation such as:

$$2FeO \cdot OH + Zn^{2+} \rightarrow ZnFe_2O_4 + 2H^+$$

to explain a possible adsorption reaction.

Whenever a choice is possible, soils with a significant content of hydrous oxides should be selected for disposal sites. Jenne [78] presents an attractive argument that sorption and fixation by certain hydrous oxides in soil are the basic mechanisms for heavy metal immobilization. However, he also states, 'the extent and rate of reversibility of heavy metal sorption by the hydrous oxides, as well as the exact nature of the "pH effect" on the reversibility is not entirely clear'. The available data indicate that heavy metal adsorption by the hydrous oxides is at least partially competitive with and reversible to exchange with other heavy metals. Jenne further proposes that Fe and Mn supply the principal matrix into which less abundant heavy metals are adsorbed, coprecipitated, or occluded.

In addition to the Fe and Mn hydrous oxides, Al hydroxides play a part in attenuation. Gibbsite ($Al(OH)_3$) is a common mineral of the more heavily weathered soils. Al and Si undoubtedly are important factors in the mobility of some of the chemical species of pollutants [94], particularly where pH is low (<4.0). Also Al hydroxide polymers present on clay surfaces (of general occurrence in soil systems according to Jackson [75]) may also significantly immobilize pollutants in soil systems.

Organic matter

Reactions between heavy metals and organic matter are thought to involve (a) the formation of complexes and (b) chelation. These are chemical reactions. Physico–chemical adsorption is often alluded to in the literature but not defined or quantized. The extent to which organic matter plays a part in physico–chemical adsorption of the trace contaminants directly is not well known. Chelation and movement of hazardous metals such as ^{89}SR–^{90}Sr(^{90}Y), for example, are real occurrences in soils and can be documented through the use of radioisotope techniques [55, 64]. Some metals, such as ^{45}Ca and ^{90}Sr, have been shown [66, 53] to move in soil as microbial residues of colloidal size.

Lime

Retention of trace and heavy metals by lime is more likely to be surface adsorption or surface precipitation by the carbonate ion. Jurinak & Bauer [82] believe that zinc, for example, is adsorbed on the crystal surface of dolomite and magnetite at lattice sites that normally are occupied by magnesium. Later, Brown & Jurinak [17] found that additions of $CaCO_3$ (limestone) to Yolo fine sandy loam beyond 5% (w/w) did not influence either Cu or Zn uptake by plants. The carbonate interaction was considered to be primarily a pH effect and secondly an effect of carbonate–bicarbonate ion on heavy metal solubility.

The composition of some municipal and industrial leachates

Classification

There is no limit to the variability in the composition of leachates, regardless of origin, whether municipal or industrial. Some order must be established, however, concerning hazardous waste classification that can be useful for orientation during this discussion. For the sake of convenience, therefore, and to better design management practices to limit the dispersal of pollutants into sensitive environments, the following classifications based on the nature of the transport vehicle are suggested:

(1) mild aqueous fluids;
(2) strong acid fluids;
(3) strong base fluids;
(4) organic solvents.

Obviously, each of the four major classes may be further subdivided into well-recognized subclasses and so on, and on. This classification scheme does not include solids or gases until, of course, they are leached or dissolved in rainwater yielding noxious aqueous fluids, for example, acid rain, or sewage sludge leachates. Almost all industrial production yields waste of some specific nature. Therefore, it is readily apparent that all of these cannot be identified and discussed in this short chapter. Environmental protection agencies of most highly industrialized countries have published lists and compositions of many potentially hazardous waste streams.

The objective here is to briefly provide some information of some hazardous constituents found in a few typical municipal and industrial wastes that are most prominently being studied today. It is recognized that the omissions far exceed the admissions. Moreover, even if one could classify all of the pollutants in the numerous waste streams, geochemical reactions with earth materials and other components of their environment are still obscure. The area of knowledge, however, has spread rapidly since the early 1970s from agricultural wastes (manures, canning wastes, and fertilizers) to municipal sewage sludges, sewage and wastewater reuse, and municipal solid waste leachates. All fit into category (1), dilute aqueous solutions and soil solutions. Pesticide degradation and movement through the soil, as a part of the agricultural industry effort, has been studied throughout the world. Only limited research attention has been given to permanent disposal of concentrated and highly toxic wastes from industrial synthetic processes. The movement of hazardous pollutants through soils from categories (2), (3), and (4) remained virtually unstudied before 1980. Many organic solvents (polar, nonpolar, etc.) move readily through soil and the vadose zone. Failure of the soil to retain strong acids and bases is now just being recognized. Research from our University of Arizona laboratories indicate rapid dissolution and solubilization of the aluminosilicates of soils exposed to strong acids and bases.

Municipal solid waste leachates

Factors in composition variability

Leachates from land disposal of solid waste varies widely in chemical composition from one location to another. Climate, age, and extent of inclusion of industrial wastes cause most of these variations. Leachate from MSW landfills is generated by infiltration of water from rain and snow and sometimes from fluctuating underground water tables. These leachates often contain polluting inorganic metals and organic constituents in sufficient concentrations to contaminate groundwaters used for domestic purposes, or seepwaters that collect on land surfaces or flow into subsurface aquifers that link to food chains: Walker [131]; Garland & Mosher [58]; Matrecon, Inc. [98] and Brown [19].

The rate of movement of heavy metals through soils depends on certain measurable chemical characteristics of the MSW leachate [44, 50, 51]. The most prominent fall into broad categories such as pH value, concentration of the soluble common salts (EC or TDS), total soluble organic carbon compounds (TOC), and certain naturally occurring metals, notably Fe (but also Al and Mn). Wide variations in the composition of these and other constituents in MSW leachates are indicated by surveys made throughout the United States and Europe: Korte et al. [84]; Garland & Mosher [58]; Hoeks [73] and Chian & DeWalle [22] (Tables 24 and 25). Great changes take place in the composition of the soluble constituents in MSW leachates with time [20, 22, 49]. Also, see Table 26 as an example of the changes in soluble constituents with age. Biodegradation of carbon compounds to CH_4, H_2, and CO_2 gases and H_2O accounts for losses of carbonaceous materials. Precipitation, complexation, and adsorption also occur with formation of sludge that collect at the bottom of the disposal excavation or tank. Some hydrous oxides (mostly Fe) precipitate on the surface of the solid waste as the level of leachate fluctuates somewhat with the rainfall distribution patterns.

A number of other factors contribute to the variability of leachate composition from different landfills [42, 43, 84, 73, 22, 60, 98, 34]. Some factors have been identified and discussed [48] such as:

(1) solid waste characteristics;
(2) physical subdivision of waste;
(3) compaction;
(4) moisture content;
(5) temperature;
(6) landfill geometry;
(7) interaction of leachate with environment;
(8) age of leachate;
(9) sampling, storage, and analytical techniques.

Organic constituents

The most prominent and most studied organic constituents in MSW landfill leachates are: total organic carbon components (TOC), volatile fatty acids, phenolic compounds, and humic and fulvic acids. These lend themselves to ready identification and quantification. Other methods for characterization consider reactive groups as carboxyl, methoxyl, ketonic, aldehydic, alcoholic, phenolic, and heterocyclic units. The total organic carbon content varies as much within a single disposal source as between sites. The trend is for a rapid drop from high values above

Table 24. *Ranges of constituents detected in the natural leachates generated from typical municipal solid waste from Tucson, AZ*

	Leachate I range		Leachate II range		Leachate III range	
Constituent	Overall	Used in study	Overall	Used in study	Overall	Used in study
			mg l^{-1}			
Time span	11/29/73–1/15/79		10/23/75–1/15/79		10/23/75–1/15/79	
COD	50–500	100–200	—	—	50–200	200
TOC	123–1155	200–900	1400–25 000	700–10 000	250–1000	250
pH	6.4–6.8	6.4–6.8	5.3–6.8	5.4	5.4–6.2	6.2
EC (mmhos cm^{-1})	1.2–4.2	2.6–3.5	5–17	9.0–11.5	1.9–2.5	1.9
TDS	768–2680	1660–2240	3200–11 000	5000–7000	1200–1600	1220
Total P	0.8–7.9	2.0–4.0	2–33	12	2–20	nd
NH_4–N	70–190	125–150	nd*	nd	nd	nd
Cl	93–350	~3900	300–600	nd	150	150
Ca	90–275	160–225	100–1750	200–1000	150–500	730
Mg	14–106	25–60	40–450	60–360	85–180	85

Na	55–150	55–150	135–750	150–640	200–300	193
K	108–2050	850–950	500–1600	600–700	40–400	40
Si	12–31	20–25	22–44	32–33	25–28	15
Cd	<0.02	<0.02	<0.02–0.45	<0.02	0.02	0.02
Co	<0.10	<0.10	<0.05–0.6	<0.05	0.01	0.01
Cr	bdl**–0.15	bdl	<0.05–2.4	<0.05	0.10	0.10
Cu	bdl–0.30	bdl	<0.05–0.7	<0.05	0.20	0.20
Fe	31–120	70–100	44–1380	900–1000	9.00	9.00
Mn	0.5–2.30	0.6–1.8	0.14–16	12	32.00	32.00
Pb	bdl	bdl	<0.5–3.3	<0.5	0.50	0.50
Ni	bdl–0.20	bdl	0.05–0.9	0.05–0.25	0.08	0.08
Zn	0.10–2.20	0.10–2.10	0.40–165	9–13	0.20	0.20

*nd – means not determined.

**bdl – means below detectable limits; bdl for the atomic absorption equipment used in $\mu g\,l^{-1}$ are: Cd = 0.005, Cr = 0.05, Co = 0.05, Cu = 0.05, Pb = 0.5, Ni = 0.05, Mn = 0.05, Zn = 0.005, Al = 0.5, and Fe = 0.05.

Leachates I and II were generated in 4000 l tanks charged with typical municipal solid waste. Leachate III is from an active Tucson city solid waste landfill.

Source: Reference [2].

Table 25. *Composition of MSW landfill leachates from three locations in the United States*

Constituent (ppm)	Location I	Location II	Location III
BOD	—	13 400	—
COD	42 000	18 100	1340
TOC	—	5000	—
Total solids	36 250	12 500	—
Volatile suspended solids	—	76	—
Total suspended solids	—	85	—
Total volatile acids as acetic acid	—	9300	333
Acetic acid	—	5160	—
Propionic acid	—	2840	—
Butyric acid	—	1830	—
Valeric acid	—	1000	—
Organic nitrogen as N	—	107	—
Ammonia nitrogen as N	950	117	862
Kjeldahl nitrogen as N	1240	—	—
pH	6.2	5.1	6.9
Electrical conductivity (μmho cm^{-1})	16 000	—	—
Total alkalinity as $CaCO_3$	8965	2480	—
Total acidity as $CaCO_3$	5060	3460	—
Total hardness as $CaCO_3$	6700	5555	—
Chemicals and metals:			
Arsenic	—	—	0.11
Boron	—	—	29.9
Cadmium	—	—	1.95
Calcium	2300	1250	354.1
Chloride	2260	180	1.95
Chromium	—	—	<0.1
Copper	—	—	<0.1
Iron	1185	185	4.2
Lead	—	—	4.46
Magnesium	410	260	233
Manganese	58	18	0.04
Mercury	—	—	0.008
Nickel	—	—	0.3
Phosphate	82	1.3	—
Potassium	1890	500	—
Silica	—	—	14.9
Sodium	1375	160	748
Sulfate	1280	—	<0.01
Zinc	67	—	18.8

Source: Reference [98].

20 000 ppm at the beginning to low levels of 10–100 ppm after three or four years of biodigestion. Volatile fatty acids followed the same trend. In contrast, the humic–carbohydrate-like and fulvic-like substances become more prominent with age. The pH level rises generally from about 4 to near neutral 7–7.5.

The presence of *phenols* in MSW landfill leachates is readily demonstrated (Artiola-Fortuny & Fuller [5]). Data in Table 27 show concentrations of phenols that far exceed the permissible levels for drinking water of 1 ppb. In another publication by the same authors [6], adsorption onto soil and polymerization with other organics of monohydroxybenzene derivatives (phenols) were found to take place. The biological synthesis of phenols in MSW leachates occurs quite readily during biodegradation of municipal solid waste. Soil properties that influence phenols adsorption the most, as reported in Table 28, are iron oxide content, surface area, pH, clay content, and CEC, oriented in descending order of importance.

Humic and fulvic acid constituents also were found [7] in MSW landfill leachates (Table 29). Although these macromolecules have just very recently been recognized as being generated in MSW leachates, they have

Table 26. *Effect of time (age) on the solubility of constituents of MSW leachate*

Constituent	MSW leachate Fresh	Old
	mg l^{-1}	
BOD	14 950	—
COD	22 650	81
TDS	12 620	1144
TSS	327	266
Total N	989	7.51
pH	5.2	7.3
Elec. Conductivity (μmhos cm^{-1})	9200	1400
Calcium (Ca)	2136	254
Chloride (Cl)	742	197
Copper (Cu)	0.5	0.1
Iron (Fe)	500	1.5
Magnesium (Mg)	277	81
Manganese (Mn)	49	—
Phosphate (P)	7.4	5.0
Zinc (Zn)	45	0.16

Source: after Reference [20].

been known to be a part of the organic matter of municipal sewage sludge since before the early 1960s.

Stage of biodegradation

MSW leachates degrade rapidly accompanied by radical changes during the early period of establishment (0–5 years). The kind and magnitude of changes, as a result of ageing, were found [2] to be approximately as follows:

(1) pH *rises* from about 3.5 to neutral 7 and slightly above;
(2) oxidation-reduction potential *rises* from a low Eh of 60 to about 150;
(3) sulphide *rises* several-fold;
(4) sludge formation tendency *rises* greatly with time;
(5) free, volatile acid *decreases* about ten-fold from a maximum;
(6) total organic carbon *decreases* about 100-fold, COD decreases many-fold with time, roughly following TOC content;
(7) electrical conductivity *decreases* several-fold;
(8) soluble heavy metal concentration *decreases* rapidly with time (often disappearing from detection);

Table 27. *The concentration of phenol and other constituents in three MSW landfill leachates*

Leachate	Time of dumping	Time of sampling	pH	EC	Fe*	TOC*	Phenols**
number	date	date		mmhos cm^{-1}	ppm	ppm	ppb
I	8/72	1/19/79	7.0	3.8	33	1080	27
		4/04/79	7.1	4.3	38	560	48
		6/15/79	6.7	4.7	40	590	40
II	8/75	3/03/78	6.7	5.3	367	1863	648
		3/11/78	6.7	5.3	350	1890	730
		1/19/79	6.9	4.7	43	1470	80
		4/04/79	6.9	5.1	40	540	63
		6/15/79	6.8	6.0	35	570	60
III	8/78	1/19/79	4.4	8.0	590	9400	2026
		4/04/79	4.9	10.0	410	8300	3450
		6/15/79	4.9	13.0	350	7800	3580

*ppm – parts per million.
**ppb – parts per billion.

(9) soluble Fe *decreases* as much as five-fold;
(10) soluble P *decreases* several-fold;
(11) common soluble ions (Ca, Mg, Na, K, Cl, SO_4, for example) *decrease* markedly.

In old MSW landfills (ten+ years) free from industrial disposals, the leachates often become little more than dirty water. If this were not the case, the thousands of landfills from every neighbourhood, from coast to coast long since would have contaminated every underground water source except for the deepest wells. Although we are fortunate for this 'built-in',

Table 28. *Regression analysis of six soil properties on capacity for adsorption of phenols*

Phenols	Added variable (stepwise 1–4)	r^2	$Mean^2$ change*
Phenol	1 – Iron oxides	.64	68.1
	2 – Surface area	.73	53.7
	3 – pH	.93	15.3
	4 – Clay	.95	11.2
o-cresol	1 – Iron oxides	.84	486.9
	2 – pH	.94	192.7
	3 – Surface area	.99	37.6
	4 – Clay	.99	26.9
p-cresol	1 – Iron oxides	.34	2026.9
	2 – Surface area	.50	1748.4
	3 – CEC	.91	324.0
	4 – pH	.99	60.1
2,6-dimethyl-phenol	1 – Iron oxides	.96	160.7
	2 – Surface area	.98	98.0
	3 – Clay	1.00	11.8
	4 – CEC	1.00	12.9
2,4-dichloro-phenol	1 – Iron oxides	.80	126.2
	2 – pH	.87	90.3
	3 – Silt	.99	8.7
	4 – CEC	.99	7.7
p-nitro-phenol	1 – Iron oxides	.64	9.8
	2 – CEC	.68	9.3
	3 – Clay	.75	8.1
	4 – Silt	.75	8.7

*High numbers and large changes give most significant correlations. Small numbers and small changes indicate poor correlation between the soil variables and the adsorption of the phenol.
Source: Reference [5].

Table 29. *Humic and fulvic acid concentrations in natural MSW landfill leachates*

		Organic carbon					
		Total sources		Humic acid		Fulvic acid	
Leachate	pH	Average*	Range	Average*	Range	Average*	Range
				ppm			
I	6.3–6.4	359	320–398	136	116–156	99	52–146
II	6.3–6.4	481	440–522	210	166–264	176	122–230
III	5.2–5.4	8535	8312–8758	144	43–256	6193	5724–6662

*Average of at least three replicates: TOC represents 5-month averages and humic and fulvic acids 2-month averages.
Source: Reference [7].

self-limiting pollution control, some serious pollution problems still persist throughout the USA originating from old municipal landfills [58, 100, 101].

Industrial waste streams

The USEPA [127] provided generalized data for Table 30 listing hazardous substances that may be found in a variety of industrial waste streams. Because of the great number of waste streams, the wide variety of constituents, and the wide variations in concentration of any one pollutant, the chapter must of necessity of space and purpose limit the discussion to a few examples. It is hoped that these will provide some concept of the magnitude of industrial hazardous waste disposal, some characteristics that must be considered, and the need for solutions as diverse as may be required. In-depth information on the characteristics of the many waste streams may be found in a large volume of literature some of which appear referenced in the text and Tables of this section.

A symposium on the management of petroleum refinery wastewaters, sponsored by the US Environmental Protection Agency and University of Tulsa, OK. [127], clarifies the demand by the public and private sectors for guidelines for acceptable disposal and/or reuse of wastes and wastewaters from petroleum refineries. Various waste treatments of the different wastes generated are addressed. The amount of wastewater generated is significant, particularly since the petroleum industry most often is located in water-deficient climates.

Research on the movement of organic solvents through soils initiated by Brown & Anderson [18] and Schramm et al. [in press] shows a clear distinction between water and solvents regarding permeability.

The prediction of pollutant attenuation

Models

The ultimate aim for identifying attenuation reactions in soils is to provide a base for predicting pollutant migration *rates*. These rates then serve as a tool for disposal site selection and modification to minimize pollution of groundwaters and food chain entrance. A model provides a simplified analogy for a natural phenomenon, according to Maisol & Gnugnoli [92].

The most common prediction method available to us at this time depends strictly on *personal judgment*. In the selection of a suitable site for disposal, the operator must apply a conceptual 'model' for the evaluation of the many factors he keeps in mind as a result of experience. A conceptual 'model' is the same as a theoretical and/or computerized model minus

Table 30. *Representative hazardous substances within industrial waste streams*

Industry	Hazardous substances										
	Arsenic	Cadmium	Chlorinated hydrocarbons[a]	Chromium	Copper	Cyanides	Lead	Mercury	Misc. organics[b]	Selenium	Zinc
Battery		X		X	X						X
Chemical manufacturing			X	X	X			X	X		
Electrical and electronic			X		X	X	X	X		X	
Electroplating and metal finishing		X		X	X	X					X
Explosives	X				X		X	X	X		
Leather				X					X		
Mining and metallurgy	X	X		X	X	X	X	X		X	X
Paint and dye		X		X	X	X	X	X	X	X	
Pesticide	X		X			X	X	X	X		X
Petroleum and coal	X		X				X				
Pharmaceutical	X							X	X		
Printing and duplicating	X			X	X		X		X	X	
Pulp and paper								X	X		
Textile				X	X				X		

[a]Including polychlorinated biphenyls.
[b]For example: acrolein, chloropicrin, dimethyl sulfate, dinitrobenzene, dinitrophenol, nitroaniline, and pentachlorophenol.
Source: Reference [98].

quantitative data and is subject to errors of judgment based on qualitative information. Thus, a model is a simplified representation of an actual waste disposal system whether it exists in the mind from practical experience or is computerized using measured parameters and complicated equations. Therefore, there may be conceptual models, physical models, mathematical models, and a host of others, both qualitative and quantitative, some of which will be described.

Ranking pollutants with soils

Another means of predicting pollutant migration rate is to rank the pollutants according to soil property (Tables 31 and 32) interactions with leachate as illustrated by Figs. 29 and 30. Although this soil-column leachate method is more quantitative than the conceptual 'model', it too is a highly qualitative procedure. Metal–soil interactions were ranked using MSW leachates containing toxic metals. The soil environment was strictly anaerobic and the leachate flow saturated, like that of conditions usually found under a landfill operation. Cations (Fig. 29) and anions (Fig. 30) were ranked separately because of differences in migration behavior in different soils. Changes in ordering of soils when going from cations to anions in Figs. 29, 30 involve a higher rank in attenuation for soils having lower pH values and/or higher free iron oxide content.

Predictions using soils and pollutant breakthrough curves

Predicted relationships for similar soils and any situation can be found by using another set of figures as typical attenuation curves presented for each soil and each selected element. One example of this is illustrated by heavy metal attenuation in aqueous solid waste leachates. Breakthrough curves (e.g. C/C_0 at increasing pore volume displacements) for different soils with a single metal (Fig. 31 (e.g. Ni with four soils)) and different metals with a single soil (Fig. 32 (e.g. Ava si c l with Ni, Be, Se)) are developed and used to construct a set of typical curves (Fig. 33), again using the soil-column technique [42]. The types of breakthrough curves for each soil and each element are given an identification letter A–E (Fig. 33). The values of C/C_0 obtained from any one soil column correspond to one of the generalized curves in Fig. 33. Weakly retained metals are represented by curves A and B where breakthrough ($C/C_0 = 1$) occurs rapidly. The rise in pollutant concentration in the effluent of A begins earlier than B, C begins before D, and so on. Curves C and D represent more of a steady-state condition. Curve E is an example of an extreme situation of a clay soil that

Table 31. *Selected characteristics of soils used in Arizona experiments*

Soil series*	Soil order	Clay %	Silt %	Total Mn ppm	Free iron oxides %	Soil paste pH	Cation exch. capacity meq 100 g^{-1}	Elec. cond. of extract μmhos cm^{-1}	Soil surface area m^2 g^{-1}	Column bulk density g cm^{-3}	Column porosity	Predominant clay minerals †
Davidson	Ultisol	52	23	4100	17	6.4	9	169	151.3	1.40	0.476	KK
Molokai	Oxisol	52	25	7400	23	6.2	14	1262	167.2	1.44	0.429	KK, G
Nicholson	Alfisol	49	47	950	5.6	6.7	37	176	120.5	1.53	0.460	V
Fanno	Alfisol	45	19	280	3.7	7.0	33	392	122.1	1.48	0.484	Mt, Ml
Mohave (Ca)	Aridisol	40	28	770	2.5	7.8	12	510	127.5	1.54	0.446	Mt, Ml
Chalmers	Mollisol	31	52	330	3.1	6.6	22	288	95.6	1.60	0.453	Mt, V
Ava	Alfisol	31	60	360	4.0	4.5	19	157	61.5	1.45	0.478	V, KK
Anthony	Entisol	15	14	275	1.8	7.8	10	328	49.8	1.87	0.360	Mt, Ml
Mohave	Aridisol	11	37	825	1.7	7.3	10	615	38.3	1.78	0.365	Ml, KK
Kalkaska	Spodosol	5	4	890	1.8	4.7	6	237	8.9	1.53	0.404	CL, KK
Wagram	Ultisol	4	8	50	0.6	4.2	2	225	8.0	1.89	0.378	KK, CL
River alluvium	Entisol	1	2	30	0.5	7.2	2	210	3.6	1.73	0.336	KK, Ml

*Oriented on basis of clay content.

†Listed in order of importance: KK = kaolinite; G = gibbsite; Ml = mica; Mt = montmorillonite; V = vermiculite; CL = chlorite.

Source: Reference [42].

did not permit the pollutant (metal) to exceed $C/C_0 = 0.1$ in the number of pore volume displacements taking place.

Elution curves obtained for each soil column are coded in Table 33. Clay soils represent slow leakage and high retention; sandy soils show sharp breakthroughs. This is useful for comparative purposes. For example, one

Table 32. *Total analysis of soils for metals and free iron oxides*

Soil series	Trace metals ($\mu g\ g^{-1}$)						Fe oxides
	Mn	Co	Zn	Ni	Cu	Cr	(%)
Wagram l s	50	bdl*	40	80	62	bdl	0.6
Ava si c l	360	50	77	110	80	55	4.0
Kalkaska s	80	25	45	50	46	15	1.8
Davidson c	4100	120	110	120	160	90	17.0
Molokai c	7400	310	320	600	260	410	23.0
Chalmers si c l	330	60	100	130	83	68	3.1
Nicholson si c	950	50	130	135	65	68	5.6
Fanno c	280	45	70	100	60	38	3.7
Mohave s l	825	50	85	100	265	18	1.7
Mohave (Ca) c l	770	50	120	120	200	40	2.5
Anthony s l	275	50	55	80	200	25	1.8

*Below detectable limit of <5 ppm.
Source: Reference [45].

Fig. 29. Relative mobility of cations studied.

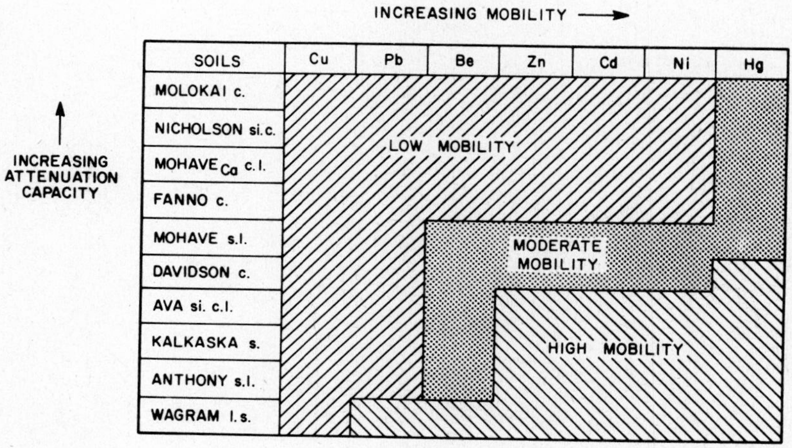

Fig. 30. Relative mobility of anions studied.

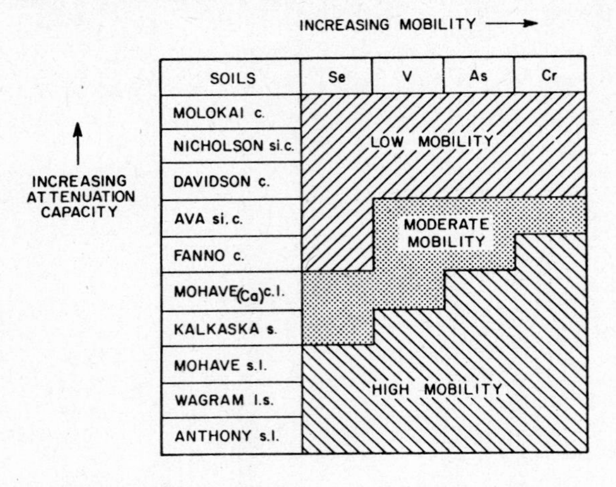

Fig. 31. Relative retention of Ni in four diverse soils.

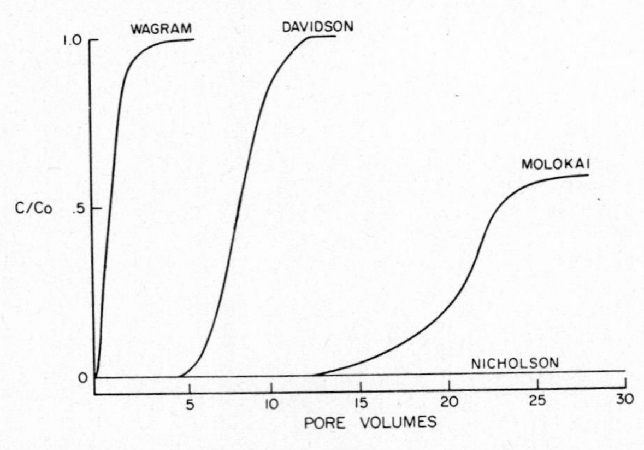

Fig. 32. Relative retention of Ni, Be and Se in Ava si c l.

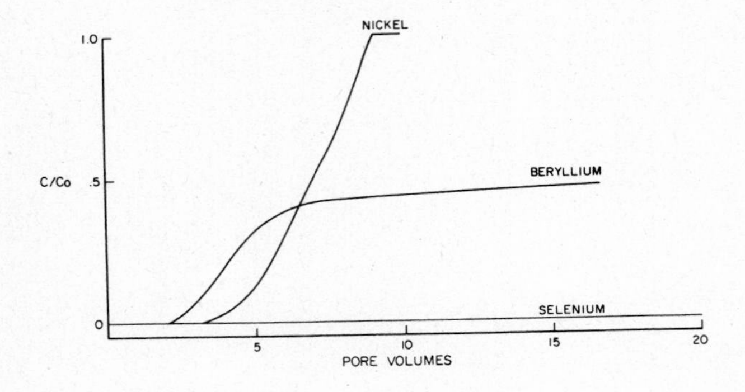

may characterize a soil from a particular location and depth and then determine the most similar soil used in this study. An estimate of the attenuation behavior can then be made.

Predictions using equations

There are a number of mathematical models suggested to describe pollutant movement through soils. A review of simulation models and their application to landfill siting is presented by van Genuchten [128]. Any of these models can be used to describe metal transport or retention in soil. All

Table 33. *Designation showing type of curve generated from each column*

Soil	As	Be	Cd	Cr	Cu	Hg	Ni	Pb	Se	V	Zn
Davidson c	C	D	C	E	E	C	B	E	E	E	D
Molokai c	E	E	E	E	E	D	D	E	E	E	E
Nicholson si c	E	E	E	D	E	B–C	E	E	E	E	D
Fanno c	C	E	E	C	E	C	E	E	E	D	C
Mohave Ca c l	E	E	E	A	E	D	E	E	C	E	E
Chalmers si c l	C	D	D	C	E	D	D–E	E	E	C	D
Ava si c l	C	D	A	C	D	B–C	A	D	E	B	A–B
Anthony s l	A	D	A	A	D	A	A	D	C	A	B
Mohave s l	B	D	D	A	E	B	C	E	C	B	B
Kalkaska s	C	D	C	C	E	C	B	E	C	D	B
Wagram l s	A	C	A	A	E	A	A	B	D	A	A

Source: Reference [84].

Fig. 33. Types of metal retention curves generated by the soil-column technique.

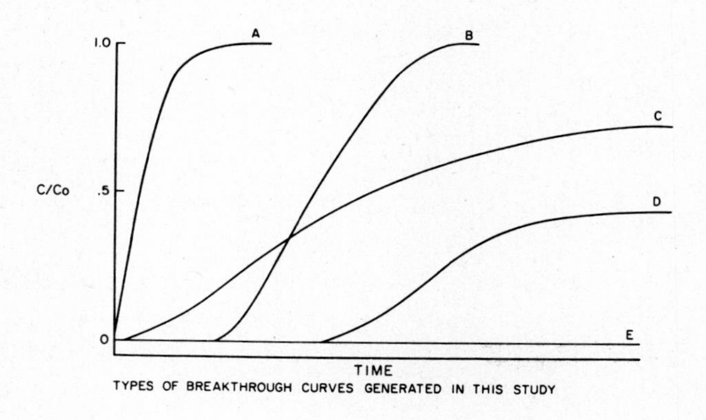

have some unknown parameters which must be evaluated from experimental data. This involves tedious computer time if solutions are complicated. The model and its solution proposed by Lapidus & Amundson [86] has an advantage in that it is relatively simple and has an analytical representation in contrast to others. The Lapidus–Amundson model (L–A) combines all unknown and immeasurable parameters into three unknowns, each of which can be determined from experimental data. The error function model is even simpler than the L–A model and requires much less computer time. Thus we will discuss these two contrasting models along with supporting equations as examples leading to attenuation predictions.

Miscible displacement theories and experimental data indicate that the solute profile is not one of piston displacement but rather is a smooth distribution of concentrations. In addition, different relative concentrations appear to travel at different rates. Since the velocity of a particular concentration ratio is a function of distance (z), it follows that the shape of the experimental breakthrough curve will also be a function of z. However, the model predicts that for $z > 10$ cm the velocity of a particular relative concentration approaches an asymptotic value; i.e., a steady-state (or near-steady-state) velocity for any relative concentration has been achieved by a depth of 10 cm.

Multiple regression analyses

The results of many soil-column tests indicate that there is a significant correlation (0.01) between (a) soil clay content, (b) particle surface area, (c) 'free' iron content and (d) pH and retention of heavy metals as pollutants (Table 19). Because of the lack of strong dominance of any clay (< 2 μm) the type of clay mineral (montmorillonite-like, mica-like, or kaolin-like) did not improve the statistical r^2 (Table 20).

The properties of the fluid or solvents carrying the pollutant also influence soil attenuation (Fig. 34) [1]. The soil and leachate properties having most effect on migration rates are regressed with the migration rates in the following way: the regression equation is written one variable at a time with all variables available to bring into the equation at each step. The first variable in the equation is the one which does the most to explain the variation in the migration rates. The second variable to enter the equation is the one which does the most to explain the variation in the migration rates not already explained by the first variable. Variables continue to be added in this way until either all variables are in the equation or until none of the unused variables are statistically significant in explaining the remaining variation in the migration rates.

Once the parameters for a specific soil and a specific element are determined, the relative concentration (i.e., $C/C_0 = C$) can be identified. Our attempt, however, is to determine the migration rate of a particular relative concentration. The question is, knowing all the parameters (diffusion coefficient, adsorption parameters, porosity, flux) for a given soil and a given element, how long it will take for a particular relative concentration, C, to reach a given depth, say groundwater. For this solution we go to the computer and the mathematical models exemplified by Lapidus & Amundson [86] and the error function equations.

Lapidus–Amundson model

By applying the Lapidus–Amundson model, the net effect of the chemical nature of the leachate or solvent on the forward and backward reaction rates at a multitude of different adsorption sites is combined into

Fig. 34. Influence of total organic carbon concentration, salt, and iron on the migration rate of Fe from Leachate II through Kalkasa sand. (Source: Reference [1].)

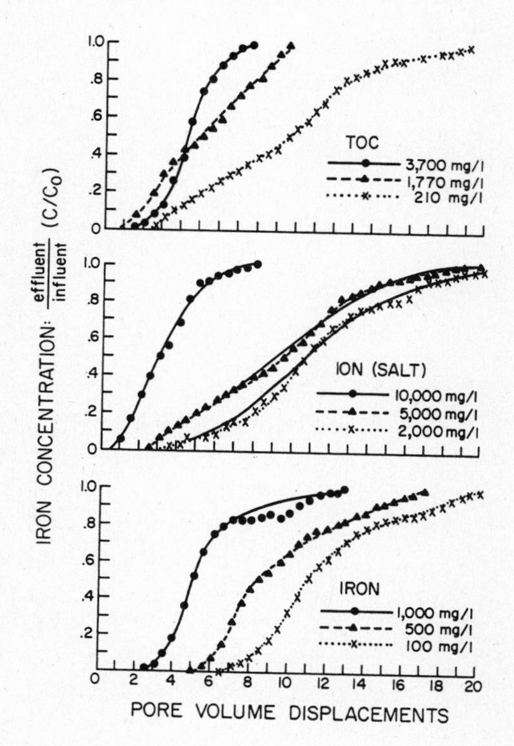

an effective forward and backward reaction rate for the particular conditions of the soil–fluid system in the experiment. Similarly, an effective diffusion coefficient is determined, or a diffusion–dispersion coefficient is determined since the effects of diffusion–dispersion combine in one variable.

The Lapidus–Amundson mathematics

$$\frac{\delta C}{\delta t} + \frac{1}{\alpha}\frac{\delta n}{\delta t} = D\frac{\delta^2 C}{\delta z^2} - v\frac{\delta C}{\delta z}, \tag{1}$$

where:

C = concentration in soil solution (M/L^3)
v = convective (pore water) velocity (L/T)
D = diffusion coefficient (L^2/T)
α = fractional pore volume of soil (L^3/L^3)
n = amount adsorbed per unit volume of soil (M/L^3)
z = vertical distance (L)
t = time (T)

The term $\delta n/\delta t$ describes a linear, nonequilibrium adsorption and is assumed to be

$$\delta n/\delta t = K_1 C - K_2 n \tag{2}$$

where

K_1 and K_2 are forward and backward reaction rates (L/T), respectively.

The boundary and initial conditions for a steady input concentration (C_0) are:

$$C = C_0,\ z = 0, t > 0 \tag{3a}$$

$$\left.\begin{matrix} C = 0 \\ n = 0 \end{matrix}\right\}, t = 0,\ z > 0 \tag{3b}$$

The solution to Eqs. (1) and (2) subject to (3a) and (3b) is

$$C/C_0 = \exp(vz/2D)[F(t) + K_2\textstyle\int_0^t F(t)dt] \tag{4}$$

where

$$F(t) = [\exp(-K_2 t)]\int_0^t \Big\{ I_0\,[2\sqrt{(K_1 K_2 \beta(t-\beta)/\alpha)}]$$
$$\exp[-(z^2/4D\beta + \beta d)]\cdot[z/2\sqrt{(\pi D\beta^3)}]\Big\} d\beta]$$

with $d = v^2/4D + K_1/\alpha - K_2 \cdot I_0$ as the modified Bessel function of the first kind of order zero. The term C/C_0 is referred to as relative concentration, C, and is a function of z and t, and contains the parameters v, α, D, K_1 and K_2.

Lapidus–Amundson equations

The resulting equations developed from this model have highly statistically significant r values for explaining the variations in the migration rates through soils of certain heavy metals found in aqueous leachates. Some examples for Cd, Ni, and Zn, where $C/C_0 = 0.5$, are:

Cd

$$\text{V.5} = \frac{v}{25}[29.9085/\text{CLAY}^* + 0.001\,085 \times (\text{CLAY} + \text{SILT})^2 + 8.532/\text{FeO} + 0.010\,15 \times \text{FeO}^2 + 84.132 \times \text{SALTS} - 205.127 \times \text{SALTS}^2 + 0.4422 \times \text{TOC} - 12.6058].$$

$r^2 = 0.84.$

Ni

$$\text{V.5} = \frac{v}{25}[13.460/\text{CLAY} + 23.8496/\text{SAND} + 6.2006/\text{FeO} + 0.303\,43 \times \text{FeO} + 69.321 \times \text{SALTS} - 222.939 \times \text{SALTS}^2 - 34.711 \times \text{TOC} + 114.796 \times \text{TOC}^2 - 5.5341].$$

$r^2 = 0.89.$

Zn

$$\text{V.5} = \frac{v}{25}[0.0136 \times \text{SAND} + 0.000\,526\,4 \times \text{SILT}^2 + 2.6371/\text{FeO} + 0.002\,88 \times \text{FeO}^2 - 14.7461 \times \text{TOC} + 77.7905 \times \text{TOC}^2 + 54.6561 \times \text{SALTS} - 148.2266 \times \text{SALTS}^2 - 5.2667].$$

$r^2 = 0.85.$

* Variables, such as, sand, silt, clay, FeO, salts are expressed as percentages.

Error function model

Error function mathematics

Another set of equations, much simpler, called *error functions* [31], has been found to provide similar predictive mathematical solutions at a much lower computer cost. The error function equation (E–F) is represented by:

$$C/C_0 = 0.5\ erfC\left[\frac{z - Avt}{\sqrt{(4D't)}}\right]$$

where

z is depth;
v is pore water velocity;

t is time;
A is a retardation factor;

and

D' is modified diffusion coefficient.

The factor A explains the position of $C/C_0 = 0.5$ in the profile and could be used to estimate the position of the breakthrough curve. The factor D' can give an estimate of the spread of the breakthrough curve. The E–F equation has the advantage of being simple and the calculations can be performed with accuracy. The adsorption–desorption is described implicitly by the factors A and D'. The parameters A and D' could be estimated for each case and then regressed against the soil and leachate properties.

Error function equations

The resulting equations developed from the error function model yields the following for Ni, Cd, Cr, and Zn:

Ni

$$\text{V.5}^* = \frac{v}{25}[19.353/\text{CLAY} + 27.785/\text{SAND} + 6.932/\text{FeO} + 0.331\,423 \times \text{FeO} + 73.6605 \times \text{SALTS} - 235.662 \times \text{SALTS}^2 - 34.1875 \times \text{TOC} + 121.32 \times \text{TOC}^2 - 6.7302].$$

$r^2 = 0.92.$

Cd

$$\text{V.5} = \frac{v}{25}[31.16/\text{CLAY} + .001\,145\,(\text{CLAY} + \text{SILT})^2 + 9.8047/\text{FeO} + .011\,635 \times \text{FeO}^2 + 83.4078 \times \text{SALTS} - 197.277 \times \text{SALTS}^2 + 0.470 \times \text{TOC} - 13.3655].$$

$r^2 = 0.84.$

Cr

$$\text{V.5} = \frac{v}{25}[52.762/\text{SAND} + .8898 \times (\text{CLAY} + \text{SILT}) + 19.915 \times (\text{SAND} \times \text{TOC}) + 1.491\,17 \times (\text{SAND} \times \text{SALTS}) + 0.465\,97 \times (\text{SILT} \times \text{SALTS}) - 8.104 \times \text{SALTS}^2 - 33.925].$$

$r^2 = 0.74.$

Zn

$$V.5 = \frac{v}{25}[0.020\,42 \times SAND + 0.000\,65 \times SILT^2 + 2.595\,08/FeO + 0.004\,11 \times FeO^2 - 13.9566 \times TOC + 73.7746 \times TOC^2 + 53.3002 \times SALTS - 142.6462 \times SALTS^2 - 5.429\,13].$$

$r^2 = 0.85.$

* V.5 is C/C_0 at 0.5.

Application of models

Predictions of metal attenuation

To use the Lapidus–Amundson and error function equations:

(1) estimate Clay, Silt, & Sand (%);
(2) estimate TOC of leachate (%);
(3) estimate total ion content of the leachate (%): this is the sum of soluble salts and iron in solution;
(4) estimate free iron oxide content (%);
(5) estimate the pore velocity of the leachate (cm/day) (the Darcian velocity divided by the leachate-filled porosity);
(6) determine the particular relative concentration;
(7) substitute the quantities from steps 1–5 into the equation determined from step 6.

For example, suppose the soil has the characteristics: Clay content = 15%, silt content = 20%, free iron oxide content = 1.5%. Suppose the leachate has total soluble ions 0.08% and total organic carbon 1.1%. If the porosity of the soil was 0.38 and the infiltration rate was 0.6 cm per day then the pore velocity would be 0.6/.38 = 1.6 cm per day. Assuming an initial Cd concentration in the leachate is 2 ppm and the concentraction limit is 1 ppm, then the relative concentration of interest is 0.5. Substitution into the L–A equation for V.5 yields

$$V.5 = \frac{1.6}{25}[29.9085/(15) + .001\,085(35^2) + 8.532/(1.5) + .01015(1.5^2) + 84.131(.08) - 205.127(0.08^2) + 0.4421(1.1) - 12.6058] = 0.149 \text{ cm per day.}$$

Using the error function model,

$V.5 = 0.104$ cm per day.

To summarize up to this point, we have plotted the data from all of the column data and have matched theoretical curves to the observed curves to get individual values for D, K_1 K_2 for each column. At this point the only data that have been used are values of α, v, C_0, and the experimental breakthrough curve from the column. And for selected relative concentration we have migration rates for each column. To broaden these results from the laboratory columns to a more general setting and also to display the results in an easily accessible form, we make use of multiple regression equations: one regression equation for each of the selected relative concentration.

Limitations and advantages of the mathematical models

Limitations: The L–A and E–F model equations must be used with caution. First, these approximations were developed on the basis of homogeneous soil columns and not under field conditions where heterogeneity dominates the soil profile and vadose area as a whole. Second, although the regression of velocity with both the soil and leachate properties have an r^2 value of high order, considerable variation in the actual velocity values and those predicted by the regression equations are still present. Here caution should be taken when these regression equations are required for highly accurate predictions of metal movement. Rather, they should be used to gain some knowledge as to which soil or leachate properties contribute or detract from metal transport velocity. Third, the models presume attenuation or migration rates of soluble constituents through soils contained primarily in *aqueous* solutions such as MSW leachates from landfills. Additional investigations are necessary to verify their use with pollutants in other solvents, although there is no good reason to believe they can not apply equally well to most soluble solutes regardless of solvents.

Advantages: Real advantages in the use of these models exist. First, both mathematical models provide the first and only real attempt to quantify attenuation of hazardous pollutants by soil. Therefore, they satisfy, in part at least, a great need as a field-oriented tool of predictive value. Thus we are put much closer to a sound management program for safe and secure waste disposal. Second, the models do not require infinite information of attenuation mechanisms which would need many years to discover and evaluate. Third, the values in the equations (e.g. D, K_1, and K_2) have been derived as real values from research experimental data making it possible to apply the models to actual land disposal problems for attenuation of potentially hazardous metals. The values represented in the models (e.g. D, K_1, and K_2) are real, beyond arm-chair theory. Fourth, these models will be applied in their present form to the evaluation of disposal sites because they:

are independent of computers, as far as the ultimate user is concerned;

yield products in the form of mathematical equations or tables which are much easier to use;
are less demanding for data than other prominent simulation model approaches;
are no more limited by the quality of input data than other approaches and (for a given site) much less limited by quantity of input data than other approaches;
are most likely to be used in evaluating relative migration of a single metal at several alternative sites. (Because of extrapolation problems, no claim of predicting exact movement rate in heterogeneous soil is made. However, the model is very likely to estimate accurately metal movement rate in homogeneous soils as might be found when landfills are lined with clay-soil mixtures or with native soils.)

Fifth, the models are concerned with nonconservative solutes (such as metals) and nonconventional leachates as well as raw, unenriched MSW leachates. Sixth, they do not conflict with the large hydrology modeling programs of the United States Geological Survey. Seventh, they are conveniently and accurately related to easily measurable physical and chemical properties of the soil–leachate system.

Desorption of attenuated pollutants

Quantitative information on desorption of soil-attenuated pollutants is conspicuous by its absence. The discussion in this chapter concentrates on the retention of pollutants up to the breakthrough point (i.e., effluent (C)/influent (C_0) = 1). The backside of the breakthrough curve, desorption, has yet to be characterized. Rouston & Wildung [109] suggested that desorption reactions of pollutants limit the permanency of land disposal, but quantitative data are still lacking. Using extraction methods, Hodgson [72] showed that heavy metal sorption is not completely reversible. He believes the nonextractability of sorbed cobalt with which he was working was due to interlattice penetration of clay minerals. Whatever the mechanism(s) involved, depending upon the capacity and thermodynamic favorability of competing sorbed reactions, an extractant would remove different percentages of sorbed material from varying depths in a soil column or profile.

One way, therefore, to evaluate the permanence of ion attenuation is by relating it to the efficiency of extraction techniques. The availability of trace metals for plant growth has long been estimated by extraction [133, 85]. Mild extractants used reflect plant availability better than total analyses [74]. To determine what portion of attenuated metal ions are subject to leaching by rainwater, extraction with water would be useful.

Extractions with harsher reagents also can provide important information. Simple 'input–output' data do not completely describe a pollutant's migration. A major shortcoming is encountered if none of the element passes through the soil column during the experiment or breakthrough cannot be achieved. It is then impossible to determine how the soil's characteristics are affecting the pollutant's movement. In any case, knowledge of the distribution of a sorbed element in a profile or soil column yields greater insight into the soil's capacity for retention of the element. Quantitative interpretation of extraction data, however, must be made cautiously.

To obtain some information on desorption, experimentation was undertaken using the homogeneous soil-column technique (Fuller [42]) with MSW landfill leachates enriched with As, Cd, Cr, Cu, Hg, Pb, Ni, Se, V and Zn. Extraction with water yields water-soluble, very loosely retained trace metal and may be expected to simulate natural leaching with rainwater. The acid solution enables sufficient extraction from each segment to obtain a profile showing migration of the element in the column. Since natural soils have been shown by Fuller & Korte [54] to contain indigenous extractable heavy and trace metals, all soils were extracted with water and 0.1 N HCl similar to those receiving landfill leachate to establish a background tare for deduction from the amounts of metals removed from the column segments.

The general appearance of the results obtained from the 0.1 N HCl extraction is illustrated in Figs. 35 and 36, showing the amount of metal extracted from each segment as a fraction of the greatest amount of metal extracted from any segment (usually the top segment). These curves reflect the type of data obtained from the leaching experiment. Cadmium was not detected in any of the effluent from the Molokai soil as evidenced by no Cd being extracted from the last half of the Molokai column (Fig. 35). The breakthrough (effluent conc. = influent conc.) of Cd through Wagram is typified by the plateau shown by the extraction profile. A third case is illustrated by Davidson c where the concentration in the effluent had leveled off or was slowly increasing. The extraction profile shows a slow decrease in extractable Cd as depth increases. This range of behavior was followed for nearly all of the elements and could be observed not only for a particular element but for an individual soil (Fig. 36). Arsenic was fully retained by Davidson clay, Zn was quite mobile, and Cr reached a steady state.

Cu and Pb were strongly retained by the soils in this study. Usually, these elements were not detected in the soil-column effluents over the course of the experiment. Extraction profiles can be used to observe differences in behavior of the individual soils toward this element. The data indicate that soil texture is the dominant feature controlling attenuation of Cu and Pb.

Fig. 35 Profile distribution of Cd extracted with 0.1 N HCl from soils receiving MSW landfill leachate enriched with Cd.

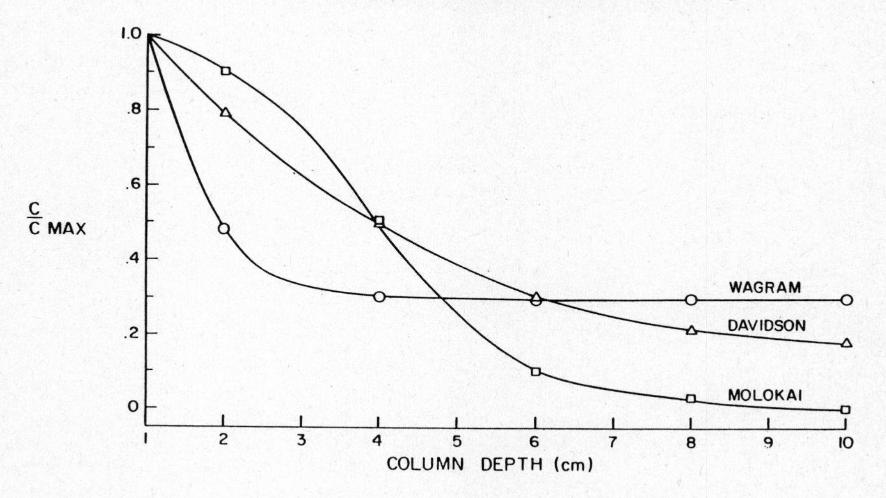

Fig. 36. Profile distribution of As, Cr, Zn, extracted with 0.1 N HCl from Davidson clay receiving MSW landfill leachate enriched with As, Cr and Zn.

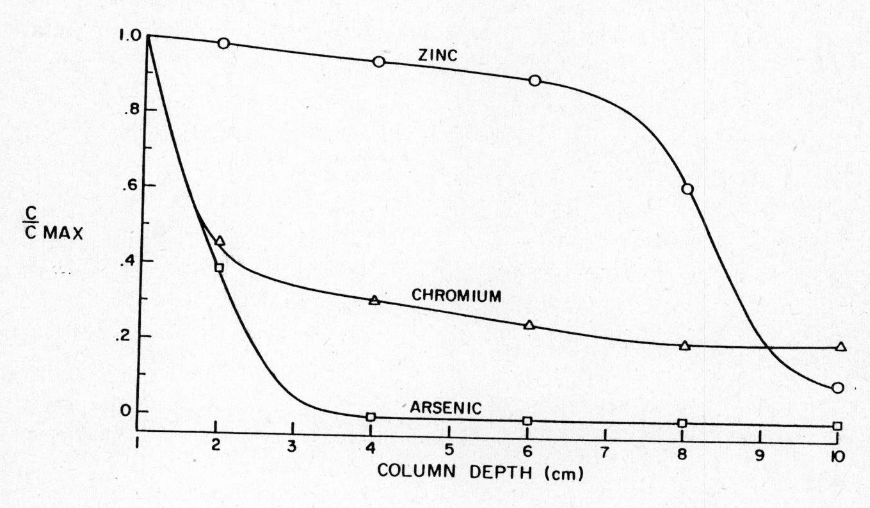

Fig. 37 shows representative extraction data for Cu. Migration was deeper in the column for those soils coarsest in texture. Copper was extracted as deep as 6 cm from Wagram, 4 cm from Mohave, and 3 cm from Fanno (Fig. 37). These data provide qualitative evidence that Cu follows trends similar to the other divalent cations as described in the discussion of the leaching phase of this experiment.

Fig. 37. Profile distribution of Cu extracted with 0.1 N HCl from three soils receiving MSW landfill leachate enriched with Cu.

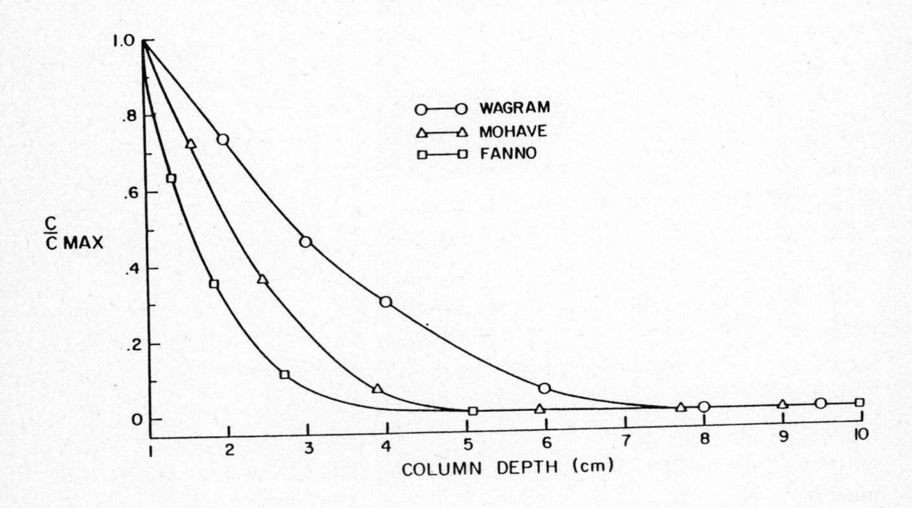

Fig. 38. Illustration of adsorption and desorption of Fe of natural MSW landfill leachate by Anthony sandy loam.

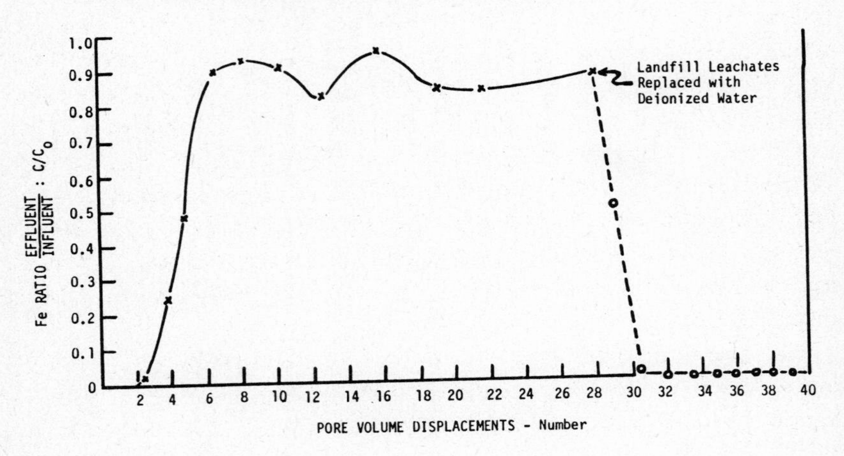

The percentages extracted by both water (Table 34) and acid (Table 35) are based on the total amount of metal retained by the column during the leaching study. The order of extractability for the elements is generally V > Se > As > Cr > Zn > Ni > Cd > Hg > Cu > Pb. Copper and Pb are the least mobile elements and the least likely to be extracted by water once they are adsorbed. The elements which are most water soluble are those sorbed as anions: V, Se, As and Cr. Unfortunately, data of this sort are difficult to relate to concentrations which would be desorbed by leaching. It can be emphasized, however, that of the total amount extracted by water from these soil columns, virtually all was taken from the top half or higher in the column. Much of this material would be readsorbed as it traveled down the column.

The order of extractability by 0.1 $\underline{N}$ HCl is Cd > Zn > Ni > V ⩾ Pb > Cu > Hg > Cr > As. There is no apparent trend. Of more importance is the large amount of material which could be extracted. In many instances all sorbed metals were extracted by the acid. This acid is stronger than what would be seen in the field but one can conclude that the more acidic the leaching solution, the more trace metal will be desorbed.

Data presented thus far are not adequate to accurately predict the potential for desorption of sorbed trace metals under all disposal site locations. This may be due in part to variations in the amount adsorbed. This creates great difficulty in comparing separate experiments. A better method for studying potential for desorption would be to leach columns with other leachate extractants after passage of solutions containing trace metals.

An example of this technique is seen in Fig. 38 which is a plot of the attenuation of Fe from MSW landfill leachate by Anthony sandy loam followed by a plot of C/C_0 after replacing the leachate with deionized water. Iron was not removed from the soil by water. Only dilution of the soil-saturated MSW leachate Fe occurred as observed in the dotted line during the 28–30 pore volume displacements. Thereafter, none of the attenuated Fe could be removed from the soil by the additional ten pore volume displacements. The retention of Fe was therefore complete, 100%. One must conclude that, in this soil, the naturally occurring landfill leachate Fe that was attenuated has a great affinity for the soil and that Fe has the capacity to resist desorption by rainwater for many pore volume displacements.

The potential for desorption of trace metals is quite variable. The amount extracted with water for an element among this group of soils can vary by one order of magnitude. The amount of material leached by water

Table 34. *The amount of sorbed metal extracted from 11 different soils by water*

Soil series	Metal desorbed									
	As	Cd	Cr	Cu	Hg	Ni	Pb	Se	V	Zn
	%*									
Davidson c	0.5	1	4	1.5	2.2	1	0	3	0.4	3
Molokai c	0	0.5	1	0.1	0	1.1	0	3	0.7	2
Nicholson si c	2	1	10	0	0	0.2	0.2	0.5	1.5	1
Fanno c	4.5	1	1	3	0.5	2	0.2	0	8	1.2
Mohave (Ca) c l	11.5	0.2	6	0.2	0	0.1	0.1	6	23	0.1
Chalmers si c l	—	—	—	—	0.5	3	—	—	3.8	—
Ava si c l	3	5	2	0.3	0	6	0.7	5	0	5
Anthony s l	0	3.5	0	3	9	18	0	1.8	28	2.8
Mohave s l	0	2	0	0.5	0	4	0.4	3	0.3	1.5
Kalkaska s	2.5	2	0.5	0.5	0	18	0.3	9	7.5	5
Wagram l s	0	9	0	0.5	1.6	7	0	1.7	9	8

Source: Reference [84].

Table 35. *The amount of sorbed metal extracted from 11 different soils by 0.1 N HCl*

Soils	Metals desorbed								
	As	Cd	Cr	Cu	Hg	Ni	Pb	V	Zn
	%*								
Davidson c	5	77	7	67	49	43	70	36	42
Molokai c	0	87	5	57	59	78	59	35	100
Nicholson si c	11	75	13	27	24	63	47	55	64
Fanno c	24	80	8	44	17	58	58	89	66
Mohave (Ca) c l	20	73	17	51	49	21	47	77	41
Chalmers si c l	—	—	—	—	22	57	—	80	—
Ava si c l	11	85	7	40	16	72	78	75	64
Anthony s l	12	100	100	68	23	100	93	63	100
Mohave s l	11	89	7	55	30	80	77	93	100
Kalkaska s	18	100	16	67	31	100	70	48	100
Wagram l s	11	90	8	60	14	100	41	100	100

*Extraction percentages are based on the total amount of metal retained by the soil in the column.
Source: Reference [84].

is not expected to be a serious migration problem. On the other hand, acidic leachates applied to a disposal site can mobilize large quantities of sorbed trace metals, even when preceded by a complete drying out of the site.

This desorption research, however, is a beginning in a possible characterization of a given soil and a single element for 'leaching' adsorbed (retained) hazardous polluting trace elements to underground water sources. It may be used for broad predictive purposes. Like the early research attempting to characterize soil P as to its 'availability' to plants, this extraction study needs considerably more attention before accurate predictions of solubility and mobility relationships can be fitted into a universally practical program. Indeed, some of the same factors influencing P fixation (retention) such as lime content in the West and iron and aluminium fixation in the East must come into play with the trace elements. Organic matter association from leachates, waste streams, soil, etc., must also be evaluated. A fruitful study would involve more dilute acid extraction, buffered salt solutions, bicarbonates, organic acids, and others.

Before embarking on trace-element desorption and solubilization soils research, one must be realistic enough to recognize that answers will come even more slowly than those for the plant nutritionist and soil fertility expert. The vehicle in which waste disposal constituents enter soils, in contrast to the purity of rainwater facing the grower, is confounded with a wide variety of leachates, wastewaters, industrial waste streams, etc., which in themselves confound further the myriad of chemical, biological, and physical reactions that may take place on land and in soils.

References

[1] Alesii, B.A. & W.H. Fuller. 1976. 'The mobility of three cyanide forms in soil', *in*: W.H., Fuller (ed.), *Residual management by land disposal*. Hazardous Environmental Protection Agency, Cincinnati, OH 45268.

[2] Alesii, B.A., W.H. Fullter & M.V. Boyle. 1980. 'Effect of leachate flow rate on metal migration through soil', *J. Environ. Qual.*, **9**, 119–26.

[3] Anderson, D. & K.W. Brown. 1981. 'Organic leachate effects on the permeability of clay liners', *in*: Proc. Seventh Ann. Res. Symp, USEPA–600/9–81–002b. US Environmental Protection Agency, Cincinnati, OH 45268, pp. 119–30.

[4] Artiola-Fortuny, J. & W.H. Fuller. 1980. 'Trace metals in feedlot manure', *Compost Sci.*, vol. 21, no. 3, pp. 30–4.

[5] Artiola-Fortuny, J. & W.H. Fuller. 1982. 'Phenols in municipal solid waste leachates and their attenuation by clay soils', *Soil Sci.*, **131**, 18–26.

[6] Artiola-Fortuny, J. & W.H. Fuller. 1982. 'Adsorption of some monohydroxybenzene derivatives by soil', *Soil Sci.* (in press).

[7] Artiola-Fortuny, J. & W.H. Fuller. 1982. 'Humic acids in municipal solid waste leachates', *J. Environ. Qual.* (in press).

[8] Aubet, H. & M. Pinta. 1977. *Trace elements in soil*. Elsevier Scientific Publishing Co., New York.

[9] Baham, J., N.B. Ball & G. Sposito. 1978. 'Gel filtration studies of trace metal–fulvic acid solutions extracted from sewage sludges', *J. Environ. Qual.*, **7**, 181–8.

[10] Banerjee, Dilip K., R.H. Bray & S.W. Melsted. 1953. 'Some aspects of the chemistry of cobalt in soils', *Soil Sci.*, **75**, 421–31.

[11] Barsdate, R.J. & W.R. Matson. 1966. 'Trace metal in arctic and subarctic lakes with reference to the organic complexes of metals', Proc. Intnat's Symp. Radioecoelogical Concen. Process Stockholm, Pergamon Press, Oxford.

[12] Bitterli, Ronda. 1981. 'Freezing and drying: effects on the solubility of municipal solid waste leachate constituents', MS Thesis. University of Arizona Library, University of Arizona, Tucson, AZ 85721.

[13] Bohn, H.L., B.L. McNeal & G.A. O'Connor. 1979. *Soil Chemistry*. John Wiley, Somerset, NJ 08873 329 pp.

[14] Bolt, G.H. & M.G.M. Bruggenwert (eds.). 1978. *Soil chemistry A: Basic elements*. Elsevier Scientific Publishing Co., New York.

[15] Bremner, J.M. 1965. 'Organic nitrogen in soils', pp. 93–149, *in*: W.V. Bartholomew & F.E. Clark (eds.), *Soil Nitrogen*, Am. Soc. of Agron., Madison, W1.

[16] Broadbent, F.E. & G.R. Bradford. 1952. 'Cation-exchange groupings in the soil organic fraction', *Soil Sci.* **74**, 447–57.

[17] Brown, A.L. & J.J. Jurinak. 1964. 'Effect of liming on the availability of zinc and copper', *Soil Sci.* **98**, 170–3.

[18] Brown, K.W. & David Anderson. 1980. 'Effect of organic chemicals on clay liner permeability', *in*: D.W. Schultz (ed.), Proc. Sixth Ann. Res. Symp. Chicago, IL, USEPA–600/9–80–010. US Environmental Protection Agency, Cincinnati, OH 45268, pp. 123–34. PB 80–175086.

[19] Brown, K.W. and Associates, Inc. 1980. 'Hazardous waste land treatment', US Environmental Protection Agency. SW–874. OWWM. Washington, DC 20460.

[20] Brunner, D.R. & R.H. Carnes. 1974. 'Characteristics of percolate of solid and hazardous wastes deposits', presented at American Water Works Association, 4th Ann. Conf., Boston, MA, 23 pp.

[21] Budzynski, James. 1980. 'Metal extractability and barley seedling metal accumulation from four municipal sewage sludges', MS Thesis. University of Arizona Library, University of Arizona, Tucson, AZ 85721.

[22] Chian, E.S.K. & F.B. DeWalle. 1977. 'Evaluation of leachate treatment', vol. 1, 'Characterization of leachate', USEPA–600/2–77–186*a*. US Environmental Protection Agency, Cincinnati, OH 45268. 210 pp.

[23] Chian, E.S.K. & F.B. DeWalle. 1977. 'Evaluation of leachate', vol. 2. 'Biological and physical–chemical process', USEPA–600/2/77–186*b*. US Environmental Protection Agency, Cincinnati, OH 45268. 245 pp.

[24] Christman, R.F. 1970. 'Chemical structures of color producing organic substances in water', *in*: D.W. Hood (ed.), Symposium on Organic Matter in Natural Waters, University of Alaska, Inst. Marine Sci. Accos. Bul. 1.

[25] Clark, F.E. 1967. 'Bacteria', pp. 1–49. *in*: A. Burges and F. Raw (ed.) *Soil biology*, Academic Press, Inc., New York, NY.

[26] Cleveland, J.M. & T.F. Rees. 1981. 'Characterization of plutonium in Maxy Flats, radioactive trench leachate', *Sci.*, **212**, 1506–9.

[27] Crook, W.M. 1956. 'Effect of soil reaction on uptake of nickel from a serpentine soil', *Soil Sci.*, **9**, 113.

[28] DeWalle, F.B. & Chian, E.S.K. 1974. 'Kinetics of formation of humic substances in activated sludge systems and their effect on flocculation', *Biotechnology and Bioengineering*, **16**, 739–61.
[29] Diamond, S. 1970. 'Pore size distribution in clays', *Clays, Clay Minerals*, **18**, 7–24.
[30] Dixon, J.B. & S.B. Weed (eds.). 1977. *Minerals in soil environment*. Soil Sci. Soc. Am., Madison, WI.
[31] Dwight, H.B. 1961. *Tables of intergers and other mathematical data*, 4th edn. The Macmillan Co., New York, NY.
[32] Ellis, B.G. & B.D. Knezek. 1972. 'Adsorption reactions of micronutrients in soils', pp. 59–68, *in*: J.J. Mortvedt, P.M.G. Giordano & W.L. Lindsay (eds.), *Micronutrients in agriculture*, Soil Sci. Soc. Am., Madison, WI.
[33] Ellis, B.G. 1973. 'The soil as a chemical filter', pp. 46–70, *in*: W.E. Sopper & L.T. Kardos (eds.), *Recycling treated municipal waste water and sludge through forest and cropland*. Penn. State Press, University Park, PA.
[34] Farquar, G.J. & F.A. Rovers. 1976. 'Leachate attenuation in undisturbed and remolded soils', *in*: E.J. Genetello & J. Cirello (eds.), 'Gas and leachate from landfills: formation, collection, and treatments', Res. Symp., March 1975. New Brunswick, NJ. USEPA–600/9–76–004. US Environmental Protection Agency, Cincinnati, OH 45268, pp. 54–70.
[35] Federal Register. 1978. 'Hazardous waste-proposed guidelines and regulations and proposal on identification and listing', vol. 43, no. 243. Washington, DC.
[36] Fields, M. & R.K. Schofield. 1960. 'Mechanism of ion adsorption by inorganic soil colloids', *New Zealand J. Sci.*, **3**, 563–79.
[37] Fraser, D.C. 1961. 'A syngenetic copper deposit of recent age', *Econ. Geol.*, **56**, 951–62.
[38] Fuller, W.H. 1975. *Soils of the desert southwest*. University of Arizona Press, Tucson, AZ 85721. 102pp.
[39] Fuller, W.H. 1975. *Management of soils of the southwest desert*. Univeristy of Arizona Press, Tucson, AZ 85621. 195 pp.
[40] Fuller, W.H. & A.D. Halderman. 1975. 'Management for the control of salts in irrigated soils', Ext. Serv. A–42 Bul., University of Arizona Agr. Expt. St. and Coop. Ext. Serv., Tucson, AZ.
[41] Fuller, W.H. 1977. 'Movement of selected metals, asbestos, and cyanide in soil: application to waste disposal problems', USEPA–600/2–77–020. US Environmental Protection Agency, Cincinnati, OH 45268. 257 pp.
[42] Fuller, W.H. 1978. 'Investigation of landfill leachate pollutant attenuation by soils', USEPA–600/2–78–158. US Environmental Protection Agency, Cincinnati OH 45268, 219 pp.
[43] Fuller, W.H. 1978. 'Soil-waste interactions', pp. 79–102, *in*: *Disposal of industrial and oily sludges by land cultivation*. Res. and Systems Management Assoc., Northfield, NJ 08225.
[44] Fuller, W.H. 1980. 'Premonitoring waste disposal sites', 1978. *in*: Am. Water Res. Assoc. Symp. Proc., 'Establishment of water quality monitoring programs', San Francisco, CA, June 1978. Minneapolis, MI 55414, 1980.
[45] Fuller, W.H. 1980. 'Soil modification to minimize movement of pollutant from solid waste operation', CRC *Critical Rev. Environ. Control*, **9**, 213–70. CRC Press, Inc., Boca Raton, FL 33431.
[46] Fuller, W.H. 1981. 'Liners of natural porous materials to minimize pollutant migration', USEPA–600/S2–81–122. US Environmental Protection Agency, Cincinnati, OH 45268. PB 81–221 863.
[47] Fuller, W.H. 1982. (in press).

[48] Fuller, W.H., B.A. Alesii & G.E. Carter. 1979. 'Behavior of municipal solid waste leachate: I. Composition variations', *J. Environ. Sci. Health*, vol. A14, no. 6, pp. 461–85.

[49] Fuller, W.H. & B.A. Alesii. 1979. 'Behavior of municipal solids waste leachate: II. In soil', *J. Environ. Sci. Health*, vol. A14, no. 7, pp. 559–92.

[50] Fuller, W.H., A. Amoozegar-Fard, E.E. Niebla & M. Boyle. 1980. 'Influence of leachate quality on soil attenuation of metals', *In* Proc. Sixth Ann. Res. Symp., March 17–20, 1980, Chicago, IL. David Schultz (ed.), USEPA 600/9–80–010, US Environmental Protection Agency, Cincinnati, OH, 291 pp. (pp. 108–117).

[51] Fuller, W.H., Colleen McCarthy, B.A. Alesii, & Elvia Niebla. 1976. 'Liners for disposal sites to retard migration of pollutants', *in*: 'Residual Management by Land Disposal', Proc. Hazardous Waste Res. Symp., Feb. 1976, Tucson, AZ. W.H. Fuller (ed.), USEPA–600/9–76–015. US Environmental Protection Agency, Cincinnati, OH 45268. 212–26 pp.

[52] Fuller, W.H. & J.E. Hardcastle. 1967. 'Relative absorption of strontium and calcium by certain algae', *Soil Sci. Soc. Am. Proc.*, **31**, 772–4, Nov.–Dec.

[53] Fuller, W.H., J.E. Hardcastle, R.J. Hannapel & Shirley Bosma. 1966. 'Calcium-45 and strontium-89 movement in soils, and uptake by barley plants as affected by $Ca(Ac)_2$ and $Sr(Ac)_2$ treatment of the soil', *Soil Sci.*, **101**, 472–85.

[54] Fuller, W.H., N.E. Korte, E.E. Niebla, & B.A. Alesii. 1976. 'Contribution of the soil to the migration of certain common and trace elements', *Soil Sci.*, vol. 122, no. 4, pp. 223–35.

[55] Fuller, W.H., & M. L'Annunziata. 1968. 'Movement of algal- and fungal-bound radiostrontium as chelate complexes in a calcareous soil', *Soil Sci.*, **107**, 223–30.

[56] Gadde, R.R. & H.A. Laitinen. 1974. 'Studies of heavy metal adsorption by hydrous iron and manganese oxides', *Anal. Chem*, **46**, 2022–26.

[57] Gamble, D., M. Schnitzer & I. Hoffman. 1970. 'Cu^{2+} – fulvic acid chelation equilibration in 0.1M KCl at 25°C', *Can. J. Chem.*, **48**, 3197–204.

[58] Garland, G.A. & D.C. Mosher. 1974. 'Leachate effects of improper land disposal', *Waste Age*, **5**, 11.

[59] Geering, H.R. & J.F. Hodgson. 1969. 'Micronutrient cation complexes in soil solution: Vol. III. Characterization of soil solution ligands and their complex with Zn^{2+} and Cu^{2+}', *Soil Sci. Soc. Am. Proc.*, **33**, 54–9.

[60] Griffin, R.A. & N.F. Shimp. 1978. 'Attenuation of pollutants in municipal landfill leachate by clay minerals', USEPA–600/78–157. US Environmental Protection Agency, Cincinnati, OH 45268.

[61] Grimm, R. E. 1953. *Clay minerology*. McGraw–Hill Book Co., Inc., New York, NY.

[62] Griffin, R.H., R.R. Frost & N.F. Shimp. 1976. 'Effect of pH on removal of heavy metals from leachate by clay minerals', pp. 259–68, *in*: W.H. Fuller (ed.), 'Residual management by land disposal', Proc. Hazardous Waste Res. Symp., Feb. 1976, Tucson, AZ, USEPA–600/9–76–015. US Environmental Protection Agency, Cincinnati, OH 45268.

[63] Hamaker, J.W. 1971. 'Decomposition: quantitative aspects', *in*: C.A.I. Goring & J.W. Hamaker (eds.), 'Organic chemicals in the soil environment', vol. 2, no. 2, pp. 432–4. Marcel Dekker, Inc., New York, NY.

[64] Hannapel, R.J., W.H. Fuller, S. Bosma & J.S. Bullock. 1964. 'Phosphorus movement in a calcareous soil: I. Predominance of organic forms of phosphorus in phosphorus movement', *Soil Sci.*, **97**, 350–7.

[65] Hannapel, R.J., W.H. Fuller & R.H. Fox. 1964. 'Phosphorus movement in a calcareous soil: II. Soil microbial activity and organic phosphorus movement', *Soil Sci.*, vol. 97, no. 6, 421–7.
[66] Hardcastle, J.E. & W.H. Fuller. 1974. 'Relative absorption of calcium and strontium by some desert soil fungi', *Chemosphere*, **2**, 59–64. Oxford, England.
[67] Silverman, M.P. & H.L. Ehrlich. 1964. *Advances in applied microbiology*, vol. 6, pp. 153–206, Academic Press Inc., New York, N.Y.
[68] Harter, R.D. 1977. 'Reactions of minerals with organic compounds in soils', *in*: Dixon & Weed (eds.), *Minerals in soil environment*. Soil Sci. Soc. Am., Madison, WI 53711 USA.
[69] Hepler, L.G. & G. Olofsson. 1975. 'Mercury: thermodynamics properties, chemical equilibria, and standard potentials', *Chem. Rev.*, **75**, 585–602.
[70] Hillel, D. 1980. *Soil and water physical principles and processes*. Academic Press, Inc., New York, NY 10003.
[71] Hobson, P.N., S. Bousfield & R. Summers. 1966. 'Anaerobic digestion of organic matter', CRC *Critical Rev. Environ. Control*, CRC Press, **4**, 131–91. CRC Press Inc., Boca Raton, FL 33431.
[72] Hodgson, J.F. 1960. 'Cobalt reactions with montmorillonite', *Soil Sci. Soc. Am. Proc.*, **24**, 165–8.
[73] Hoeks, J. 1976. 'Pollution of soil and groundwater for land disposal of solid wastes', Tech. Bul. 96, 1976. Inst. Land and Water Management. Wageningen, Netherlands.
[74] Ivanov, D.N. & V.A. Bolshakov. 1969. 'Extracting available forms of trace elements from soils', *Khim. Sel. Khoz.*, **7**, 229–32.
[75] Jackson, M.L. 1963. 'Soil clay mineralogy analysis', *in*: C.I. Rich & G.W. Kunze (eds.), *Soil clay mineralogy*. University of North Carolina Press, Chapel Hill, NC. pp. 245–94.
[76] James, R.O. & T.M. Healy, 1972. 'Adsorption of hydrolysable metal ions at the oxide water interface: I. Co(II) adsorption on SiO_2 and TiO_2 as model systems', *J. Colloid and Interface Sci.*, **40**, 42–52.
[77] Jenne, E.A. 1979. 'Chemical modeling in aqueous systems', Am. Chem. Soc. Symp. ser. 93, 897 pp.
[78] Jenne, E.A. 1968. 'Control of Mn, Fe, Co, Ni, Cu, and Zn concentrations in soil and water: The significant role of hydrous Mn and Fe oxides', *Adr. Chem. Ser.*, **73**, 337–387.
[79] Jenny, Hans. 1980. *The soil resource, origin and behavior*. Springer-Verlag, New York, NY, pp. 377.
[80] Jensen, S. & A. Jernelov. 1969. 'Biological methylation of mercury in aquatic organisms', *Nature*, **223**, 753–4.
[81] Jones, H.R. 1973. 'Waste disposal control in the fruit and vegetable industry', *Poll. Tech. Rev.*, no. 1. Noyes Data Corporation, Park Ridge, NJ. 261 pp.
[82] Jurinak, J.J. & N. Bauer. 1956. 'Thermodynamics of zinc adsorption on calcite, dolomite, and magnesite-type minerals', *Soil Sci. Soc. Amer. Proc.*, **20**, 466–71.
[83] Kenney, D.R. & R.E. Wildung. 1977. 'Chemical properties of soil', *in*: Elliott & Stevenson (eds.), *Soils for management of organic wastes and waste waters*. Am. Soc. Agron., Madison, WI, pp. 75–97.
[84] Korte, N.E., W.H. Fuller, E.E. Niebla, J. Skopp & B.A. Alesii. 1976. 'Trace element migration in soils: Desorption of attenuated ions and effects of solution flux', pp. 243–58, *in*: W.H. Fuller (ed.), 'Residual management by land disposal', USEPA–600/9–76–015. US Environmental Protection Agency, Cincinnati, OH 45268.

[85] Krauskopf, K.B. 1972. 'Geochemistry of micronutrients', pp. 240, *in*: J.J. Mortvedt, P.M. Giordano & W.L. Lindsay (eds.), *Micronutrients in agriculture*. Soil Sci. Soc. Am., Madison, WI.

[86] Lapidus, L. & N.R. Amundson. 1952. 'Mathematics of absorption in beds. VI: The effect of longitudinal diffusion in ion exchange and chromatographic columns', *J. Phys. Chem.*, **56**, 984–8.

[87] L'Annunziata, M.F. & W.H. Fuller. 1968. 'The chelation and movement of Sr^{89}–Sr^{98} (Y^{90}) in a calcareous soil', *Soil Sci.*, vol. 105, no. 5, pp. 311–19.

[88] Leeper, G.W. 1972. 'Reactions of heavy metals with soil and special regard to their application of sewage wastes', Department of Army, Corps of Engineers, Washington, DC. Contract no. DACW 73–73–C–0026, 70 pp.

[89] Leeper, G.W. 1978. *Managing the heavy metals on the land.* Marcel Dekker Inc., New York, NY.

[90] Letey, J. 1977. 'Physical properties of soils', pp. 101–12, *in*: Elliott & Stevenson (eds.), *Soils for management of organic wastes and waste waters.* Soil Sci. Soc. Am. Madison, WI.

[91] Lindsay, Willard L. 1979. *Chemical equilibria in soils.* Wiley-Interscience Publication. John Wiley and Sons, New York, NY.

[92] Maisol, H. & G. Gnugnoli. 1972. *Simulation of discrete stochastic systems.* Sci. Res. Assoc., Inc., Chicago, IL, 465 pp.

[93] Marbut, C.F., 1935. *Atlas of American agriculture. Part III: Soils of the United States.* US Department of Agriculture, Government printing office, Washington, DC.

[94] Marion, G.M., D.M. Hendricks, G.R. Dutt & W.H. Fuller. 1976. 'Aluminum and silica solubility in soils', *Soil Sci.*, **121**, 76–85.

[95] Martin, J.P. 1971. 'Side effects of pesticides on soil properties and plant growth', pp. 733–92, *in*: C.A.I. Goring & J.W. Hamaker (eds.) *Organic chemicals in the soil environment*, vol. 2. Marcel Dekker, Inc., New York, NY.

[96] Doner, H.E. 1978. 'Chlorides as a factor in mobilities of Ni(II) and Cd(II) in soil', *Soil Sci. Soc. J.*, **42**, 882–5.

[97] Martin, J.P. 1946. 'Microorganisms and soil aggregation: II. Influence of bacterial polysaccharides on soil structure', *Soil Sci.*, **61**, 157–166.

[98] Matrecon, Inc. 1980. 'Lining of waste impoundment and disposal facilities', Municipal Environmental Research Laboratory, Office of Research and Development. US Environmental Protection Agency, Cincinnati, OH 45268.

[99] Meeks, B., L. Chesnin, W.H. Fuller, R. Miller & D. Turner. 1975. 'Guidelines for manure utilization in the western region USA', Rept. WRRC. W–124 Comm. Wash. State University, Col. Agr. Res. Center. Bul 814, Pullman, WA.

[100] Merz, R.C. 1954. 'Final report on the investigation of leaching of a sanitary landfill', Pub. 10. State Water Pollution Control Board, Sacramento, CA., pp. 1–54.

[101] Merz, R.C. & R. Stone. 1965. 'Factors controlling utilization of sanitary landfill site', US HEW and University So. Cal. Rept., Los Angeles, CA. 77 pp.

[102] Murrman, R.P. & F.R. Koutz. 1972. 'Role of soil chemical processes in reclamation of waste water applied to land', pp. 48–74, *in*: 'Water management by disposal on land'. Spec. Rept. No. 171. US Army Cold Regions Research Engineering Lab., Hanover, NH.

[103] Novozamsky, I., J. Beek & G.H. Bolt. 1978. 'Chemical equilibria', *in*: '*Soil Chemistry: A. Basic elements*', Bolt & Bruggenwert (eds.), Elsevier Scientific Publishing Company, New York, Oxford. pp. 13–42.

[104] O'Neal, A.M. 1949. 'Soil characteristics significant in evaluating permeability', *Soil Sci.*, **67**, 403–9.

[105] Mortensen, J.L. 1963. 'Complexing of metals by soil organic matter', *Soil Sci. Soc. Am. Proc.*, **27**, 179–186.
[106] Rao, P.S.C. (ed.). 1978. *Pentachlorophenol: chemistry, pharmacology and environment toxicity*. Plenum Press, New York, NY. 402 pp.
[107] Rebhun, M. & J. Manka. 1971. 'Classification of organics in secondary effluents', *Env. Sci. and Tech.*, **5**, 606.
[108] Rogers, R.D. 1976. 'Methlation of mercury in soils'. *J. Environ.*, **5**, 454–8.
[109] Rouston, R.C. & R.E Wildung. 1969. 'Ultimate disposal of wastes to soils', *in*: L.K. Cecil (ed.), *Water. Chem. Eng. Prog. Symp.*, ser. 65, **97**, 19–25.
[110] Schnitzer, M. 1969. 'Reactions between fulvic acid, a soil humic compound and inorganic soil constituents', *Soil Sci. Soc. Am. Proc.*, **33**, 75–81.
[111] Schnitzer, M. & H. Kodama. 1977. 'Reactions of minerals with soil humic substances', pp. 741–70, *in*: Dixon and Weed (eds.), 'Minerals in soil environment', *Soil Sci. Soc. Am.*, Madison, WI 53711 USA.
[112] Schnitzer, M. & S.I.M. Skinner. 1965. 'Organo-metal interactions in soils: 4. Carboxyl and hydroxyl groups in organic matter and metal retention', *Soil Sci.*, **99**, 278–84.
[113] Schoen, U. 1953. 'Kennzeichung von tonen durch phosphatbindung und kationenumtausch', *Z. Pflanzenernahr. Dungung u. Bodenk.*, **63**, 1–17.
[114] Shapiro, J. 1964. 'Effect of yellow organic acids on iron and other metals in water', *J. Am. Water Works Assoc.*, **55**, 1062.
[115] Siegel, A. 1971. 'Metal organic interaction in the marine environment', *in*: S.D. Faust & J.V. Hunder (eds.), *Organic compounds in aquatic environments*. Marcel Dekker, Inc., New York, NY.
[116] Slowey, J.F., L.M. Jeffrey & D.W. Hood. 1967. 'Evidence for organic complexed copper in sea water', *Nature*, **214**, 377.
[117] Sommers, L.E. 1977. 'Chemical composition of sewage sludges and analysis of their potential use as fertilizers', *J. Environ. Qual.*, **6**, 225–31.
[118] Sommers, L.E., D.W. Nelson & D.J. Silveira. 1979. 'Transformations of carbon, nitrogen, and metals in soils treated with waste materials', *J. Environ. Qual.*, **8**, 287–94.
[119] Steengjerg, F.T. 1933, 'The manganese content of Danish soil', *Tidsskr. Planteavl*, **39**, 401–36.
[120] Stevenson, F.J. & M.S. Ardakani. 1972. 'Organic matter reactions involving micronutrients', pp. 79–114, *in*: J.J. Mortvedt, P.M. Giordano & W.L. Lindsay (eds.), *Micronutrients in agriculture*, Soil Sci. Soc. Am., Madison, WI.
[121] Stumm, W. & H. Bilinsky. 1973. 'The apparent supersaturation', Proc. Intnat. Conf. on Water Poll. Res., Jerusalem, 1972. *Adv. in Water Poll. Res.*, S.H. Jenkins (ed.). Pergamon Press, Oxford.
[122] Summers, A.O. & S. Silver. 1978. 'Microbial transformations of metals', *Ann. Rev. Microbiol.*, **32**, 637–72.
[123] Soil Survey Staff. 1951. *Soil Survey Manual*. USDA Handbook 18, pp. 189–234. Government Printing Office. Washington, DC.
[124] Thibodeau, Louis J. 1979. *Chemodynamics: Environment movement of chemicals in air, water, and soil*. Wiley-Interscience Publication. John Wiley and Sons, Inc., New York, NY.
[125] Thomas, C.W. & A.R. Swoboda. 1970. 'Anion exclusion effects on chloride movement in soils', *Soil Sci.*, **110**, 163–6.
[126] Turjoman, Abdul Mannan. 1978. 'The behavior of lead as a migrating pollutant in six Saudi Arabian soils'. Ph.D. Thesis, University of Arizona Library, Univ. of Arizona, Tucson, AZ 85721.
[127] US Environment Protection Agency. 1976. 'Proceedings of the open forum on

management of petroleum refinery waste waters', F.S. Manning (ed.), University of Tulsa, OK, pp. 512.

[128] van Genuchten, M.T. 1978. 'Simulation models and their application to landfill disposal siting: A review of current technology', *in*: D.W. Schultz (ed.), 'Land disposal of hazardous waste', Proc. Fourth Ann. Res. Symp. USEPA–600/9–78–016. US Environmental Protection Agency, MERL, Cincinnati, OH 45268. 425 pp.

[129] Veith, J.A. & G. Sposito. 1977. 'On the use of the Langmuir equation in the interpretation of "adsorption phenomena"', *Soil Sci. Soc. Am. J.*, vol. 41, no. 4, pp. 697–702.

[130] Vomicil, J.A. 1965. 'Porosity', *in*: C.A. Black (ed.), *Methods of soil analysis*, Pt. 1. Amer. Soc. Agron., Madison, WI.

[131] Walker, W.H. 1969. 'Illinois groundwater pollution', *J. Am. Water. Assoc.*, **61**, 31–40.

[132] Walker, J.M. 1974. 'Trench incorporation of sewage sludge', Proc. Nat. Conf. on Municipal Sludge Management, 11–13 June, 1974. Pittsburg, PA.

[133] Walsh, L.M. & J.D. Beaton (eds.). 1973. 'Soil testing and plant analysis', *Soil Sci. Am.*, Inc., Madison, WI.

[134] Wallace, A. 1962. 'A decade of synthetic chelating agents in inorganic plant nutrition', Arthur Wallace, Los Angeles, 195 pp.

[135] Weiss, A.J., & P. Colombo. 1980. 'Evaluation of isotopic migration in land burial', BNL/NUREG–51143.288. Edwards Brothers, Inc., Ann Arbor, MI, pp. 195.

[136] Schnitzer, M. & S.U. Kahn. 1972. *Humic substances in the environment*. Marcel Dekker Inc., New York, NY.

9

Toxicological assessments and their relevance to hazardous waste management

Terminology

Toxicity is the ability of a chemical molecule or compound to produce injury once it reaches a susceptible site in or on the body [1]. 'Acute toxicity', as applied to materials which are inhaled or absorbed through the skin, refers to a single exposure of duration measured in seconds, minutes or hours. For ingested materials it refers to a single dose. The term 'chronic toxicity' is used in contrast to acute toxicity and means 'of long duration'. When applied to materials which are inhaled or absorbed through the skin it refers to prolonged or repeated doses over periods of days, months or years. Absorption is considered to be the entry of a molecule into the blood stream so that it may be carried into any part of the body. Systemic effects are damage effects resulting from absorption which occur at sites in the body other than those where local contact has taken place.

Toxicology is defined as the study of the action of poisons on the living organism. Toxicology is not an exact science when compared with chemistry, physics or mathematics since toxicological phenomena can not always be predicted with accuracy or explained on the basis of physical or chemical laws. The term poison is used to define a substance which causes harm to living tissue when applied in a relatively small dose.

As we shall see, it is not always easy to make a clear-cut distinction between poisonous and non-poisonous substances.

The determination of health responses

Classical considerations of the health response of an organism to an applied dose of a study material would be expected to produce a relationship such as that shown in Fig. 1.

This characteristic curve would be obtained from a study of a large number of individuals (human beings and other mammals) from a material which exhibits the property of being beneficial to health at low concentrations yet harmful at high concentrations. This concept allows us to identify the inherent phases labelled survival, deficiency, normal health, toxicity and lethality. Classical theory of toxicology suggests that the relationship between dose and response is that shown in Fig. 2.

This theory allows us to define the RD_{50} figure, which is the response obtained for 50% of the organisms for a particular concentration of material in the nutrient. When death is the response being measured, this is expressed as LD_{50}.

Experimental methods of assessing toxic effects

Human beings may be exposed to toxic chemicals in many ways ranging from occupational exposure during manufacture or use to digestion of foods contaminated by residuals from agricultural or water pollution sources. Toxicological studies have therefore to be designed with the method of exposure in mind. Occupational exposure generally considers contact with the skin and mucous membranes or inhalation, whereas feeding studies need to be employed to assess risks from ingestion. Toxicological studies of drugs have a somewhat different objective since they need to determine the therapeutic effects of short-term administration of low doses compared to the toxic effects of high doses or long-term administration. Pesticides warrant different considerations. In this case toxicological investigations are needed to determine the maximum dose

Fig. 1. An ideal dose–response relationship.

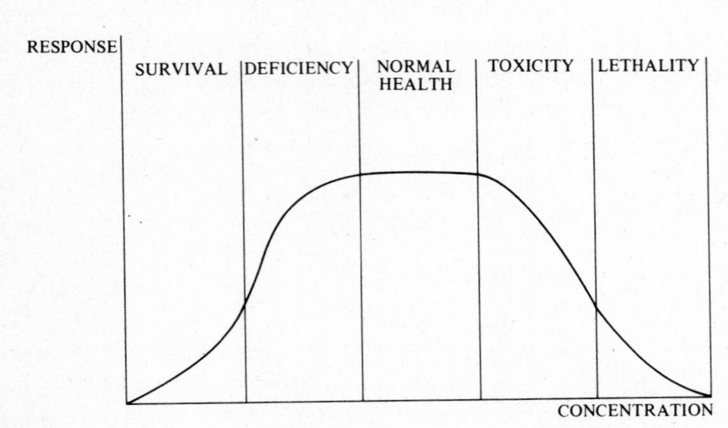

that may be tolerated by animals (and men) throughout their lifetime without manifesting any adverse effects.

The susceptibility of different species of animals to toxic agents varies considerably, yet, clearly, the prospects of undertaking toxicity studies on men (and women) is very limited. Consequently, several different species of animal have to be used for evaluations and it is generally assumed that man is at least as sensitive as the most sensitive species tested. The problems encountered in extrapolating laboratory studies to 'man scale' are examined later but, first, consideration must be given to some practicalities.

Non-human primates

Studies employing non-human primates have contributed considerable knowledge to the assessment of public health hazards and diseases affecting man [2]. Rhesus, stumptail and bonnet monkeys (Macaca mulatta, Macaca arctoides and Macaca radiata respectively) have been used to study inhalation effects of pollutants since they have a well-developed bronchial tree similar to humans [3]. Although monkeys and man (unlike rats) have similarly well developed respiratory bronchioles Schmidt [4] has argued that it is presumptuous to assume that monkeys are ideal models of man. In any event, research with primates in recent years has suffered from an ever-decreasing availability of animals. Trapping and exportation of animals is constrained by legal restrictions in many countries.

Fig. 2. An ideal dose–response curving showing determination of RD_{50}.

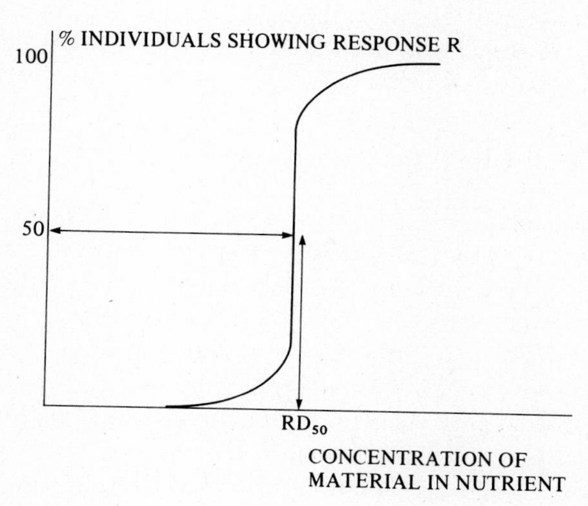

Rodents and small animals

Most acute toxicity studies are carried on under controlled laboratory conditions on mice, rats, dogs, marmosets, pigeons, quail, rabbits, miniature pigs and goats. Reliable data can only be obtained if the species used is disease free, of similar age and weight, equal proportions of the sexes, fed with the same diet and maintained in identical laboratory conditions. Even then large numbers of test animals are required and statistical analysis of results is essential. The selected animals may be outbred or inbred and randomly or non-randomly mated. Over recent years the tendency has been to increase the longevity of the tests from ten days, to two weeks, to three months, to two years and even to a lifespan. Tests which are confined to measurement of the number of dead specimens are now recognised to be of little value. It is necessary to monitor blood and urine specimens and examine body organs histologically for the detection of carcinogenesis, mutagenesis and teratogenesis.

Dermal toxicity

Dermal toxicity tests are often carried out on rabbits by clipping the skin free of hair, causing a shallow abrasion of the skin and introducing liquids to the area by means of a sleeve. Single exposures of increasing dosage levels are studied. Damage to mucous membranes and eyes is generally studied using rabbits and toxicants are applied to the conjunctiva of the eye, the vaginal vault and the tip of the penis. It must be recognised that the eye of the rabbit differs functionally and anatomically from the human eye and some researchers prefer to employ monkeys since their lacrimating glands resemble those of humans. Typically, six albino rabbits are used for each test and care must be taken to eliminate sawdust, wood chips and other extraneous materials from the animals' cages. Tests are limited to the use of 0.1 ml or 1 mg and one eye of the test animal is exposed; the other eye serves as a control.

Inhalation toxicity

Inhalation toxicity studies vary according to whether gases, dusts, mists or vapours are being studied. Generally, animals are exposed under conditions in which they breathe the atmosphere containing known concentrations (expressed in mg m^{-3}) for various periods of time. Groups of white rats (body weight 200–300 g) are frequently used for inhalation studies. Recently it has been recognised that the particle size of the droplet as well as concentration affects inhalation toxicity since particles greater than 10 μm diameter are effectively prevented from reaching the lung areas, whereas those of 3 μm diameter and less are inhaled into the alveolar region. Consequently, inhalation test chambers require sophisticated dispersing equipment and it is necessary to avoid conditions in which condensed test materials can be orally ingested [5, 6].

Acute oral toxicity

Measurement of oral toxicity can be made by either adding test materials to the diet or intragastrically by gavage. For dosage through diet, it is necessary to carefully control the quantity and quality of normal food in relation to the test animal's body weight. Though rats and mice are frequently chosen for such studies it is necessary to take into consideration differences in metabolic processes. Biotransformations within the test animal may serve to detoxify the test material or, alternatively, to increase its toxicity. For example, there is a tendency to detoxify aromatic amines in rats and humans but the response in dogs is the reverse. Acute, short-term and long-term studies of oral toxicity are necessary. The protocol is to determine the nature of toxic effects induced by massive dosages and gradually work down to minimal or no-effect levels. To compensate for some of the uncertainties in extrapolating animal studies to man it is advisable to use at least two different animal species, one of which is a non-rodent.

Acute oral toxicity studies are designed to permit estimation of the lethal dose for 50% (LD_{50}) of a group of test animals, expressed in terms per unit of body weight. Test substances are administered in an inert vehicle in solution or suspension after an overnight (16–18 hour) fast followed by the offer of food. Between 24 and 120 test animals are required to provide a reasonable estimate of LD_{50}, its standard error and the slope of the dose–response curve.

Short-term (sub-acute) toxicity

Short-term (sub-acute) oral toxicity involves the daily ingestion of the test substance for a period of one to six months at levels ranging from the maximum tolerable to the minimum effective. A period of 90 days in a young rat covers the span from infancy to full sexual maturity and, therefore, many toxicologists argue that this test has considerable value and may, in certain cases, obviate the need for chronic testing. The major objective is to estimate the minimum daily intake that will induce an adverse effect in animals and the maximum daily intake that they will tolerate throughout a lifetime. Although the rat has been widely used for such studies it should be recognised that the rat differs from man in its natural capacity to synthesize vitamin C and, anatomically, in that it lacks a gall bladder. The most popular non-rodent species used is the dog but dogs must receive immunological treatment against leptospirosis and hepatitis and must be deparasitised before and during all experimental tests. Dogs are seldom found to be completely free of intestinal parasites at autopsy even when bred and maintained under laboratory conditions. Mongrels of unknown history should not be used and beagles are the breed most frequently chosen by toxicologists. For short term oral toxicity studies groups of the test animals are fed a series of three to five graded doses of the test material and a control group is fed the basal diet dose. Each group consists of 20 animals, ten of each sex although the control group should be somewhat larger. Rats are started on test shortly after weaning (55 ± 5 g average weight) and dogs are placed on test about six months of age after preliminary conditioning and acclimatisation have been completed. The doses of test materials are expressed on a body weight basis; the diet is fed *ad libitum* and a record of food consumption is kept as an index of appetite. In the case of young rats it is necessary to make frequent adjustments of dietary dose levels to compensate for the marked diminution of food intake relative to body weight as maturity approaches. Difficulties arise where the test material affects the palatability of the diet. The assessment of critical dosages in short-term feeding studies demands experience, judgement and intuition on the part of the investigator. Oser [5] points out that it is theoretically impossible to establish the minimum 'effective' dose since the animal organism is too complex a biological system to permit drawing a sharp line between positive and negative responses. To establish the threshold dose at which minimal or no adverse effects are induced, much effort has to be expended to look for very little. The subtle effect of marginal doses are only revealed by means of a battery of clinical, biochemical and physiological tests and exhaustive post-mortem examinations. The observations on living animals should include growth, appearance, and behaviour; chemical examination of body fluids, tissues and excreta; physiological tests of organ function and biopses of accessible target organs (see Table 1). Post-mortem examinations must include gross and histomorphological examinations of all major organs and tissues.

Chronic toxicity

Chronic toxicity tests refer to the major fraction of the lifetime of the species. For studies employing rats this requires experiments lasting two years. The testing protocol is generally an extension of the short-term tests but they are often applied to a greater number of test animals.

Table 1. *Typical observations to be made during short-term toxicological studies of small animals*

Physical appearance	Conditions of fur, skin, eyes, ears, extremities, genitals, anal region, tail (d)*
Behaviour	Signs of lethargy, posture, erratic movements (d)
Growth response	Body weight, length, body surface area (w)**
Food consumption	Appetite check, check to determine dose actually ingested (w)
Water consumption	Relevant to urine volume (w)
Hemocytology	(a) erythrocyte count or haematocrit, leukocyte count, differential leukocyte count (pt) (b) erythrocyte morphology (pt)*** (c) reticulocyte count (pt) (d) platelet count (pt)
Hemochemistry	(a) blood glucose, urea nitrogen (pt) (b) serum proteins (c) triglycerides and cholesterol (d) alkaline phosphatase, glutamic hydrogenase, lactic dehydrogenase isoenzymes (e) red cell cholinesterase (f) prothrombin and/or clotting time (g) sedimentation rate (h) serum bilirubin
Urinanalysis	Volume, specific gravity, reducing sugars, albumin, occult blood, sediment (pt)
Special metabolic and functional tests	Carbohydrate metabolism – glucose tolerance, Liver function – bromosulpholein retention, Renal function – phenolsulphonephthalein excretion, creatinine clearance
Mortality	Expressed as % mortality of each group, median survival time
Necropsy	Macroscopic inspection of all vital organs. Recording absolute and relative weights of liver, kidneys, spleen, heart, thyroids, adrenals, pituitary and gonads. Fixing in 10% formalin these organs and the stomach, large and small intestines, pancreas, bladder, gonads, thymus, salivary gland, lymph nodes, lungs, marrow skin, muscle, spinal cord and brain
Histomorphology	Section and stain major organs and tissues in control animals and highest dosage groups and liver and kidneys of all groups

*d – Examined daily.
**w – Examined weekly.
***pt – Examined before the test and 1, 3, 6, 12 and 24 weeks after. Other tests are measured once or twice during the study. It is not necessary to carry out all these tests in each study.
Source: after Reference [5].

Typically, 35 rats of each sex would be used so that ten may be sacrificed at the termination of the short-term test. The assessment of observations over the test period is usually based on a 95% probability factor.

Aquatic toxicity

Aquatic toxicity determinations are generally confined to fish and water fleas (Daphnia).

Fish toxicity

The study of fish toxicity begins with an assessment of exposure concentrations and exposure saturations. Duthie [7] has suggested the protocol shown in Fig. 3.

From this it can be seen that the toxicologist needs access to a wide data base before he can begin to study a particular chemical toxicant. The next test is the study of acute toxicity and the selection of the fish has to be made. The choice of fish depends not only on the kind of material and the expected exposure area but also on the availability of data from related compounds. Fathead minnow (Pimephales promelas), bluegill (Lepomis macrochirus) and rainbow trout (Salmo gairdherii) are popular choices.

Four criteria have been used

(1) the time which passes before the first clear indications of poisoning are evident;

(2) the time taken, for example, to depress the respiration rate to some fraction of normal, or
the time at which the fish lose their sense of balance or swim upside down;

(3) the minimum time of exposure from which recovery is not possible;

(4) the time of immersion necessary to kill the fish.

Dourdoff [8] recommends that the test animals should be from the same species and obtained from the same body of water. The largest specimens must not be more than one and a half times the length of the smallest and they should be acclimatised to test conditions for at least one week and not fed for two days before use. The choice of water source poses difficulties since it must be chemically and 'environmentally' similar to the unpolluted water used in the test and be customary for the species of fish chosen. Test temperatures, containers, water depth, dissolved oxygen, duration of test and feeding arrangements all need standardising. Lloyd [9] describes the establishment of concentration–response curves and the determination of LC_{50} from 3, 6, 24, 48 and 96 hour observations. Fig. 4 shows characteristic responses.

Fig. 3. Aquatic hazard assessment and evaluation. (Source: Reference [7].)

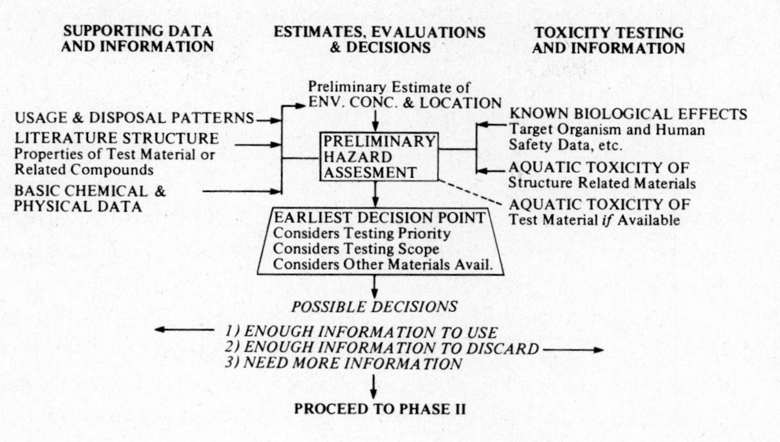

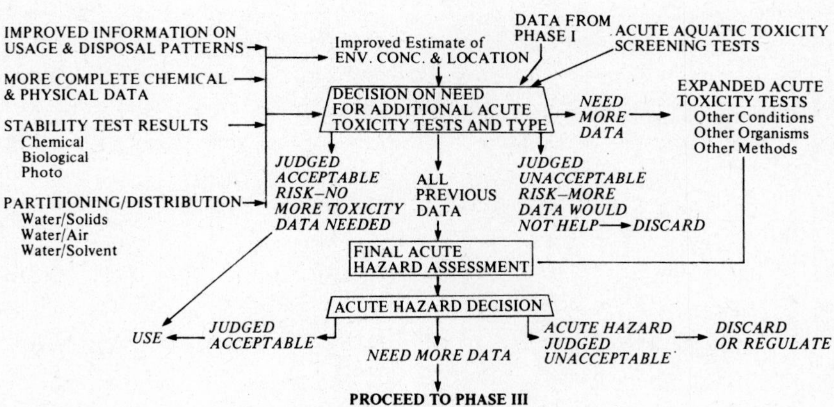

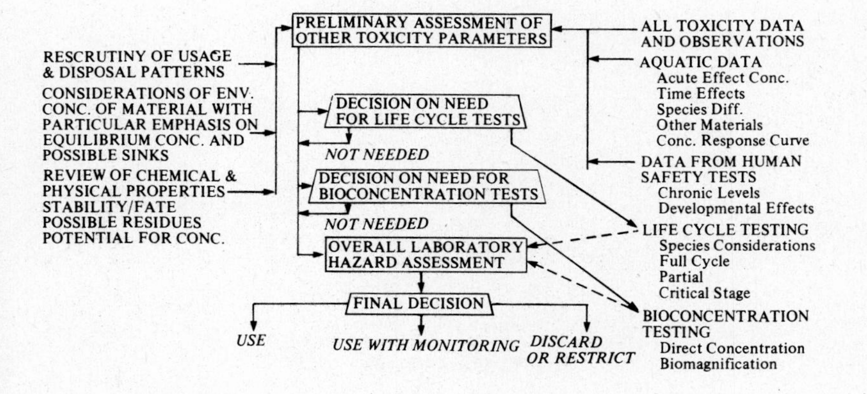

For Curve A, which becomes parallel with the time axis after a short exposure LC_{50} can be easily established, indicating that the fish can detoxify the chemical or become acclimatised to it. Curve B indicates that the chemical can not readily be detoxified and Curve C shows low tolerance. The International Standards Organisation has recently proposed a standard testing procedure [10].

At least ten fish should be used in any one experiment and as many as 100 have been recommended [11]. Klein [12] has reviewed acute fish toxicity studies and reported lethal concentrations and exposures for a wide variety of fish. Kenega [13] reported the data shown in Table 2 which includes LC_{50} and the maximum allowable toxicant concentration (MATC) and illustrates the observed differences between fish of different species in chronic lifecycle tests and the calculation of application factors.

Mehrle et al. [14] reported toxic effects of polychlorinated biphenyls on fish and pointed out that changes in the fish diet had a marked effect on the results. Katz [15] has reported that the aquatic toxicity of heavy metals is affected by the ammonia and chlorine content of the water. Marking [16] reported that mixtures of chemicals produced unexpected effects; some beneficial and some hazardous. Though mixtures of copper, zinc and nickel were additive in toxicity to rainbow trout, commercially available mixtures of insecticides produced synergistic toxic effects.

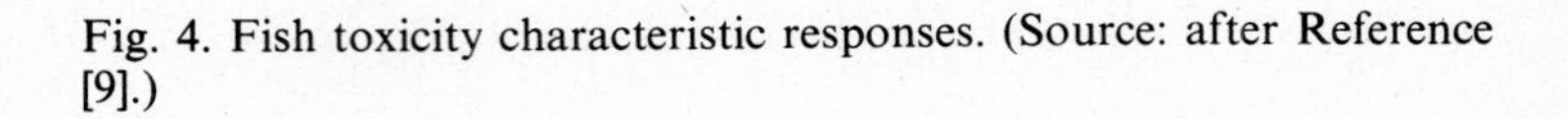

Fig. 4. Fish toxicity characteristic responses. (Source: after Reference [9].)

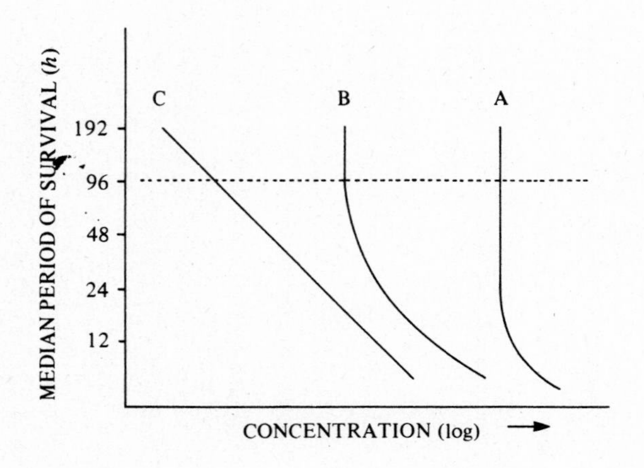

Table 2. *Toxicity of selected toxicants to fish*

Chemical	Species	LC50[a] or LTC[a, b] (ppb) (4 days)	Range of MATC's (ppb)	Type of chronic lifecycle test C = complete P = partial	Range of application factors: MATC/LC50 or MATC/LTC
Chlorinated hydrocarbon insecticides and PCB's					
Lindane	Fathead minnow	69[d]	9.1–23.5	C	0.13–0.34
	Bluegill	30[d]	9.1–12.5	C	0.30–0.42
	Brook trout	26[d]	8.8–16.6	P	0.34–0.64
Chlordane	Fathead minnow	36.9	0.32	C	0.009
	Bluegill	59	0.54–1.2	P	0.009–0.02
	Brook trout	47	0.32–0.66	P	0.007–0.014
	Sheepshead minnow	12.5	0.5–0.8	C	0.04–0.06
Heptachlor	Fathead minnow	7[d]	0.86–1.84	C	0.12–0.26
	Sheepshead minnow	10.5	0.97–1.9	P	0.09–0.18
Endrin	Sheepshead minnow	0.38	0.12–0.31	C	0.35–0.91
Pentachlorophenol	Sheepshead minnow	442	47–88	C	0.11–0.20
Endosulfan	Fathead minnow	0.86[d]	0.2–0.4	P	0.23–0.47
Toxaphene	Fathead minnow	5.3[b]	0.025–0.054	P	0.005–0.01
	Channel catfish	15[b]	0.129–0.299	P	0.009–0.020
	Brook trout	4.1[b]	<0.039	P	<0.010
DDT	Fathead minnow	48	0.09	C	0.019
Methoxychlor	Fathead minnow	8.63	<0.125	P	<0.014
	Yellow perch	22.2	<0.625	P	<0.028
	Sheepshead minnow	49	12–23	P	0.24–0.47

Table 2. (*cont.*)

Chemical	Species	LC50[a] or LTC[a, b] (ppb) (4 days)	Range of MATC's (ppb)	Type of chronic lifecycle test C = complete P = partial	Range of application factors: MATC/LC50 or MATC/LTC
Phosphate and carbamate insecticides					
Malathion	Fathead minnow	10 750[b]	200–580	C	0.019–0.053
	Fathead minnow	9000(S)	200–580	P	0.022–0.064
	Bluegill	82[b]	3.6–7.4	P	0.043–0.090
	Bluegill	108	3.6–7.4	P	0.033–0.069
	Flagfish	349	8.6–10.9	C	0.019–0.054
	Sheepshead minnow	51	4.0–9	P	0.08–0.18
Fenthron	Fathead minnow			C	0.07–0.12
Diazinon	Fathead minnow	7800(S)	6.8–13.5	P	0.0009–0.0017
	Brook trout	770	< 0.8–2.4	P	0.001–0.003
Azinphosmethyl	Fathead minnow	1900(S)	0.51–1.8	C	0.000 09–0.000 27
Carbaryl	Fathead minnow	9	0.21–0.68	C	0.023–0.075
Carbofuran	Sheepshead minnow	386	15–23	C	0.04–0.06
Miscellaneous organic chemicals					
LAS (linear alkyl benzene sulfonate)	Fathead minnow	4350	630–1200	P	0.14–0.28
2,4-D(BE)	Fathead minnow	5600(S)	300–1500	P	0.054–0.268
Propachlor	Fathead minnow			C	0.17–0.36
Dinoseb	Lake trout	77(S)	< 0.5	P	< 0.006
Trifluralin	Fathead minnow	115[d]	1.9–5.1	P	0.017–0.044
	Sheepshead minnow	190	1.3–4.8	C	0.007–0.025

Butralin	Fathead minnow			C	0.2–0.42
Picloram	Lake trout	2700(S)	< 35	P	< 0.012
Atrazine	Fathead minnow	15 000(S)[d]	210–520	P	0.014–0.035
	Bluegill	6700[d]	90–500	P	0.014–0.075
	Brook trout	4900[d]	60–120	P	0.013–0.024
Acrolein	Fathead minnow	84[d]	11.4–41.7	P	0.14–0.50
Captan	Fathead minnow	64[b]	16.5–39.5	P	0.25–0.61
Aroclor 1242	Fathead minnow	15[e]	0.86–5.4	C	0.057–0.14
	Fathead minnow	234–300[f]			0.0028–0.018
Aroclor 1254	Fathead minnow	7.7[e]	0.52	C	0.064
	Fathead minnow	> 33[f]			< 0.016
Metals[g]					
Copper (sulfate)	Fathead minnow	75	10.6–18.4	P	0.14–0.24[h]
	Fathead minnow	470	14.5–33.0	P	0.031–0.07[i]
	Bluegill	1100	21–40	P	0.019–0.036
	Brook trout	100	9.5–17.4	P, C	0.095–0.17
Cadmium (sulfate)	Fathead minnow	7200	37–57	P	0.0051–0.0079
	Bluegill	20 400	31–80	P	0.0015–0.0039
Nickel (chloride)	Fathead minnow	27 000	380–730	P	0.014–0.027
Zinc (sulfate)	Fathead minnow	9200	32–180	P	0.003–0.02

[a]Tests are flowing-water, except (S) denotes static test.
[b]Mean lethal threshold concentration (LTC).
[d]Incipient LC50 used.
[e]Newly hatched fry.
[f]Two-month-old fish.
[g]LC50's are on a metal ion basis.
[h]Soft water.
[i]Hard water.
Source: Reference [13].

Daphnia

The waterflea (Daphnia magna) is reported to be more sensitive to toxicants than fish [17], and, of course, is cheaper. It has a typical lifecycle of three weeks and this permits determination of LC_{50}'s within 48 hours. Pearson [18] has shown close relationships between the toxic effects to Daphnia and Fathead minnows in the case of some 30 organic compounds. The toxicities were 4–7 fold greater to Daphnia which is regarded as a good correlation. Table 3 reports LC_{50} and MATC data for Daphnia in the presence of metals.

Bioluminescence

A relatively new development which appears to offer considerable potential in toxicity studies is the use of luminescent bacteria. Bulich [20] describes a method of assay based on the changes of light output of particular luminescent bacteria as measured by a photometric device. On exposure to toxicants the light output is quickly diminished in proportion to the concentration of the toxicant. The test is completed in 2–5 minutes and shows acceptable correlation with 96-hour acute fish toxicity data. On-

Table 3. *Application factors for Daphnia magna calculated from acute LC_{50} and MATC data for metals*

	Daphnia LC_{50} (ppb)			
Metal ion	48-hour	3-week	MATC[e] (ppb)	Application factor (MATC/48-hour LC_{50})
Sodium	1 820 000	1 480 000	680 000	0.37
Calcium	464 000	330 000	116 000	0.25
Magnesium	322 000	190 000	82 000	0.25
Potassium	166 000	97 000	53 000	0.32
Iron	9600	5900	4380	0.47
Tin	55 000	42 000	350	0.006
Zinc	280	158	70	0.25
Nickel	1120	130	30	0.027
Lead	450	300	30	0.067
Copper	60	44	22	0.37
Cobalt	1620	21	10	0.006
Cadmium	65[d]	5	0.17	0.003

[e]Criteria – <16% reproductive impairment.
[d]No food; all others with food.
Source: Reference [19].

going studies in the USA indicate potential for the monitoring of leachates from waste disposal sites and the technique may also have application in the initial screening of waste consignments to determine whether they should be regarded as hazardous. The equipment has recently been introduced into Britain [21].

The toxic effects of inorganic materials

All minerals are toxic when administered in excess, yet at lower doses in trace quantities some minerals are essential to human life and others are used with no apparent toxic effects. At minute doses some minerals have a stimulatory effect on animals. Metals such as sodium, potassium and magnesium are essential to growth, health and survival of animals. Micro quantities of about ten other metals are also essential to health, probably due to their influence on enzymes and hormones. A metal in trace amounts (less than 0.01% of the weight of an organism) is said to be essential when an organism fails to grow or complete its life cycle in the absence of that metal. Conversely, the same essential metal becomes toxic at a higher concentration.

Some of the harmful effects of excess metals in mammals are listed in Table 4.

The toxic effects of organic materials

The dichotomy of opposing effects related to inorganic materials is also evident in the case of synthetic organic chemicals but the problems are compounded by the realisation that many organic products have only been manufactured in recent decades and there is little or no historical evidence of analogous situations which provides much comfort. Many halogenated organic materials are proving to be persistent and bioaccumulative. For example, Muller & Korte [23] have shown that the herbicide, monolinuron and the insecticide dieldrin, are resistant to biodegradation. DDT residues are now found in very nearly all fish and the harvesting of mackerel in Californian waters has been banned as a result [24].

The interaction of organic and inorganic materials

The prediction of the behaviour of most inorganic species (*per se*) in the environment is usually possible since access is available to geological and mineralogical data over many thousands of years. Similarly, naturally occurring organic materials are well understood and their biodegradation processes are comprehensible. However, the interaction between these processes is far less clear. We are aware that several interactive mechanisms exist. For example, the solubilising of inorganic materials by combination with certain organic compounds (known as chelation) is an important biological process. Insoluble iron compounds found in soil form soluble chelants with organic compounds secreted by vegetation [25]. Such chelation processes explain the ability of moss and lichen to extract trace metals from barren rocks. Metals also react with proteins, containing COO- and SH-groups. Nucleic acids form soluble metal compounds by reaction with phosphate, nitrogenous base and hydroxyl groups. In this way aluminium, strontium, nickel, barium, chromium and manganese are solubilised.

Again, it must be recognised that the complexity of reactions between inorganic and organic materials is magnified by man's synthesis of new organic compounds and polymers in the recent past. The behaviour of these products under waste management conditions can not be predicted

Table 4. *Some effects of elemental excesses on mammals*

X	Effects of excess
As	Stomach pain: convulsions: goitre*
Be	Lung cancer by inhalation †
Ca	Atherosclerosis: cataract: gall stones*
Cd	Hypertension?: nephritis †
Co	Heart failure: polycythemia
Cr	Lung cancer by inhalation †
Cu	Jaundice: Wilson's disease †
F	Mottled teeth: bone sclerosis*
Fe	Haemochromatosis: siderosis*
Hg	Encephalitis: neuritis †
I	Hyperthyroidism*
Mg	Anaesthesia
Mn	Ataxia †
Mo	Growth depression
Ni	Dermatitis: lung cancer by inhalation †
Pb	Anaemia: brain damage, neuritis: kidney cancer †
Sb	Heart disease
Se	Cancer: deformed nails and hair*
Si	Kidney stones: lung disease †
V	Reduced growth
Zn	Anaemia

*Found in some human populations.
†Caused by industrial overexposure.
Source: Reference [22].

with certainty. Yet there is evidence that inorganic–organic reactions are environmentally significant.

The mobility of organic mercury compounds is now well established [26] and has caused health defects in many thousands of people in Japan [27]. Research has shown that cadmium is also complexed and solubilised by chelants [28].

Polio viruses adsorbed on landfill soil have been shown to be solubilised by chelants [29].

Concentration mechanisms and food chains

One of the key characteristics of living cells is their ability to take up ions from a solution against a concentration gradient. This is most clearly demonstrated by free living aquatic organisms such as algae which obtain all their nutrients from solution. Indeed, Mauchline & Templeton [30] have shown that marine organisms have the ability to concentrate ions by as much as 100 000 times. Table 5 shows their data.

This feature is important because microorganisms form the beginning of the natural food chain and as one predator consumes another the toxicants are passed on to the higher forms of life. Plant life is also capable of concentrating toxicants. It has been demonstrated [31] that multicellular plants concentrate elements by factors which range from 150 to 7500 by either absorption through the soil or from rainfall falling on their leaves.

As man digests food from both animal and vegetable sources he retains some of its components in the body and excretes some in urine, sweat and faeces. Retention data for metals in the human body is shown in Table 6 but corresponding data for most synthetic organic compounds is not yet available.

Table 5. *Concentration of elements in sea water by marine organisms*

Concentration factor	Elements
<1	Ce, Na
1–10	B, Br, Ca, F, Li, Mg, S
10–100	Cs, K, Mo, Rb, Sb, Sc, Sr, U
100–1k	As, Ba, Be, Hg, Ra, Si, V
1–10k	Ag, Al, Cd, Co, Cr, Cu, Ga, I, Ni, Re, Zn
10–100k	Mn, N, P, Sc, Sn, Th, Ti, Zr
>–100k	Fe, La, Pb

Source: Reference [30].

Health responses

Classical considerations of the health responses of an organism to an applied dose of a study material were originally thought to be as shown in Fig. 1 in which survival, deficient, normal health, toxic and lethal phases can be identified. At this point it is important to stress the differences between toxic and lethal effects. As we all know, death is somewhat absolute and easy to measure and, for this reason, reference to lethal effects

Table 6. *Percentage of daily intake of reference man which is retained within the human body*

Element	Anions	Element	Cations
As	95–75	Ag	92.8–87
B	23	Al	99.78
Br	7	Ba	92.3
Cl	6	Be	91.7
Cr (VI)	54	Bi	92
F	45	Ca	83
Ge	7	Cd	95.5–33
I	15	Co	99.3–78
Mo	50	Cr (III)	99.6
Ni	3	Cs	10
Nb	42	Cu	98.6
P	36	Fe	98.4
S	6	Hg	93–66
Sb	97–20	K	15
Se	67–20	Li	60
Si	50	Mg	61
Te	22	Mn	99.994
V	99.25	Na	25
		Ni	97.3
		Pb	90–15
		Ra	96.5
		Rb	14
		Sc	95
		Sn	99.5
		Sr	82
		Th	96.7
		Ti	61
		Tl	66
		U (VI)	84
		Zn	96.2
		Zr	96.4

Source: Reference [22].

is often made in arguments about pollution control. However, society's main concern should be with toxic effects. Toxic effects include all the harmful effects resulting from exposure (both acute and chronic) to a poison. Toxic effects include growth retardation, decreased fullness of health and intellectual capability, detrimental changes in reproductive cycle with mortality of offspring, increased morbidity, pathological changes, appearance of tumours, chronic disease symptoms, and decreased longevity.

Toxicologists remind us that laboratory studies are influenced by:

(a) the chemical nature of the material in the nutrient;
(b) the method of feed of the nutrient;
(c) the age or state of development of the organism;
(d) the time period between the applied dose and the measured response.

In studies related to human beings, it has also been reported [32] that the sex of the species, the stress conditions and the imbalance of other biological conditions in the subject being studied also affect the responses obtained. Additionally, there are other complicating factors. For example, the response can be in two phases. In the first phase, a metal (for example) can be absorbed without causing damage but then, in a second phase, it seems to be released and causes damage to the tissue concerned. The accumulation of toxic metals occurs as a function of ageing and the gradual deterioration of the excretory mechanisms in the kidney, the lung or the prostate. The same effect may result from chronic accidental exposure of humans to toxicants.

Extrapolation of laboratory results to man

The results of studies on carcinogenicity according to Jones (Medicines Division, DHSS, London) [33] are difficult to interpret and extrapolate. Analysis of comparative pharmokinetics in the species concerned and knowledge of the proximate carcinogen is necessary. In practice, knowledge in these areas is almost always incomplete. Almost all carcinogenesis studies are conducted in rats and mice and yet, for numerous compounds, their pharmokinetics differ from man. Jones concludes that

> the quantification of carcinogenic risk is much more difficult; it involves the use of various mathematical models to extrapolate from the incidence of tumours found in animal studies at the various dose levels. Even if the latter extrapolation were valid, decisions based on risk–benefit assessment would still involve

value judgements rather than objective scientific comparisons of different courses of action. We need a great deal more knowledge to improve the quality of our decision making in this area.

Mutagenic tests to determine whether a material has the potential to cause carcinogenic effects or genetic alterations in the somatic cells and/or germinal cells are limited to new drugs. Few tests have been conducted on other chemical products and even for drugs, which warrant high priority, there are no established guidelines.

Prior to the thalidomide episode, it was neither customary nor expected that drugs and pesticides should be subjected to testing for teratogenic properties. Since then various rodent species, primates and chick embryos have been used for teratogenicity studies but, according to Oser (USFDA Research Laboratories) [5], it is not possible to gauge the predictive value of these procedures. Notwithstanding this reservation only a very limited range of chemicals have been subjected to such tests.

Throughout the field of toxicology there exists divergent views about the shape of the dose–response curve. Classical theory viewed the relationship as shown in Fig. 2. but at the present time this is in question. Fig. 5 illustrates the dilemma faced by research workers. ·

Fig. 5. The problems in estimating human risks from animal bioassays. (Source: Reference [34].)

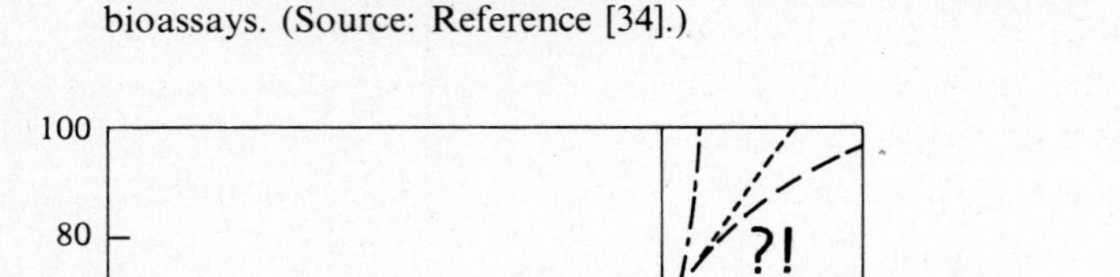

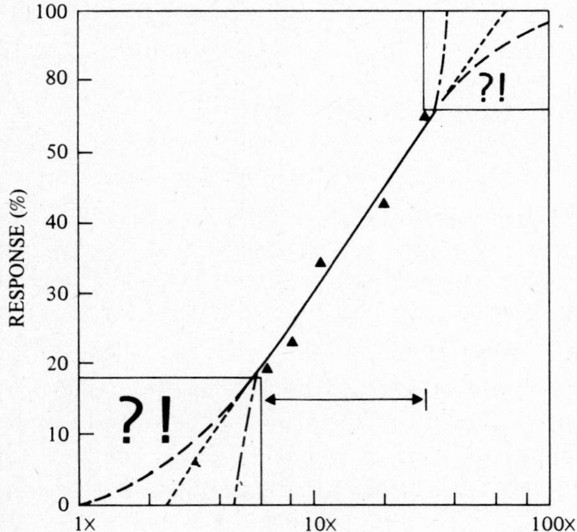

Experimental data in the centre portion of this figure is determined by laboratory studies. Straight-line extrapolation suggests that there is no response at low doses (i.e. a threshold). Scientists continue to disagree about this feature particularly in connection with carcinogens. It is argued that carcinogens are metabolised in a different way to other chemicals and give unrestrained growth of cells and irreversibility. Some argue that every molecule of a chemical carcinogen reacts with a DNA receptor in a cell of the body. Though there is limited evidence to support either of the opposing views, most government regulations such as the US Food and Drug Act and OSHA Regulations lean towards the conservative view that carcinogens do not exhibit thresholds.

Scientific research has identified more than 4.3 million chemical compounds of which some 63 000 are thought to be in common use [35]. Additionally, about 1000 new compounds enter the world market each year [36]. It is therefore necessary to consider the expenditure and resources required to conduct an acceptable testing programme. The present cost of a comprehensive, extensive toxicological test battery to assess satisfactorily human risk (including carcinogenic, teratogenic, mutagenic) may run as high as $800 000 for one chemcial [37]. A less complete test may cost $200 000–300 000 and a current average cost is around $500 000 per chemical.

These tests are almost exclusively confined to single chemicals and, since waste materials are inevitably mixtures of unknown and varying composition, the enormity of the hazard assessment task which confronts waste management specialists can be seen.

It is valuable to ponder the reservations expressed by eminent researchers in the field of toxicology:

D.I. Mount, Director of USEPA Environmental Research Laboratory at Duluth said in 1976 [38]

> We are recognizing more and more problem chemicals in the environment that are not directly toxic but instead bioaccumulate, producing residues in the body. Certainly, toxicity tests designed to measure this end point must be different from those to measure direct toxicity. We also can see now that many of the problem contaminants in the environment are persistent ones. However, we are in the very infancy of knowing how to test persistence of materials effectively and realistically, and a method to measure persistence is an important and current need in our field.

H.L. Falk reported in 1977 [39]

> Many toxic substances reach the environment which have not been tested for their toxicity whatsoever; and if we assume that they all must be tested, there may not be enough mice or rats or even investigators around to do the job. . . . The amount of work needed is staggering and related costs prohibitive. . . . The question of which tests are necessary for a conclusive (satisfactory) answer on product safety for mankind can not be answered with present knowledge and consequently should not be asked.

J.A. Zapp commented that [40]

> We must always keep a measure of skepticism about any extrapolation until it has been verified in human experience . . . which of course is out of the question.

In the waste management field there is no opportunity to control the reactions occurring beneath a landfill site particularly where hazardous liquid wastes have been codisposed with other materials. The composition of the waste materials is unknown to us and, since the composition changes with time as biodegradation occurs, prediction of the precise chemical nature is currently beyond our capability. Knowing the intrinsic difficulties of assessing the harmful effects of individual chemicals we must recognise the impossibility of predicting the toxicity of landfilled waste mixtures. Consequently, in the interests of public health and the maintenance of lowest attainable costs, we must be sure to manage waste residuals so that they can not, at any time, escape into our environment. Our responsibility must be to concentrate and contain hazardous wastes rather than dilute and disperse them into the biosphere.

References

[1] Goldwater, L.J., 'Toxicology', *in* N. Irving (ed.), *Dangerous Properties of Industrial Materials*, Van Nostrand Reinhold Co., New York, 1963.

[2] Castleman, W. et al., 'The Role of Non-Human Primates in Environmental Pollution Research', 174th American Chemical Society Meeting, Chicago, as reported in S.D. Lec & J.B. Mudd, *Assessing Toxic Effects of Environmental Pollutants*, Ann Arbor Science Publishers Inc., Michigan, pp. 15–42, 1980.

[3] Von Hayck, H. (translated by V.E. Krahl), *The Human Lung*, Hafner Publishing Co., New York, 1960.

[4] Schmidt, L.H., *in* W.I.B. Beveredge (ed.), *Selection of Species for Various Uses, Primates and Medicine, Vol 2, Using Primates in Medical Research*, Karger, New York, 1969.

[5] Oser, B.L., 'Toxicology of Pesticides to Establish Proof of Safety', *in* R.

White-Stevens (ed.), *Pesticides in The Environment*, Vol 1, Part 2, Marcel Dekker Publishers, New York, 1971.

[6] US Army Corps. Research and Development Community, 'Respiratory Challenge of Animals with Microorganisms by Means of the Henderson Apparatus', US Army Biological Warfare Labs. Report, Frederick, Maryland, 1957.

[7] Duthie, J.R., 'The Importance of Sequential Assessment in Test Programs for Estimating Hazard to Aquatic Life, *in* F.M. Mayer & J.L. Hamelink (eds.), *Aquatic Toxicology and Hazard Evaluation*, ASTM STP 634, American Society for Testing and Materials, pp. 17–35, 1977.

[8] Dourdoff, P. et al, 'Bio-Assay Methods for the Evaluation of Acute Toxicity of Industrial Wastes in Fish', *Sewage Industr. Wastes*, **23**, 1380–97, 1951.

[9] Lloyd, R., 'The Use of Concentration–Response Relationship in Assessing Acute Fish Toxicity', *in* K.L. Dickson, A.W. Maki & J. Cairns Jnr (eds.), *Analyzing the Hazard Evaluation Process*, American Fisheries Society, Washington DC, 79–92254, 1979.

[10] International Standards Organisation, 'Recommended Changes to ISO Proposed Flow Through Test Procedures', ISO/TC 147/SC5 WG3 DOC 18, May 1978.

[11] Wuhrmann, K. & H. Woker, 'Die Giftigkeit von Phenol für vershiedene Fischarten', *Schweiz Z. Hydrol*, **12**, 271–87, 1950.

[12] Klein, L., *River Pollution II. Causes and Effects*, Butterworths, London, 1962.

[13] Kenega, E.F., 'Aquatic Test Organisms and Methods Useful for Assessment of Chronic Toxicity of Chemicals', *in* K.L. Dickson, A.W. Mali & J. Cairns Jnr (eds.), *Analyzing the Hazard Evaluation Process*, American Fisheries Society, Washington DC, 79–92254, pp. 101–111, 1979.

[14] Mehrle, P.M., F.L. Mayer & W.W. Johnson, 'Diet Quality in Fish Toxicology', *in* F.L. Mayer & J.L. Hamelink (eds.), *Aquatic Toxicology and Hazard Evaluation*, ASTM STP 634, American Society for Testing and Materials, pp. 269–80, 1977.

[15] Katz, B., 'Relationship of The Physiology of Aquatic Organisms to the Lethality of Toxicants', *in* L.L. Marking & R.A. Kimmerle (eds.), *Aquatic Toxicology*, ASTM STP 667, American Society for Testing and Materials, pp. 62–76, 1979.

[16] Marking, L.L., 'Method for Assessing Additive Toxicity of Chemical Mixtures', *in* F.L. Mayer & J.L. Hamelink (eds.), *Aquatic Toxicology and Hazard Evaluation*, ASTM STP 634, American Society for Testing and Materials, pp. 99–108, 1977.

[17] Ellis, M.M., 'Detection and Measurement of Stream Pollution', *Bull. US Bur. Fish*, **48**, 365–437, 1937.

[18] Pearson, J.G. et al., 'An Approach to The Toxicological Evaluation of a Complex Industrial Wastewater', *in* L.L. Marking & R.A. Kimmerle (eds.), *Aquatic Toxicology*, ASTM STP 677, Association of Testing and Materials, 1979.

[19] Dickson, K.L., A.W. Maki & J. Cairns Jnr (eds.), *Analyzing the Hazard Evaluation Process*, American Fisheries Society, Washington DC, 79–92254.

[20] Bulich, A.A., 'Use of Luminescent Bacteria for Determining Toxicity in Aquatic Environments', *in* L.L. Marking & R.A. Kimmerle (eds.), *Aquatic Toxicology*, ASTM STP 667, American Society for Testing and Materials, pp. 98–106, 1979.

[21] Available from Beckman–RIIC Ltd, 6 Stapleton Road, Orton Southgate, Peterborough, PE2 OTB.

[22] Bowen, H.J.M., *Environmental Chemistry of The Elements*, Academic Press, London, 1979.
[23] Muller, W.P. & F. Korte, 'Microbiological Degradation of Benzo-A-pyrene, Monolinuron and Dieldrin in Waste Composting', *Chemosphere*, Vol. 4, No. 3, pp. 195–8, June 1975.
[24] Kelly, M.G. & J.C. McGrath, *Biology: Evolution and Adaptation to The Environment*, Houghton Mifflin Co., 1975.
[25] Rollinson, C.L., *Coordination Chemistry*, Plenum Press, New York, 1969.
[26] Friberg, L. & J. Vostal, *Mercury in the Environment*, CRC Press Inc., 219pp. Boca Raton, FL 33431, 1972.
[27] Smith, W.E. & A.M. Smith, *Minimata, A Warning to the World*, Chatto and Windus Publishers, London, 1975.
[28] Kobayashi, J., F. Morii & S. Muramoto, 'Removal of Cadmium from Polluted Soil with the Chelating Agent EDTA', Proc. University of Missouri's Annual Conference, pp. 179–92, Columbia USA, June 1974.
[29] Sobsey, M.D., C. Wallis & J.L. Melnick, 'Development of Methods for Detecting Virus in Solid Waste Landfill Leachates', *Appl. Microbiol*, Vol. 28, No. 2, pp. 232–8, August 1974.
[30] Mauchline, J. & W.L. Templeton, *Oceanog. Mar. Biol. Annu.*, No. 2, p. 229, 1964.
[31] Timofeeva-Resovskaya, E.A. et al., *Dokl. Akad. Nank SSR*, **140**, 1437, 1961.
[32] Luckey, T.D. et al., *Heavy Metal Toxicity Safety and Harmology*, Georg Thieme Publishers, Stuttgart, 1975.
[33] Jones, G., 'Toxicity Requirements of the UK and EEC', *in* J.W. Garrod (ed.), *Testing for Toxicity*, Taylor and Francis, pp. 1–10, 1981.
[34] Gehring, P.J. & G.E. Blace, 'Mechanisms of Carcinogenesis: Dose–Response', *J. Environ, Pathol. Toxicol.*, **1**, 163, 1977.
[35] Muir, W.R., 'Prevention of Occupational Cancer (Part IV)', *Ann. NY Acad. Sci.*, **271**, 491, 1976.
[36] Council of Environmental Quality, 'Toxic Substances', Washington DC, 1971.
[37] Stara, J.F. & D. Kello, 'Relationship of Long Term Animal Studies to Human Disease' *in* S.D. Lee & J.B. Mudd (eds.), *Assessing Toxic Effects of Environmental Pollutants*, Ann Arbor Science, Michigan, p. 43–76, 1980.
[38] Mount, D.I., 'Present Approaches to Toxicity Testing – A Perspective', *in* F.L. Mayer & J.L. Hamelink, *Aquatic Toxicology and Hazard Evaluation*, ASTM STP 634, American Society for Testing and Materials, pp. 5–14, 1977.
[39] Falk, H.L., 'The Toxic Substances Control Act and In Vitro Toxicity Testing', *In Vitro*, **13**, 676, 1977.
[40] Zapp, J.A. Jnr, 'Extrapolation of Animal Studies to The Human Situation', *J. Toxicol. Environ. Health*, **2**, 1425, 1977.

10

Hazardous waste treatment and disposal options

Introduction

This discussion is concerned with the treatment and disposal of hazardous waste off-site; that is, wastes which have been transported through the factory gate. As we have seen from other chapters in this book, the most frequently used method of industrial waste disposal is landfill. The alternatives to landfill are described here and all these technologies are in current operation. Certain exotic systems such as pyrolysis and wet air oxidation are omitted as being of no practical significance at the present time but this is not to say that they will remain unavailable commercially. Land-based incineration is discussed in a separate chapter owing to its relative importance in the UK waste disposal field. The following systems are discussed in approximate order of installed capacity, to reflect, as far as possible, the relative significance in terms of the volume of waste handled:

(1) waste solidification;
(2) sea disposal;
(3) simple chemical treatment with sewer facility;
(4) oil/water separation;
(5) mine disposal;
(6) marine incineration;
(7) fuel conversion

There was promised some legislative pressure to promote proper treatment of wastes in specially designed plant during the early 1970s. The Control of Pollution Act 1974 [10] set out to impose minimum standards so far as waste disposal practice was concerned but the results of the delays in implementation of the Act are plain to see from a study of Table 1.

Waste solidification

By far the most exciting and significant development in modern waste disposal practice is that of the introduction of solidification processes into the market place.

Table 1. *Installed treatment capacity and current throughput of UK hazardous waste treatment facilities*

Operator	Location	Type of wastes processed	Capacity t/year	Current throughput t/year	Spare capacity t/year	%
DP Effluents	Runcorn, Cheshire	Acid treatment and liquid wastes	30 000	15 000	15 000	50
Polymeric Treatments Ltd	Killamarsh, North Derbyshire	Dewatering of sludges	5000	Due to commence end of 1980	5000	100
Hargreaves Clearwaste Services Ltd	Wakefield, West Yorkshire	Most inorganic wastes	10 000	2000	8000	80
Re-Chem International Ltd	Fawley, Hampshire	Most inorganic wastes	20 000	11 000	9000	45
Re-Chem International Ltd	Pontypool, Gwent	Most inorganic wastes	20000	11 000	9000	45
Re-Chem International Ltd	Roughmute, Stirlingshire	Most inorganic wastes	20 000	1000	19 000	95
Safeway Sludge Disposal	Garretts Green, Birmingham	Acid treatment and sludge dewatering	30 000	12 000	18 000	60
Totals			135 000	52 000	83 000	61
'Solidification' plants (mainly for treatment of inorganic wastes)						
Polymeric Treatments Ltd	Aldridge, West Midlands		250 000	150 000	100 000	40
Stablex	Thurrock, Essex		300 000	25 000	275 000	92
Totals			550 000	175 000	375 000	68

Incineration (mainly for treatment of organic wastes)						
Berridge Incinerators Ltd	Hucknall, Notts	Liquids including halogenated	9000	6000	3000	33
Croda Synthetic Chemicals	Four Ashes, Wolverhampton	Non-halogenated liquids	5000	3000	2000	40
Hargreaves Clearwaste Services Ltd	Wakefield, West Yorkshire	Non-halogenated liquids	15 000	1000	14 000	93
Re-Chem International Ltd	Fawley, Hampshire	Solid and liquid wastes – virtually all types	15 000	15 000	—	—
Re-Chem International Ltd	Pontypool	Solid and liquid wastes – virtually all types	20 000	12 000	8000	40
Re-Chem International Ltd	Roughmute, Stirlingshire	Solid and liquid wastes – virtually all types	20 000	10 000	10 000	50
Cleanaway Ltd (formerly Redland Purle Ltd)	Ellesmere Port, Cheshire	Bulk liquids, including halogenated	20 000	10 000	10 000	50
Cleanaway Ltd (formerly Redland Purle Ltd)	Rainham, Essex	Non-halogenated	13 000	2000	11 000	85
Totals			117 000	59 000	58 000	50

Source: Reference [11].

Undoubtedly, the greatest problem facing industry as far as waste disposal is concerned is the safe and effective disposal of its liquids and sludges. Solid wastes can generally be landfilled at suitable sites but free-flowing materials present many problems when considering this traditional outlet. Solidification processes are designed to overcome the inherent difficulties experienced in the disposal of liquids and sludges in that the material is converted to a solid mass prior to ultimate disposal.

Legislative and public pressures during the early 1970s should have encouraged the widespread development of solidification as a waste disposal method and the introduction of the Control of Pollution Act in 1974 promised to consolidate and translate pressures into practice. However, such was not the case and the piecemeal implementation of the Act resulted in loss of impetus and momentum so that solidification has had a slow start instead of the deserved mushroom. Nevertheless, there is little doubt that the UK represents the world leader in solidification technology.

The reader is referred to the paper by Chappell & Butler [12] presented at the Effluent and Treatment Convention 1978 for a resumé of the impact of legislation on the development of waste treatment solidification plant.

The solidification of industrial waste is a valuable end in itself, but, concomitant with the physical solidification, there is usually an additional chemical fixation benefit. The two words are often used synonymously and, since the two benefits are inseparable, this is acceptable although not strictly correct.

The various processes involve the more or less homogeneous fixation of toxic waste(s) in a solid matrix, the embedding of larger units of toxic wastes within a continuous three-dimensional monolithic mass of the solidification agent and the encapsulation of individual packages of waste by surrounding them with the solidification agent.

The basic objective of fixation is to isolate the toxic wastes from the environment by producing materials having low leachability and/or low permeability and possessing mechanical and structural integrity. Such materials may then be disposed in a relatively non-hazardous form as a land reclamation medium or, possibly, re-used as an engineering material.

Solidification systems available

Inorganic systems are usually based upon silicate-containing constituents. They are generally used to produce a slurry which sets into a solid material with properties which range from a soil-like material to hard, monolithic, rock-like structures. Vitrification is a rather special and expensive inorganic system which is reserved for highly dangerous wastes such as high-level radioactive wastes.

Organic systems use either thermoplastic or thermosetting polymers as solidification agents. Thermoplastics are used in the molten form while thermosetting materials are polymerised *in situ*.

Inorganic systems

The major system offered commercially is SEALOSAFE, introduced in August 1974, and being currently offered at three UK plant. The concept is one of regional treatment centres such that wastes are brought into the centre for storage and subsequent processing. The alternative strategy of solidification of a waste at a producer's premises is discussed later when the CHEMFIX process is outlined.

SEALOSAFE is operated at two plant in Brownhills, West Midlands by Polymeric Treatments Ltd and one at West Thurrock, Essex by Stablex Ltd. Planning consent is extant for a plant near Morley in West Yorkshire and, being a regional centre concept, more are planned subject to market forces.

UK Patent 1 485 625 [13] describes the SEALOSAFE principles and the principal claim is that of solidification. Subsidiary claims relating to permeability and compressive strength are presented and the examples quoted in the Patent include results of leaching tests.

The operators' claims for the resultant product, called STABLEX, after allowing to set are:

low permeability;
low leaching characteristics;
significant compressive strength;
non-biodegradable;
non-flammable;
non-odorous;
unattractive to vectors of disease.

Thus, the product is said to be inert and the pollutant species of the original wastes are not subject to significant release by any natural agency.

Basically, the process involves the mixing of a prepared waste 'stock' with various reagents to produce a slurry which sets in around three days and attains its maximum strength in about six months. The reagents used are cementitious and pozzolanic and this accounts for the ability to set. Various lime-containing cements and complex alumino-silicate pozzolans can be used.

Process description

Prior to acceptance of any waste at a SEALOSAFE plant, a sample is submitted for laboratory analysis, evaluation, formulation and testing. At the laboratory the waste sample will be thoroughly screened for its suitability for SEALOSAFE treatment and a sample product will be made and tested. The test work performed will include leaching, permeability and compressive strength studies and determinations. When the laboratory is satisfied that the chosen formulation meets the STABLEX standards of product performance the formulation sheet is signed off and a copy passed to the plant.

This formulation note is used at the plant as the basis of treatment method for the particular waste. The process involves the blending of various wastes from storage in pre-determined sequences and quantities and high-speed disintegration of solids to form a homogeneous waste 'stock'. Pre-treatment chemicals may be added at this stage.

The stock is then blended with two or more powdered reactants in a mixer and this stage of the processing is the 'polymerisation' step. Inorganic polymerisation based on silicate chains takes place and a setting reaction commences.

After a suitable contact time in the mixer, the waste is fully reacted and is safe and suitable for transport to a disposal site. At this stage, the polymerised product is in the form of a slurry which can be pumped or tankered away to a remote disposal site or poured for casting purposes.

There are no atmospheric emissions or discharge to sewer and there are no elevated temperatures or pressures involved during processing. The whole waste is solidified during reaction and no byproducts are formed requiring separate disposal.

It has been claimed that X-ray crystallographic evidence suggests that fixation of pollutants in the setting matrix has a chemical basis in addition to physical entrainment (Chappell [14]).

Plant description

A SEALOSAFE plant is of flexible design and no two need be the same. Basically, a plant comprises storage, processing and disposal units.

(a) Storage

Wastes received at a SEALOSAFE plant can be in solid, sludge or liquid form and can arrive in a range of containers and quantities varying from retail packages to bulk road tankers.

Solid wastes are usually segregated into general groups (for example, filter cakes, paint solids, lime cake, etc.) and stored in covered bunkers.

Drummed wastes are separated and segregated into cordoned-off compounds and drums are labelled with unique reference numbers related to a plant register containing analytical details and treatment method.

Liquid wastes can be handled in one of two ways. At the Thurrock STABLEX plant, liquid wastes are tested for compatibility and, if suitable, are stored pending treatment in one of the three concrete, below-ground holding basins which have a total capacity of some 4 million gallons (18 million litres). At the Brownhills SEALOSAFE plant, the liquid wastes are segregated into sulphuric, hydrochloric, nitric and chromic acids, neutral wastes, alkaline wastes and organically contaminated wastes. Storage tanks in this case are of glass-reinforced polyester construction, with or without polyvinyl chloride lining, or mild steel depending upon the contents.

(b) Stock preparation

From storage, liquid wastes are drawn upon by pumps and pipes of suitable materials and construction and are loaded according to a pre-determined programme into a 'stock preparation' (SP) tank. Each of these SP tanks sits on a weigh-cell which detects and records the weight of waste added.

The SP tanks are equipped with either a high-speed disintegrator or an homogenising device adapted from a conventional washmill. This equipment serves to mix, blend and homogenise the waste and also break up any solid matter present in the liquid wastes.

To the blended, liquid stock the solid wastes can be added for homogenisation by screw feed, conveyor feed or mechanical shovel. In the case of a washmill type SP tank, whole drums can be added and these will be disintegrated: for example, drummed cyanide is loaded whole into the SP tanks at Thurrock.

(c) Polymerisation

The plant chemist checks each SP tank's contents and when the prepared stock is ready for polymerisation, he will signal to the plant controller. From the SP tank the prepared waste mix is pumped into a holding silo at the top of the polymerisation tower. The stock waste is then fed to the reactor vessel, together with powdered reagents which are drawn from storage in high-level hoppers alongside the waste silo. The feeding of waste and of reagents is governed by electronically controlled weigh-batching apparatus that ensure the proportions of materials being delivered to the reactor conform with the original laboratory formulation.

The reactor vessel itself is a horizontal, sealed mixer fitted with doubleblade paddles.

(d) Discharge
From the reactor vessel, the treated waste is discharged via hydraulically operated doors into a holding hopper. This STABLEX can then be pumped away to a remote site or discharged into awaiting road, rail or canal transport.

Product assessment

The testing of STABLEX can include a number of important procedures designed to measure various parameters.

The principal areas of interest, especially as far as environmental significance is concerned, are:

permeability;
leaching;
compressive strength.

(A) Permeability

The degree of water pollution that might occur when industrial waste is dumped into a landfill site is strongly influenced by a parameter known as permeability; the rate at which water passes through a material. High permeability means that water passes rapidly through the waste and there is therefore a correspondingly high risk that the water will be badly polluted. Low permeability means that the water passes slowly through the waste and that the water is more likely to run off the surface of the waste or evaporate than pass through it, thus greatly reducing the possibility of water contamination.

Furthermore, the permeability of the ground surrounding and beneath the waste deposit is also important as it can affect the degree of pollution of underground water supplies which may be widely used for potable and industrial purposes.

For these reasons, the permeability of SEALOSAFED wastes is an important consideration in assessing risks of water pollution.

Permeability is measured by passing water under pressure through a specimen clamped in a specially designed holder and determining the water flow rate. The co-efficient of permeability is defined as the rate of flow of water per unit area of specimen under unit hydraulic gradient, the dimensions being cm sec^{-1}.

The STABLEX under test is cast as a slurry into a greased ring mould to produce a squat cylinder of 5 cm diameter and 1.5 cm depth. The sample is vibrated or tapped to remove entrained air. Each mould is placed in an incubated environment at $20°C \pm 2°C$ at equilibrium humidity for three

The mould should be smeared with light oil or releasing fluid. The sample should be vibrated or tapped to remove entrained air. The specimen is cured for three days at an equilibrium humidity in an incubator at 20°C ± 2°C then removed from the mould and coated with paraffin wax. Curing at 20°C ± 2°C is continued for 25 days.

After curing, the wax is removed from the specimen. If the ends are not plain they are either trimmed to a tolerance of 0.05 mm or capped with gypsum plaster or cement paste to a tolerance of 0.05 mm. The length and cross-section area of the specimen are measured and the specimen placed centrally on the lower platen of the machine.

The force is applied to both ends of the specimen such that the rate of deformation is uniform and approximately 1 mm min^{-1}. The maximum force exerted during the test is recorded.

The unit compressive strength of the specimen is calculated as maximum load per unit area.

Results

Table 7 presents some typical strength data on solidified wastes.

It can be seen that strength increases with time and maximum is usually attained after about six months. There is considerable variation in the measured strengths of various samples and the differences depend upon the nature of the waste that has been solidified.

By way of comparison, the following data are presented in Table 8.

Thus, the solidified products of the SEALOSAFE process are comparable in strength to grouts which are used in civil engineering for void filling and soil stabilisation. The SEALOSAFE products are, therefore, suitable for reclaiming disused quarries and for bulk cavity filling and/or grouting.

Table 7. *The compressive strength of SEALOSAFED waste*

Compressive strength in MN m^{-2}			
	3 days	7 days	28 days
An arsenical waste		2.69	5.17
An aqueous effluent	1.33	2.28	4.21
A chromic waste		0.74	1.52
A chromium waste		1.07	2.14

Source: Reference [15].

Practical considerations

SEALOSAFE is a practical option for waste treatment/disposal and the following brief summary describes the history of the processing capabilities:

invented:	early 1970s;
patented:	1973 in the UK;
pilot plant:	Brownhills, 1973–74;
first plant:	nominal 50 000 t/a, August 1974, Brownhills;
second plant:	nominal 250 000 t/a, July 1978, Thurrock;
third plant:	nominal 200 000 t/a, July 1978, Brownhills;
fourth plant:	planning stage, nominal 50 000 t/a, West Yorkshire;
international:	operating plant in Japan; several in advanced preparation/construction phases in North America; developments throughout Europe.

Since the first plant at Brownhills was opened, more than 1 million tonnes of liquid, sludge and solid wastes have been treated using SEALOSAFE technology.

Range of wastes treated

There are currently over 2000 different types of wastes formulated for SEALOSAFE treatment and these represent wastes produced by a large section of modern industry.

Table 8. *Comparative compressive strength of materials*

Typical compressive strength of materials	
	MN m^{-2}
Concrete (BS 12) consisting of a standard mix of Portland Cement, sand and gravel after 28 days	30.0 ± 7.0
Mortar (BS 12) consisting of a standard mix of Portland Cement and sand, after 3 days	14.5
Industrial grouts used for void filling, soil stabilisation mud jacking and general site work, after 28 days	0.5 – 4.0
STABLEX after 28 days	1.4 – 5.5

Source: Reference [15].

Compounds consisting wholly or partly of any of the following can be readily processed [14]:

- Hydrochloric acid
- Sulphuric acid
- Nitric acid
- Chromic acid
- Phosphoric acid
- Hydrofluoric acid
- Other acids
- Sodium or potassium hydroxides or oxides
- Calcium oxide
- Alkaline cleaners
- Calcium hydroxide
- Sodium or potassium carbonates
- Magnesium
- Aluminium
- Titanium
- Vanadium
- Chromium
- Manganese
- Cobalt
- Nickel
- Copper
- Zinc
- Arsenic
- Selenium
- Molybdenum
- Silver
- Cadmium
- Tin
- Antimony
- Tellurium
- Barium
- Mercury
- Thalium
- Lead
- Nitrites and nitrates
- Borates
- Arsenates and arsenides
- Sodium or potassium cyanide
- Soluble complex cyanides
- Ferrocyanides and ferricyanides
- Sulphides
- Selenides
- Tellurides
- Hypochlorites and chlorites
- Peroxides
- Chromates and chrome compounds
- Fluorides, silicofluorides, borofluorides
- Certain phenolic compounds
- Polyester resins
- Latex
- Ion exchange resins
- Filter media
- Paint wastes
- Interceptor pit wastes
- Ink manufacture wastes
- Acid tars
- Asbestos and asbestos products

The types of wastes handled and the typical pollutants present are summarised as [14]

Liquids

Source	*Pollutants*
Adhesive manufacture	Mercury
Gas purification plant	Arsenic

Gas purification process	Alkaline sulphides
Paint spray booth	Cadmium, lead, tin
Plating waste	Acid, tin
Plating liquor	Acid, chromium
Plating solution	Acid, cadmium
Metal processing	Cyanide

Solids

Source	*Pollutants*
Catalyst residues	Vanadium
Heat treatment residues	Cyanide, barium
Petrochemical catalyst	Cobalt, molybdenum, nickel
Spent catalyst	Chromium
Metal refining dust	Lead
Metal smelting dust	Fluoride
Fire extinguisher manufacture	Alkali, cadmium
Metal refining	Antimony
Metal recovery plant	Copper, nickel, zinc
Gas purification	Sulphides, cyanides
Furnace residues	Vanadium
Building products and asbestos	Asbestos and asbestos products

Sludges and filter cakes

Source	*Pollutants*
Plating industry	Cadmium, zinc, cyanide
Metal smelting	Alkali, arsenic
Tannery	Sulphide, chromium
Acid pickling	Acid, chromium, zinc
Electrical components manufacture	Carbon, cyanide
Railway washings	Alkali
Effluent treatment	Copper, tin
Printing trade	Copper, zinc
Chemical manufacture	Barium
Pharmaceutical manufacture	Zinc
Electroplating plant	Chromium, lead, copper, nickel
Chloralkali plant	Mercury
Metal finishing	Cyanide, zinc, lead, tin
Aluminium finishing	Alkali, heavy metals

Re-use capabilities

At Brownhills, the STABLEX is being used in a land reclamation project in which an abandoned clay quarry, some 30 metres deep, is being filled. The clay hole is a 'safe' site and could be used for the reception and deposition of a wide range of industrial wastes in an untreated form, together with domestic refuse. However, when fully reclaimed with STABLEX, as it will be in the near future, it will be possible to release the new landform for immediate development. There will be no time wait of 20–30 years to allow for decomposition and settlement as would be the case if refuse were being deposited. Indeed, it is envisaged that the site will be developed and provided with light industrial units and warehousing facilities within a short time after final STABLEX deposition.

The Thurrock scheme envisages the infilling and reclamation of disused chalk quarries. Many such sites exist in the south east of the UK and have remained derelict owing to the massive potable aquifer that underlies them. Traditional landfill in such sites is impossible: STABLEX can be used in such 'sensitive' sites without the same risks of pollution.

During such landfilling and reclamation schemes as are operated at Brownhills and envisaged at Thurrock, STABLEX sometimes exhibits drying and shrinkage cracks during hot weather. This phenomenon does not affect the STABLEX mass as the next deposition, in slurry form, seeks out and grouts any fissures that are present. Thus, the final landfill comprises a genuine monolithic 'plug' which, to all intents and purposes, is cemented onto the base and sides of the original hole.

Economic appraisal

Direct costs for waste disposal methods are relatively simple to compute. In its submission to the House of Lords enquiry on hazardous waste disposal, Polymeric Treatments Ltd gave the following (Table 9) cost breakdown (1980–1 prices).

These prices reflect the chemical reagent costs and the handling difficulties associated with a particular waste. For example, when considering drummed wastes, part of the price relates to emptying, cleaning and manual handling prior to processing.

An important factor in the costs of waste disposal is the contingent liability should anything go wrong with the disposal route chosen. For example, should landfilling of wastes be chosen and polluting liquid issue from the landfill site, very large costs can be incurred over many years. The principal factor to consider must be one of 'environmental impact'. In America now, each new project at the planning stage must be accompanied by the Environmental Impact Statement which must assess the various effects that the project will have on the environment. Waste disposal is an essential part of each statement and the regulatory authorities are required to consider the various effects before making a decision. The Council of the European Communities is considering adopting a Directive which will implement a similar system of 'environmental impact analysis' to be used at the planning stage in the Member States (Draft Directive 7972/80 [22]).

The cost to the community of future liabilities must be added to the direct costs. For example, if aquifers underlying tip sites are polluted then the costs of provision of alternative water supplies must be included in the estimate of waste disposal costs. Evidence to date suggests that the product of wastes solidified by SEALOSAFE maintains its integrity and so consequential future liabilities after disposal are minimal and probably zero.

Other inorganic systems

Chemfix

Operated in the UK under licence, by Wimpey Waste Management Ltd, the Chemfix process comprises a trailer-mounted rig which is able to visit industrial premises to undertake solidification of collected batches of wastes. For example, whole collecting ponds or lagoons of wastes may prove amenable to treatment. Essentially, a solution of sodium silicate (or other alkali silicate) is mixed with a setting agent such as calcium chloride, cement or calcium sulphate and with the waste, for example mineral acid, to form a gel. The product is pumped away as a slurry and allowed to set.

Obviously, for Chemfix to be suitable as a treatment option a placement

Table 9. *Costs of waste solidification*

Waste category	Proportion of total waste treated %	Cost range £/tonne
1	3.5	0–10
2	81.0	10–20
3	9.0	20–50
4	5.5	50–100

Source: Stablex Ltd: submission to the House of Lords Enquiry on Hazardous Wastes (unpublished but lodged on public register for statutory period).

area to receive the product must be available on site so that the waste lagoon can be pumped over and set on another area.

UK Patent 1 337 301 [23] describes the process which is offered by Chemfix Inc. of Kenner, Louisiana.

Others

Other UK and USA systems are offered by:

Petrifix (UK) Ltd., Foregate Street, Worcester (See US Patent 4 124 405 [24]);

ANEFCO, White Plains, New York;

Chem. Nuclear Systems Inc., Bellevue, Washington;

Ionization International Inc., Chicago III (Chem-onazone Division);

Chemical Waste Management Inc., Calumet City III;

Dow Chemical Co., Midland, Michigan;

Dravo Corp., Pittsburgh, Pennsylvania;

Environmental Technology Corp., Pittsburgh, Pennsylvania;

Industrial Resources Inc., Chicago III;

IU Conversion Systems Inc., Horsham, Pennsylvania;

John Sexton Landfill Contractors Inc., Oakbrook;

Manchak Environment Inc., Santa Barbara, California;

Nuclear Sources and Services Inc., Houston, Texas;

Solidtec Inc., Morrow, Georgia;

Stabatrol, Norristown, Pennsylvania;

Stablex Corp., Radnor, Pennsylvania;

Stock Equipment Co., Chagrin Falls, Ohio;

System Technology Corp., Zenia, Ohio;

Teledyne Corp., Timorium, Maryland;

Todd Shipyards Corp., Galveston, Texas;

TRW Inc., Cleveland, Ohio;

United Nuclear Industries Inc., Richland, Washington;

Washington State University, Pullman, Washington;

Wehran Engineering Corp., Middletown, New York;

Werner and Pfleiderer Corp., Waldwick, New Jersey.

Many of the American systems are only applicable to scrubber sludges from power stations.

Organic systems

Thermoplastic methods

These involve the mixing of bitumen, polyethylene or other thermoplastics with a dried waste at considerably elevated temperatures. The mixture solidifies on cooling.

Clearly, this is a purely physical isolation. It is a very expensive process and must be carried out with extreme care where the heating of waste gives rise to toxic or hazardous emissions. The product will, of course, be inflammable. Organics in the waste which act as a solvent or plasticiser for the thermoplastic cannot be treated in this way.

Because of the high cost of the process, it is unlikely to find general application and will not be competitive with alternative methods of dealing with organics, such as incineration. Werner and Pflederer Corp. have a process for binding nuclear waste by this method.

Thermosetting methods

In situ polymerisation is undertaken by intimately mixing monomer with the waste and then polymerising in a container in which it is to be disposed. The most common system uses urea formaldehyde condensation. Again, this is an extremely expensive system and is unlikely to find widespread use. Processes are available from Teledyne Corp. and ANEFCO. Dow and Washington State University have developed other polymer systems for use in solidification.

Sea disposal

The dumping of industrial wastes at sea is subject to its own specific legislation and this has been dealt with at some length elsewhere. As with all waste disposal practice, the modern-day sea disposal operations exist because of legislation and so there is much operational detail included in the section on the Dumping at Sea Act 1974 [25].

This method of disposal is, one could suppose, the ultimate 'dilute and disperse' option. Reliance upon the natural marine environment to accommodate industrial pollutants and to dilute them sufficiently and to alter them chemically or biochemically to non-polluting forms is the keystone of this particular philosophy. Undoubtedly, the capacity of the marine environment to accept polluting loads is without realistic measure. There are mechanisms of chemical cycling and environmental dispersion such that, in general, pollutants discharged in this way soon become indistinguishable from natural background levels.

Caution as to whether or not the capacity to absorb is infinite has to be expressed: the inter-governmental conventions reflect this and seek to restrict the option of dumping at sea to those wastes that are most likely to degrade and disperse without harm to the natural ecosystems in the seas.

Perhaps the colossal spread of persistent pesticides such as DDT has helped to engender this caution: certainly, the consequences of mercury poisoning in Minimata Bay have had their impact.

Thus, there is a significant body of opinion seeking to curtail the disposal of industrial wastes at sea so long as there are realistically available land-based alternatives. For example, heat treatment cyanide salts and contaminated rubbish were, for a long time, dumped overboard in drums at authorised dumping grounds. This, formerly, was an essential practice since no sensible alternative was available. However, the SEALOSAFE process was shown to be environmentally preferable and economically acceptable: nowdays, very little cyanide wastes are dumped at sea. The principal disposer of cyanides still maintains a small volume of sea disposal business in order, one suspects, to preserve the principle should the need ever arise in the future to extend the practice once again: one would imagine, and hope, this to be unlikely.

Chemical treatment to sewer

There is a handful of treatment plant in the UK that are able to offer chemical treatment of industrial wastes. Most involve simple neutralisation of aqueous effluents followed by coagulation or flocculation and discharge of supernatant liquors to sewer.

The plants are simple in concept and design. Acidic effluents can be taken in and reacted with either alkaline wastes, waste lime or purchased caustic salts to effect neutralisation. Concomitant precipitation of metals as their hydroxides renders the plant effluent suitable for discharge to sewer following a period of settlement or chemical coagulation of suspended solids. Thus, various industrial wastes can be treated such that normal sewer discharge consent limits can be met, namely neutral pH, limited soluble metals and limited solids in suspension. Recovered solids must be disposed to landfill separately.

In essence, these treatment plant are available in the market place as an alternative to installation of an in-house plant at an industrial premises. Cost savings for the industrialist may be considerable when making use of such regional waste treatment plant.

The discharge to sewer is, of course, controlled and consented by the appropriate water authority and the legislative provisions relating to this are discussed elsewhere.

Descriptive literature on the treatment plant is scarce and it is not, therefore, possible to give further details in this section. However, processes such as cyanide oxidation using chlorine gas or hypochlorite solutions and chromium VI reduction to chromium III are operated at some plant in addition to simple neutralisation.

Oil/water separation

As an extension to the concept of a simple plant treating industrial wastes to render them suitable for sewer discharge, various oil recovery plant in the UK have been developed. These, of course, become increasingly popular as the cost of virgin oil rises.

In principle, oily wastes and oil/water emulsions are taken in to these plant for separation and recovery of oil. The oil can be used in-house for steam generation at the plant or sold as a low-grade fuel.

As with the chemical neutralisation plant, information on oil recovery from industrial wastes is scant. Nevertheless, the basic principles are well established.

The crudest form of oil/water separation is that of gravity settlement. Wastes are taken in to large storage vessels and allowed to settle and stratify. Oil is drawn off from the top, 'dirty' water from the middle and settled sludge from the base. The water should be suitable for sewer discharge and the sludge for landfill.

As an aid to settlement and for the purposes of 'cracking' oil/water emulsions into the separate constituent phases, heat and/or acids can be used. Heat, of course, can be provided by using recovered oil to fuel on-site boilers and acids can be supplied in the form of waste acids so long as neutralisation facilities are available at the plant premises.

The 'dirty' water may require further treatment prior to sewer discharge, depending upon water authority requirements by way of consent conditions. This 'polishing' can be achieved by use of:

centrifugation;
ultrafiltration;
tilted-plate separators;
combinations of above.

Mine disposal

Reproduced here is a modified and updated version of a paper presented at the International Environment and Safety Exhibition and Conference during 1–4 September 1980 at the Wembley Conference Centre, London [26]. (By permission of Labmate Ltd, St. Albans, Hertfordshire.)

It describes the systems applicable to the UK in so far as sub-surface disposal of industrial wastes is concerned.

Historical

Disposal of liquid wastes via 'soakaways' is an age-old process which relies upon the natural filtration (physical, chemical and biological) of the earth to purify the waste.

The deliberate disposal of liquid wastes into underground voids is also ancient but, relatively recently, has become subject to legislative control. Usually, access to sub-surface disposal outlets has been private and reserved for single waste types: for instance, several private concerns have access to abandoned mineshafts for the controlled disposal of their own process wastes.

It is necessary to note that the voids from former coal workings are the property of the National Coal Board and permission must be obtained from them.

There are two facilities now operating in the UK whereby access to underground voids is offered in the market place and not reserved for singular or private use.

Legislation

Since 1963, discharge to underground strata via wells, boreholes or pipes has been subject to the authorisation and consent provisions of Section 72 of the Water Resources Act 1963 [27], now administered by the regional water authorities. Such consents usually have conditions attached thereto which serve to satisfy the regulatory authority that no undue damage to underground waters will occur.

These provisions will be replaced and repealed by Section 34 of the Control of Pollution Act 1974 and registers of consents granted by virtue of this Section will be made available to the public at a date prescribed upon implementation of Section 41 of the same Act.

At present, registers of consents are not, by virtue of law, available to the public. However, many water authorities do afford public access.

The provisions of Part I of the Control of Pollution Act 1974 relating to the licensing of activities for the disposal of controlled waste also apply in certain circumstances.

Current activities

It is impossible to paint an accurate picture at the moment as some water authorities decline to make consent information available. Nevertheless, it is certain that most consents relate to domestic septic tank discharges.

Of the industrial discharges, the following list illustrates the scope of activities that currently obtains:

coal washery discard;
paper mill sludge;
organohalogen residues;
cooling water;

mercury sludge;
coke oven liquor;
propylene (storage as liquid);
creamery washdown;
gas liquor condensate;
ammoniacal liquor;
phenolic liquor;
sulphuric acid pickle liquor;
pottery slip and drilling muds;
various industrial liquid wastes.

As indicated, all but two of these are for private use. The two commercially operated underground disposal facilities are at Chatterley Whitfield and at Walsall Wood and these are now discussed.

Chatterley Whitfield

The National Coal Board operate a disposal facility for the discharge of liquid pottery slip and bentonite drilling muds into a disused mine complex at Chatterley Whitfield, Stoke-on-Trent.

The disused mineworkings are shown in generalised section in Fig. 3. Access to the flooded workings for liquid waste is gained via one of the twin

Fig. 3. The mine complex at Chatterley Whitfield.

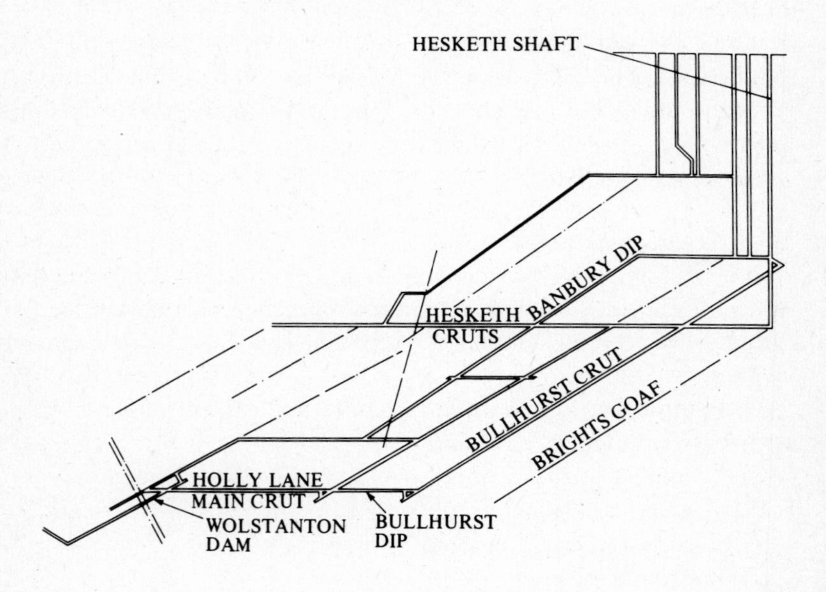

steel pipes within the old Hesketh shaft which conducts the waste into the old Brights seam.

Prior to closure, Chatterley Whitfield was connected via a series of roadways to Wolstanton Colliery, which is currently in operation. When Chatterley Whitfield closed it was planned that any make of mine waters in the Chatterley workings would migrate to Wolstanton where they would be pumped to waste. To control this a dam was erected in the Wolstanton–Chatterley connection.

Since the mine waters being pumped from Wolstanton are now subject to standard consent limits set by the water authorities, it is an essential requirement of the operation that no heavy metals from the pottery slip find their way into the Wolstanton discharge and this requirement was a main consideration in the planning of the disposal facility.

Near the base of the Hesketh shaft is an access road into the old Brights working at Chatterley Whitfield and one of the two steel pipes within the Hesketh shaft was designed to conduct the waste into this seam. At some stage it was envisaged that the pottery slip would displace mine water from the Brights workings and this mine water would then flood the pit bottom areas and the Hesketh cruts which feed the Bullhurst dip towards the Wolstanton dam. Initially, any such water would be contained within the pit bottom area by virtue of two specially built half-walls within the pit roadways, the height of these walls being coincident with the base level of the shaft which provided a means of sampling any water reaching this level.

If the pottery slip displaces clean water without heavy metal contamination, this can be allowed to overflow the half-walls and travel to Wolstanton, but if displaced water is contaminated with heavy metals the process would have to cease before being allowed to overflow the half-walls.

To ensure that the overflow is 'clean' a comprehensive control and monitoring scheme has been adopted. All wastes are initially sampled and analysed to assess suitability for the facility by the licence holder, Effluent Disposal Ltd. Only those wastes that are suitable are scheduled for disposal. At the disposal complex, each tanker is sampled and checked by NCB operatives prior to being coupled to the discharge pipe. The samples are retained and one in 30 (formerly one in ten) are subjected to full analysis by the NCB at their comprehensive laboratory complex. Additionally, the Staffordshire County Council take random samples for analysis by the County Analyst.

Weekly tests for the presence of liquid in the pit bottom are made by the NCB but none has yet appeared. Weekly analyses of this liquid are stipulated in the site licence and also of the discharge from the Wolstanton dam. Consent limits for suspended solids and toxic metals are 200 mg l^{-1} and 4 mg l^{-1} at the pit bottom and 50 mg l^{-1} and 1 mg l^{-1} at the dam.

The facility relies upon settlement of the clay fines from the slip and drilling muds. Since operations started in December 1978, the total volume of liquid waste discharged has already exceeded the estimated volume of the pit bottom area and suggests that the facility could be operating as a vast settlement chamber. However, the efficiency cannot be estimated until analysis of any liquid from the pit bottom has been obtained.

The opening of this disposal outlet has helped to reduce the volume of liquid waste being dumped onto landfill sites in the Staffordshire area. This is, of course, important, as the whole concept of disposal of liquid wastes onto landfill sites is in question as being suspect.

Walsall Wood

Many people in the waste disposal industry will be familiar with the mine disposal facility operated by Effluent Disposal Ltd in the late 1960s and early 1970s (Brownhills, West Midlands).

The underground workings in this case are completely dry and the facility has been described as a 'clay bottle' having no overflow.

The facility opened in 1966 for the reception of a variety of liquid wastes including inorganic acids, organic acids, heavy metal containing solutions, phenolic liquids, oily wastes, dairy wastes, scrubber liquors, lagoon liquors, treated cyanide wastes and many others. At its peak, it was accepting around 35 000 000 gallons of liquid waste per annum. Towards the end of 1976, however, the access became blocked by falling brickwork from the deteriorating shaft.

Again, the voidspace is owned by the NCB. Fig. 4 delineates the geographical extent of the worked mine. It will be seen that the coal measures of the Walsall Wood complex are isolated by massive geological faults which have thrown low permeability strata into juxtaposition with them and which have effectively isolated and sealed the workings as a 'bottle'.

After the collapse of the shaft in 1976, Effluent Disposal Ltd, the site operators, sought and obtained permission to drill an exploratory borehole into the workings. This intersected four workings, each of which was free of liquid, and the sampled atmosphere was typical of mine air. In January 1978, a plastic transmission line was inserted into the exploratory borehole and liquid wastes from an adjacent treatment lagoon were discharged into the workings. This particular activity is now terminated and the treatment lagoon is in the final stages of infilling and reclamation.

In December 1978, Effluent Disposal Ltd were granted planning permission, following an extensive public enquiry, to construct a new discharge borehole into the workings from the Empire site which also houses the SEALOSAFE treatment plant of Polymeric Treatments Ltd.

Fig. 4. The Walsall Wood mine workings

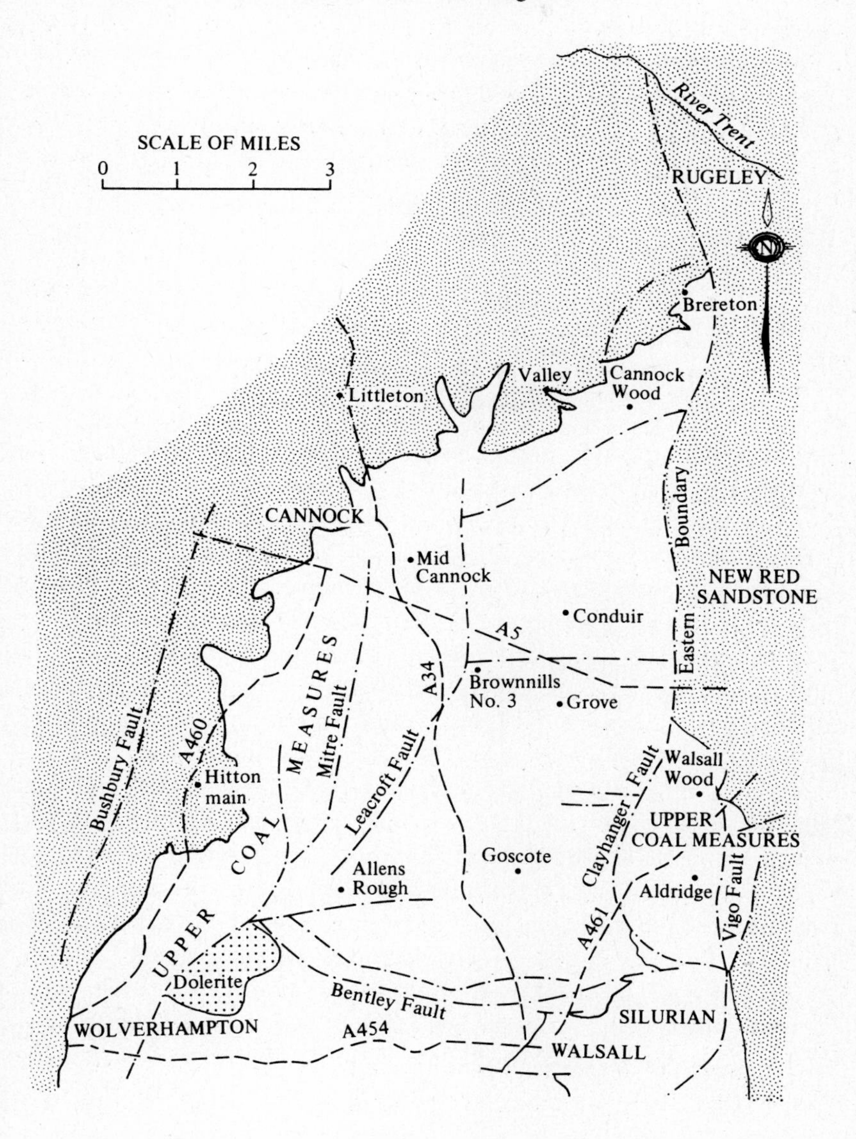

However, the range of wastes which will be discharged into the workings is markedly different from those that were discharged during the life of the original shaft. The opportunity has been taken to integrate the proposed facility into the Empire complex which operates wastes solidification and oil recovery and will shortly be operating incineration. Thus, only those wastes which are unsuitable for positive treatment or recovery will be scheduled for the discharge borehole and then only after treatment to precipitate and settle out solid matter.

There is presently a shortfall of facilities capable of safely handling a variety of liquid wastes including:

aromatic hydroxyls (e.g. phenols);
ammonia salt solutions;
polyvinyl acetates and alcohols;
organic acids and esters;
paint solutions;
printing inks, dyestuffs;
chelating agents;
stearates, soaps, aqueous synthetic lubricants;
organically contaminated acidic and aqueous wastes.

Incineration is often quoted as the panacea for waste disposal. The wastes quoted above can, undoubtedly, be incinerated but the cost of support fuel needed to evaporate the water prior to incineration of the organic constituents is, very definitely, prohibitive. Even incineration does not represent total destruction, as some people proclaim, since exhaust emissions and scrubber liquors do have an environmental impact.

As regards the safety and efficacy of the mine facility, the evidence produced at the public enquiry was very positive. Apart from the geological and structural integrity of the system, the control and monitoring procedures are rigorous.

As a planning requisite, each of the nine disused mineshafts which led to the workings will be stabilised by infilling, drilling and grouting to NCB specifications if not already done. Three monitoring boreholes are required to be drilled into the workings to enable venting and monitoring of mine air and/or gases and fumes and also liquid levels. Samples of vented gases and any liquid present in these monitoring holes will be analysed.

The control of the facility is by pre-sampling and analysis plus an on-site control laboratory. In general, control is by exclusion and wastes in the following categories are classed as unsuitable for discharge, at least in their raw state:

those having a flashpoint of less than 60°C;
spontaneously inflammable materials;
those that liberate toxic, inflammable, irritant or explosive gases under acid conditions;
radioactive materials;
highly odorous compounds;
halogenated solvents;
carcinogens.

The decision on whether or not a waste is suitable for treatment and discharge is based upon prior determination of pH, flashpoint, total dry solids at 105°C, ash on ignition, ammonia, cyanide, sulphide, total organic carbon in solution, cadmium, chromium, copper, iron, lead, nickel, zinc, odour and visual appearance.

After pre-treatment, blended wastes are held in a batch tank for sampling and on-site determination of pH, flashpoint, cyanide, sulphide, total organic carbon, odour and visual appearance prior to discharge. Daily composite samples will be taken for similar analysis.

Actual discharge limits, rather than exclusions, are pH of between 2 and 6 and temperature of less than 35°C.

'Reserved' planning matters have been the subject of protracted discussions and finalisation of these is expected in the near future. Indeed, it is anticipated that by the time this book is published the facility will be available to industrialists as a waste disposal outlet. The terms of a planning agreement enabling operations to be commenced were approved by the Walsall Metropolitan Borough Council Area Planning Committee on 17 June 1981, $2\frac{1}{2}$ years, all but two days, from the Secretary of State's decision on the project!

During the past operations of this facility, the whole system was regarded as being a disposal mechanism by way of indefinite storage. However, independent research has suggested that many reactions will occur in the mineworkings that will serve to purify the discharged wastes. The faults bounding the workings contain quartz and clays. Quartz is pure silica and is chemically unreactive; quartz vessels are often used to store acids and various organic solutions. The clays include kaolinite, illite, chlorite and montmorillonite; again, these are chemically stable and not attacked by the materials proposed for discharge. The clays have a base exchange capacity: that is, positive ions can be replaced with others of a similar size while leaving the structure unaffected. Thus, environmentally significant metals such as cadmium, copper and nickel can be exchanged for environmentally insignificant ones like sodium, potassium and calcium.

Research work indicates that waste liquids in the mineworkings will be subjected to quiescent settlement, base exchange reactions, physical filtration and other beneficial phenomena. It may, therefore, be possible in the future to abstract, from down-dip in the workings, liquids which are amenable to simple treatment and discharge to sewer. It is hoped that the Walsall Wood facility will hence prove to be useful as a treatment system rather than an ultimate disposal outlet.

Summary

The concept of disposal of liquid wastes to underground strata is not new. There are many consented discharges of sewage and of industrial effluent to underground cavities and some abandoned mineworkings. The concept of employing underground workings for treatment of wastes rather than for ultimate discharge is, however, relatively new.

There are two such systems in the UK and both offer a true waste disposal service, not being restricted to private and single waste discharges. The systems operate on different principles and are unique. The scope for further facilities along these lines is difficult to predict. Chatterley Whitfield is unique by virtue of its location to the pottery industry that it serves; Walsall Wood is unique by virtue of its geological structure. Nevertheless, the principle of using underground strata for waste treatment and/or disposal is sound and there must be possibilities for future facilities to be developed. This must represent one of the more exciting aspects in the waste disposal field.

Marine incineration

The principles of incineration have been discussed separately. Thermal destruction of wastes on the high seas is a unique application of this disposal method and much has been said about the controlling legislative mechanisms in another chapter.

During incineration of chlorinated compounds, and chlorinated hydrocarbons in particular, the chlorine is liberated as hydrogen chloride gas. The gas is hygroscopic and soon forms hydrochloric acid. For land-based incinerators, therefore, it is necessary for them to be equipped with acid exhaust removal apparatus such as wet scrubbers. The produced scrubber sludges must then be disposed of separately. The high temperatures required and the provision of control equipment to prevent release of obnoxious gases pose real problems to land-based incineration practice.

Thus it was that during the late 1960s and early 1970s two marine incineration vessels were commissioned both operating from Holland and both being principally dedicated to the incineration of vinyl chloride monomer (VCM) and other chlorinated products from its manufacture. VCM is produced by reacting ethylene with chlorine and is the raw material for the production of polyvinyl chloride plastics (PVC). VCM is particularly hazardous being a suspected carcinogen. West European manufacturing capacity of VCM is currently over 5 000 000 tonnes per annum.

Vulcanus

This incinerator vessel was first owned by the Vulcanus Shipping Co. Ltd of Singapore using Ocean Combustion Service BV of Rotterdam as agents. The vessel has recently been purchased by the USA enterprise Waste Management Inc. and it is believed that operations will shortly be transferred to America from Europe.

With a load capacity of 4000 tonnes the vessel is of double hull and double bottom construction and various design features are in excess of IMCO requirements. Saacke rotary cup atomisers fuel the twin open-bowl incinerators situated at the stern of the ship and liquid incinerable wastes are finely and evenly atomised into the chambers. Intimate mixing with air is achieved and combustion temperatures of up to 1500°C ensure a minimum of 99% thermal destruction of the wastes.

Each of the two incinerators is 4.5 metres in diameter and 8 metres high and is provided with three Saacke atomisers. Prior to injection into the incinerators the liquid wastes are milled to -2 mm to ensure that no blockages occur during combustion. Each incinerator has a capacity of 20–30 tonnes per hour.

There are 13 waste storage tanks of varying capacity from 116 to 520 m^3.

The German and Dutch authorities impose stringent controls on the vessel and its operations. An automatic camera is loaded and sealed by the authorities prior to each sailing and a photograph is taken every 30 minutes of the instruments on the centre console: incinerator temperature, time clock with date and the display co-ordinates of the Decca navigator are all photographed. Additionally, a Decca tracker-plotter draws out the ship's journey on a map. These records are then made available to the authorities when docking. The vessel operates on a three-week cycle.

The Vulcanus, with its stern incinerators, is obliged to operate its burners whilst on the move.

Matthias vessels

There have been three Matthias incinerator vessels owned and operated by Süd-Müll GmbH & Co. KG of Frankenthal-Hessheim and each equipped with a single Stahl-und Blech-Bau GmbH designed and manufactured bowl incinerator. The world agents (excluding Germany and Benelux) are TR International (Chemicals) Ltd, a wholly-owned subsidiary of Simon Engineering Ltd.

The original Matthias I was a 500-tonne coastal tanker modified to carry and incinerate wastes. In 1972, Matthias II was commissioned with a 1500-tonne capacity and in 1976 Matthias III came into service with a 15 000-tonne capacity. This latest and largest addition was scheduled to service the USA market and received limited approval from the Environmental Protection Agency (USEPA Report 68–022165 V, 11/12/76 [28]) after various trials. However, the vessel has since been scrapped owing to lack of business: this is quite an interesting facet to marine incineration in view of the recent acquisition of Vulcanus by Waste Management Inc.

Matthias I has also been scrapped. Matthias II is, therefore, the single remaining vessel in this series and this ship loads from both Holland and the UK. Rotterdam and Newcastle-upon-Tyne are the two ports of operation at present but there are plans to increase the number of UK operational bases.

As with Vulcanus, the Dutch Rijkswaterstaat and the German Hydrographisches Institut control and supervise the operations and are responsible for the automatic monitoring and recording equipment.

Matthias II has the following specifications:

length:	72.5 m;
beam:	10.8 m;
gross tonnage:	1135;
storage tanks:	12;
total capacity:	1200 m^3;
furnace weight:	180 t;
incineration capacity:	10–12 t hr^{-1}.

Unlike Vulcanus, Matthias II is stationary when burning wastes and the furnace is mounted forward. The bowl is first heated, using fuel, to 1000°C and then wastes injected progressively until the normal operating temperature of 1450°C is reached and maintained. Combustion efficiency is said to be better than 99.9% and air is provided by 16 blowers having a throughput of 10 000 m^3 $hr-^1$ at standard temperature and pressure, well in excess of stoichiometric requirements.

The incineration area is some 200 miles from both English and Dutch coasts and well south of oil rigs. UK operations are supervised and governed by the Ministry of Agriculture, Fisheries and Food who are responsible for the issuing of certificates of technical acceptance for individual wastes proposed for marine incineration and for the authorisation and licensing of each and every bulk load taken on board for incineration.

Marine incineration has taken the following pattern [29] (tonnages of wastes ex. Europe):

1969	4000 t	(Matthias I);
1970	8000 t	(Matthias I);
1971	28 000 t	(Matthias I and Vulcanus);
1972	66 000 t	(Matthias I and II and Vulcanus);
1973	87 000 t	(Matthias I and II and Vulcanus);
1974	100 000 t (15 000 in USA)	(Matthias I and II and Vulcanus);
1975	85 000 t	(Matthias I and II and Vulcanus);
1976	85 000 t	(Matthias I and II and Vulcanus); (Matthias III built and scrapped).

It seems that the tonnages handled may be levelling out but the removal of Vulcanus to America may affect later figures.

UK vinyl chloride monomers

The residues from VCM manufacture in Germany, Holland, Scandinavia and Belgium are all incinerated at sea. UK production by BP and by ICI (approximately 40 000 tonnes per annum) is disposed to spent salt mines in Cheshire. The North West Water Authority has issued three consents for the disposal and storage of organohalogen residues in pressure-tight brine cavities.

Fuel conversion

The mechanical conversion of domestic refuse and light commercial/trade wastes into a usable fuel is becoming a reality. Several large-scale installations are now working in the UK with varying degrees of success. Also, biological methods of conversion of sugar and starch wastes to ethanol are being developed.

Industrial wastes in general, and 'chemical' wastes in particular, are not so easy to process into a useful fuel or fuel additive. Nevertheless, attempts are being made and there is at least one commercially available process for the conversion of organic wastes. 'Leigh fuel' is one such system. It is said to convert paint residues, tars, latex suspensions, thick oily sludges and other organic compounds into a useful fuel. The wastes are blended together in admixture and then converted into a solid material in pellet, briquette or pulverised form. Being a solid fuel, the product can be easily stored and fed to conventional mechanical and pneumatic solid fuel feed systems in boiler plants. Typically, the makers claim that the product has a calorific value of 7000 BTU/lb (3875 kcal/kg), a water content of 20% and an ash content of 40%.

Further reading

[1] National Association of Waste Disposal Contractors, 'Practical Waste Management': Training Course, 15–17 April 1980, Coventry; 7–9 October 1981, Coventry; 7–9 June 1982, Runcorn.

[2] Holmes, J.R. (Ed.). 1982, *Essays in Modern Waste Management*. Wiley, London.

[3] Beselievre, E.B. & M. Schwartz, 1976, *The Treatment of Industrial Wastes*. McGraw–Hill, New York.

[4] Bradshaw, A.D. & M.J. Chadwick, 1980, *The Restoration of Land*. Blackwell Scientific Publications, Oxford.

[5] Bridgewater, A.V. & C.J. Mumford, 1979, *Waste Recycling and Pollution Control Handbook*. George Goodwin, London.

[6] Pojasek, R.B. (Ed.). 1980, *Toxic and Hazardous Waste Disposal*: Vol. 1, 'Processes for Stabilization/Solidification'; Vol. 2, 'Options for Stabilization/Solidification'; Vol. 3, 'Impact of Legislation and Implementation on Disposal Management Practices'; Vol. 4, 'New and Promising Ultimate Disposal Options'; Vol. 5, 'Perspectives on the Management of Hazardous Waste Disposal'; Vol. 6, 'Development of State Hazardous Waste Management Programs'. Ann Arbor, Michigan 48106.

[7] Sell, N.J. 1981, *Industrial Pollution Control*. Van Nostrand Reinhold, New York.

[8] Walters, J.K. & A. Wint, 1981, *Industrial Effluent Treatment:* Vol. 1, 'Water and Solid Wastes'. Applied Science Publishers, London.

[9] Wilson, D.G. (Ed.). 1977, *Handbook of Solid Waste Management*. Van Nostrand Reinhold, New York.

References

[10] *Control of Pollution Act 1974*, Elizabeth II, Ch. 40. HMSO, London.

[11] Porter, C. 1981, 'Lawmakers Could Insist on Regional Treatment', *Surveyor*, 11 June 1981.

[12] Chappel, C.L. & P.L. Butler, 1978, 'The Changing Face of Toxic Waste Treatment and Disposal: Impact on Treatment Plant Development' *Effluent and Water Treatment Convention*, ENPOCON, 16 November 1978. Brintex Exhibitions Ltd. London.

[13] *UK Patent 1 485 625*, 'Improvements in and Relating to the Conversion of Liquid Hazardous Wastes to Solid Form' (Application No. 26201/73, 1 June 1973).

[14] Chappell, C.L., 1980, 'Il Trattamento e lo Smaltimento Dei Rifiuti Industriali Mediante Solidificazione con il Processo "SEALOSAFE"', *SEP/Pollution Congress*, Padua, Italy, 22 April 1980.

[15] Chappell, C.L., 1979, 'The Disposal of Wastes from the Galvanizing Industry "Intergalva '79"', *Twelfth International Galvanizing Conference*, Paris, 17–23 May 1979. European Galvanizers' Association/Zinc Development Association, London.
[16] Chappell, C.L. & S.L. Willetts, 1979, 'Isolation of Heavy Metals from the Environment', *International Conference: Management and Control of Heavy Metals in the Environment*. Imperial College, London, September 1979. CEP Consultants Ltd, Edinburgh.
[17] Chappell, C.L. & S.L. Willetts, 1980, 'Some Independent Assessments of the SEALOSAFE/STABLEX Method for Toxic Waste Treatment', *Journal of Hazardous Materials*, Vol. 3, pp. 285–91. Elsevier, Amsterdam.
[18] National Sanitation Foundation, 1979. *Leachate Testing of Hazardous Chemicals from Stabilised Automotive Wastes*, January 1979. Ann Arbor, Michigan 48106.
[19] Willetts S.L. 1981, 'Chemical Fixation', *Symposium: The Handling and Treatment of Chemical Wastes*: Inetrsymp Ltd., Manchester, 10 September 1981.
[20] American Society for Testing and Materials. *Proposed Methods for Leaching of Waste Materials*, May 1978. ASTM, Philadelphia 19103.
[21] US Environmental Protection Agency, 1980, *Test Methods for the Evaluation of Solid Waste, Physical/Chemical Methods*. SW–846. USEPA Office of Solid Waste, Washington DC 20460.
[22] *CEC Draft Directive (7972/80)*, Concerning the Assessment of the Environmental Effects of Certain Public and Private Projects. *OJ* C169/15–22, 9 July 1980.
[23] *UK Patent 1 337 301*, 'Land Improvement with Waste Materials' (Application No. 802/71, 7 January 1971).
[24] *US Patent 4 124 405*, 'Process for Solidifying Aqueous Wastes and Products Thereof' (Application No. 710 666, 2 August 1976).
[25] *Dumping at Sea Act 1974*, Elizabeth II, Ch. 20. HMSO, London.
[26] Willetts, S.L. 1980, 'Sub-surface Disposal of Liquid Wastes', *International Environment and Safety Exhibition and Conference*, Wembley, 1–4 September 1980. Labmate Ltd, St Albans, Herts.
[27] *Water Resources Act 1963*, Elizabeth II, Ch. 38. HMSO, London.
[28] US Environmental Protection Agency. *Status of Ocean Incineration of Organic Chlorides Aboard Matthias III as of 9th March 1976*. Report No. 68–022165 V.
[29] Fabian, H.W., 1979. 'Verbrennung chlorierter Kohlenwasserstoffe', Sonderdruck aus *UMWELT*, No. 1, 1979 (VDI–Verlag GmbH, Dusseldorf).

11

The incineration of hazardous wastes

Introduction

Thermal oxidation of wastes is probably the most widely used system for disposal, next to landfill. A popular misconception is that incineration actually constitutes a method of 'destruction' but this is not the case. Incineration or combustion techniques serve as methods of waste reduction and conversion: there remain residues in different forms in need of separate disposal. Nevertheless, incineration is a valuable tool in today's waste management industry.

Several municipal incinerators exist that are specifically designed and equipped to accommodate domestic refuse and collected trade and commercial wastes but, for the purposes of this discussion, these are not included here since the amount of 'hazardous' wastes accepted at such plant is insignificant.

As far as incineration plant for industrial wastes are concerned, there are very many in-house systems at premises devoted to catering for on-site process waste. These are not available in the general market place and cannot be considered as tools in the waste management industry.

Having removed most of the existing incinerators from the scope of the discussion it is pertinent to ask 'with what are we left?'! The answer is some 117 000 tonnes per annum installed capacity plant at a handful of locations in the United Kingdom.

Re-Chem International Ltd, part of the British Electric Traction Group, operate the three largest facilities, these being at Southampton, Pontypool in South Wales and Bonnybridge in Scotland. All three can accommodate liquid, solid and drummed wastes and are equipped with exhaust scrubbers. The significance of this will become apparent later.

Cleanaway Ltd operate two plant equipped with exhaust gas scrubbers suitable for bulk liquid wastes only: one at Ellesmere Port and one at Rainham, Essex.

S. Berridge and Son Ltd operate one facility at Hucknall near Nottingham. This has no gas scrubbing facility and is suitable for liquid wastes only.

There may be other facilities, but these are of no practical significance because of the low waste volumes involved.

It is immediately obvious that incinerator facilities for industrial wastes are few and far between and this situation is mirrored in all other industrialised countries. The reason is also obvious: the plant are sophisticated, technically complex and very expensive. Thus, each plant serves a large geographical area and wastes are very often transported over long distances for disposal by this method. Because of the high cost of incineration, the method is usually considered as the last resort except for those wastes for which incineration is the only practical method of disposal.

Suitability of wastes

Wastes exist in many physical forms: gases in cylinders or cans, volatile liquids, mobile liquids, viscous liquids, sludges, suspensions, solids and mixtures. When these different physical states are considered in terms of being a 'feedstock' then it is readily apparent that the plant designed to receive and incinerate them must either be restricted to one principal state or be designed as a 'multiple' unit and this is the case.

Chemically, not all wastes are suitable for incineration. In general, organic wastes can be thermally oxidised to yield carbon dioxide, water and minor byproducts depending upon the nature of the waste. Certain inorganic wastes can be oxidised to good effect such as azides and other nitrogen-containing species.

The efficiency of the process in carrying out complete oxidation is dependent upon temperature, turbulence and time and the design of any incinerator must take account of these parameters. These will be explored in some depth later.

Where wastes contain inorganic salts and/or halogen compounds and/or sulphur compounds then problems arise. Inorganic salts will liberate metal oxides on incineration, the most common being sodium, and these will require high-energy scrubbers, usually of the venturi type, to effect their removal from the exhaust stream. Halogen-containing wastes present particular problems in that the acid (hydrochloric and hydrofluoric most commonly) must be formed for effective removal by wet scrubbing as the halogen gases themselves are relatively insoluble. This demands an excess of hydrogen to be provided by a hydrocarbon support fuel: for example, if trichloroethylene is a major component in a waste it can be incinerated according to:

$$CHClCCl_2 + 2O_2 \rightarrow 2\ CO_2 + HCl + Cl_2.$$

The chlorine gas would be vented to atmosphere after having been passed through a wet scrubber virtually unhindered. However, if a hydrocarbon fuel is present then the reaction follows:

$$CHClCCl_2 + 7/2\ O_2 + (CH_4) \rightarrow 3\ CO_2 + 3HCl + H_2O$$

and all of the hydrogen chloride acid gas can be scrubbed out. There is a statutory limit of 460 mg m^{-3} for HCl in UK stack emissions.

Sulphur can be present in wastes as either inorganic sulphides, sulphites and sulphates or as sulphonated organic compounds. In a minimum of excess air sulphur dioxide, SO_2, will be formed and this can be completely removed using caustic scrubbing. In excess air the trioxide will be formed and this is more difficult to remove since it will readily form sulphuric acid in a wet scrubber.

Nitrogen in wastes is of little problem even though nitrogen oxides are acid-forming. The reason for this is that the combustion air will itself contain almost 80% nitrogen and so oxide emissions as NO_X are inevitable. Oxidation of the nitrogen can be kept to a minimum by ensuring combustion takes place as quickly as possible so that the combustion products remain as nitrogen gas, without further oxidation.

Plastics products, polymeric precursors and intermediates and polymeric scrap can produce their own peculiar problems. Polyurethane, for example, can liberate carbon monoxide and hydrogen cyanide in the exhaust if not efficiently incinerated. Carbon monoxide affects the ability of haemoglobin to carry oxygen in the blood and cyanide disrupts the ability of cells to use oxygen and so there is a biological synergism between the two compounds rendering them, in combination, particularly hazardous. Polyvinyl chloride is a considerable nuisance in incinerators owing to its high chlorine content of some 56%. Its combustion demands a high rate of hydrocarbon support fuel usage.

Polystyrene comprises 92% carbon and the polymer has a one-step degradation at just below 300°C. Because of this, soot generation is a problem and the carbon particle 'aerosol' so formed requires elevation to over 800°C before it will burn. Thus, high temperature, adequate residence time and good mixing are essential to avoid smoke emissions. There is a limit of 114 mg m^{-3} for particulates in stack emissions in the UK.

Polychlorinated biphenyls (PCB's) present peculiar problems in that they are temperature stable. Incineration of these compounds will only be effected at temperatures above 1100°C with at least 3% excess oxygen and with a residence time of more than 2 seconds.

Steam plume generation

As previously noted, a principal product of incineration is water. Together with water that may be picked up from a wet scrubber, this means that exhaust gases carry steam to the atmosphere which, on cooling, manifests itself visibly as a dense plume depending upon atmospheric conditions at the time. The problem is more psychological than material but is, nevertheless, very real.

Additionally, wet scrubbing systems for removal of acid gases result in a near-saturated, cooled exhaust gas and such exhaust will not respond to electrostatic precipitator methods of particulates removal. Venturi scrubbers are, however, efficient at removing particulates.

It is a paradox that efficient scrubbing out of acid gases and particulates usually serves to stabilise steam plumes and cause excessive persistence [1]. Re-heating of the cooled, saturated exhaust is one method of eliminating, or at least minimising, the steam plume but is only conceivable given cheap or recovered energy for fuelling a heat exchanger.

Design principles

(i) The furnace volume must be capable of receiving the physical bulk of the material and of providing sufficient residence time for the burnout of flying particulate matter.

(ii) The correct combustion air requirements must be met and usually must be kept in the order of 50%–150% above stoichiometric requirements.

(iii) There must be minimum use of underfire air so as to keep solids, particulates and ash on the grate(s) and prevent them from entering the gas stream.

(iv) There must be maximum use of overfire air to ensure ample oxygen and turbulence in the combustion chamber.

(v) The design temperature must be attained and maintained. Minimum combustion temperature must not fall below 800°C so as to ensure carbon particles are oxidised and not liberated as soot. Average combustion temperatures of 900°C–1100°C are needed to ensure destruction of odourous compounds and proper oxidation of heavy organic compounds. Sustained temperatures of 1100°C–1200°C are required for oxidation of PCB's and PCT's (polychlorinated terphenyls).

(vi) The high-temperature zone must have a gas path of sufficient length and volumetric capacity to enable complete combustion of volatile constituents. Theoretically, a residence time of between 0.5 and 0.75 seconds is required given efficient mixing and correct temperatures but more than 2 seconds may be required: for example, when incinerating PCB's and PCT's.

Variables affecting proper waste combustion

(1) Combustibility

The nature of the waste itself dictates its 'combustibility' or ease of incineration and any waste scheduled for disposal by this route must first be screened in the laboratory to determine the relevant properties and characteristics. The calorific value of the waste, its upper and lower flammability points, its flashpoint, its ignition and auto-ignition temperatures all characterise the combustibility of the material.

For industrial waste incinerators, of course, constancy of feedstock is unlikely and probably even impossible. Thus, compatibility of wastes for mixing, either during storage or for blending prior to injection into the incinerator, must be assessed. Some viscous liquids may require to be preheated before injection; some materials may need to be dissolved in specific solvents before incineration; some wastes may require constant agitation during storage to prevent settling out and consequent feed pump blockages; some wastes may demand immediate incineration on delivery, for example those that would self-polymerise on storage; some wastes may be corrosive and require special handling. Again, plastics, and organic polymers in general, warrant closer attention as these compounds often contain fire-retardants, plasticisers and other additives which affect the combustibility of a particular formulation. Prediction of the burning characteristics is not easy: in the case of PVC, 52 separate compounds have been identified and even wood has 19 and the interaction of these compounds together with those of other, added wastes must be complex.

(2) Temperature

The temperature in the combustion chamber must be monitored and controlled, preferably automatically. Too low a temperature will result in incomplete combustion and smoke emissions; too high a temperature will damage the refractories and may even cause fusing of the ash or a sintering with the refractories. There are four basic means of controlling the combustion temperature:

(i) Excess air control: the adiabatic flame temperature is a function of type of fuel and amount of air. Limiting orifices and nozzles and metering pumps are used to prevent overfiring during continuous feeding of liquid or gaseous fuels. For liquids, the problem is aggravated if the volatility of the feedstock fluctuates.

(ii) Radiant heat transfer: transfer of useful heat from the combustion chamber is usually the single purpose of any fuel-burning system (i.e. heating units) but not so with a waste incinerator. Some municipal installations do incorporate heat recovery systems that take away heat from the combustion zone using heat exchangers but this is not generally the case for industrial waste incinerators.

(iii) Two-stage combustion: in this system, the first chamber operates in a deficiency of air and at a reduced temperature. Incomplete combustion occurs and the process is more properly described as one of destructive distillation in that volatile matter is gasified and driven off without the benefit of complete oxidation. Combustion of these volatile elements continues at higher temperatures in the second chamber. This system provides dual control opportunities.

(iv) Direct heat transfer: heat-absorbing materials can be added to control combustion temperature. Aqueous wastes, for example, consume a lot of heat in simply evaporating the water before the organic constituents can be oxidised. Actually, this concept is somewhat inverted at modern-day industrial waste incinerator plant because aqueous wastes predominate and there is usually a need to purchase support fuel to evaporate the water. This used not to be the case only a few years ago but the price of oil and oil products has given the impetus to recovery processes where possible in preference to incineration: for example, dirty solvents are now redistilled or laundered rather than scheduled for disposal and this has resulted in the loss of high CV fuels to the incinerators.

(3) Turbulence

The degree of turbulence, or the intimacy of mixing between fuel and air, affects combustion efficiency significantly. In general, mechanical or aerodynamic systems are used to ensure good mixing. For example, reciprocating rakes and moving grates can be used to provide a mobile bed for heavy solids; tangential fans and injectors can be employed to produce cyclonic burn paths; fluid-beds can be induced using high air pressures.

(4) Residence time

Combustion chamber volume and path length are fixed at the design stage and determine the residence time realised. Convoluted chamber formations, baffles and vanes can be used to extend residence times.

Types of incinerators

There is no doubt that the rotary kiln and the multiple hearth types of incinerators provide the highest degree of flexibility and adaptability and these are outlined first in this section. The other types are then introduced in no particular order since categorisation can be achieved on many different bases. An indication of the varieties of equipment and their applications is presented diagrammatically in Fig. 1.

Rotary kiln incinerators

In recent years this type of incinerator has become established as first choice for chemical wastes.

Basically, it comprises a rotating drum in the horizontal plane in which combustion takes place. The drum is refractory-lined and the length is usually 3- to 4-times the diameter which may be up to 4 m. It is inclined some 2–5° from the horizontal and rests on rollers which can rotate the drum at a variable speed from $10\ \text{r hr}^{-1}$ to $4\text{–}5\ \text{r min}^{-1}$.

The upper, feed end can be butted into a grate feeder and sometimes even a dual grate. Solids are fed onto the first, 'cold' grate which is inclined towards the mouth of the drum on which drying occurs. From here, the solids are automatically and mechanically scraped onto the second, 'hot' grate where ignition occurs prior to actual loading into the drum.

The gradual rotation and inclination of the drum results in a progression of the burning waste down the drum to the lower, discharge end. Liquid wastes are usually counter-fired into this lower end. This ensures a gentle and continuous mixing of the burning material and has the advantage of presenting fresh surfaces for combustion which prevents the formation of baked, insulating layers which would otherwise result in incomplete combustion. The ash, slags and other unburnt materials such as metal scrap are tumbled down to the discharge end.

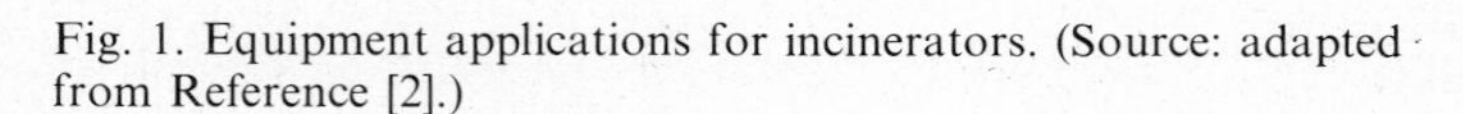

Fig. 1. Equipment applications for incinerators. (Source: adapted from Reference [2].)

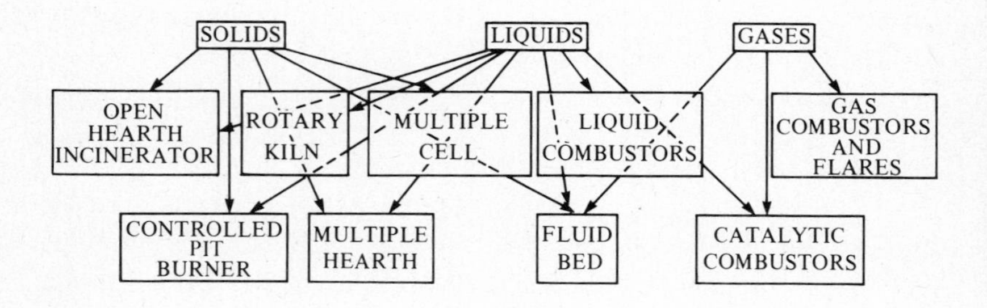

There are two major disadvantages to the rotary kiln installation. Firstly, the high-temperature air seals between moving and non-moving parts and the close tolerances required between the rotating rings and rollers mean high initial and maintenance costs. Secondly, the cascading action of the burning waste gives the fine ash an enhanced opportunity to be carried away with exhaust gases and the same action serves to abrade the refractories and prolong contact time with corrosive distillation products. These latter points are of significance because denser refractories of a higher than normal grade must therefore be specified.

A schematic installation is illustrated in Fig. 2.

Multiple hearth incinerators

This type of incinerator found much success in the metallurgical industry having been conceived in 1889 by Herreshoff for pyrite roasting for sulphuric acid manufacture. It was adapted in 1934 to the burning of sewage sludge: later developments proved successful in accommodating chemical wastes, tars, greases etc. The unit consists of a refractory-lined cylindrical shell in the vertical plane equipped with several fixed hearths at different levels. There is an air-cooled central shaft, equipped with rabble arms, which slowly rotates so as to move the solid material around each hearth in one direction. Each level has an opening such that material ploughed round by the rotating rabble arm falls through the opening onto the next lower hearth.

Solid waste is introduced onto the top hearth and progresses down

Fig. 2. Rotary kiln installation.

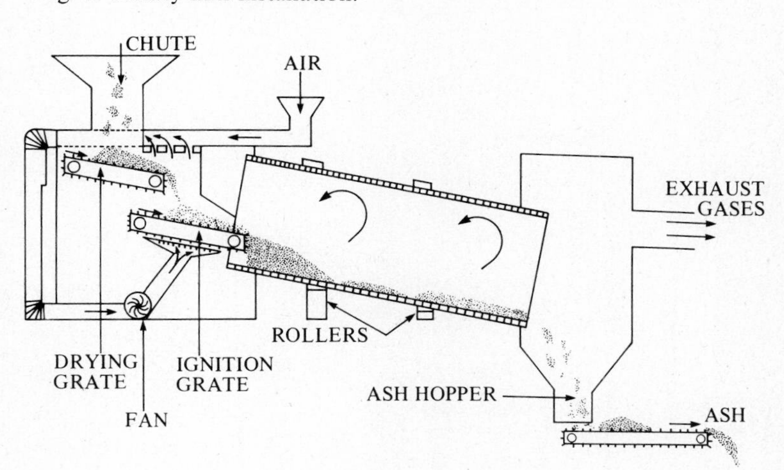

through the incinerator, being scraped from one hearth to the next until it emerges as ash through the bottom. The individual hearths and rabbles can be designed so as to vary the retention time at each level and three zones are generally recognised: the upper levels constitute the drying and preliminary combustion zone; the middle levels comprise the completion of combustion zone; and the lower levels the ash cooling zone.

Air is drawn up through the hearths and the lower zone serves to preheat this air as it passes over the hot ashes. However, combustion air is usually injected at each level to guard against any possibility of oxygen starvation that might arise from overloads on an earlier hearth.

Liquid wastes are injected into the chamber via auxiliary burner nozzles. Fig. 3 schematically illustrates the multiple hearth concept.

An alternative configuration is the 'moving' or 'reciprocating grate' furnace in which two banks of grates are set in a stepped juxtaposition as illustrated in Fig. 4. One bank is fixed and the other bank moves in and out such that the burning material is pushed over one grate onto the next lower.

Open hearth incinerators

The simplest of systems: this is a case of throw a match in and run! Nevertheless, open burning is often the only safe way of incinerating some particularly explosive wastes.

Fig. 3. Multiple hearth incinerator. (Courtesy NA Centre GSA.)

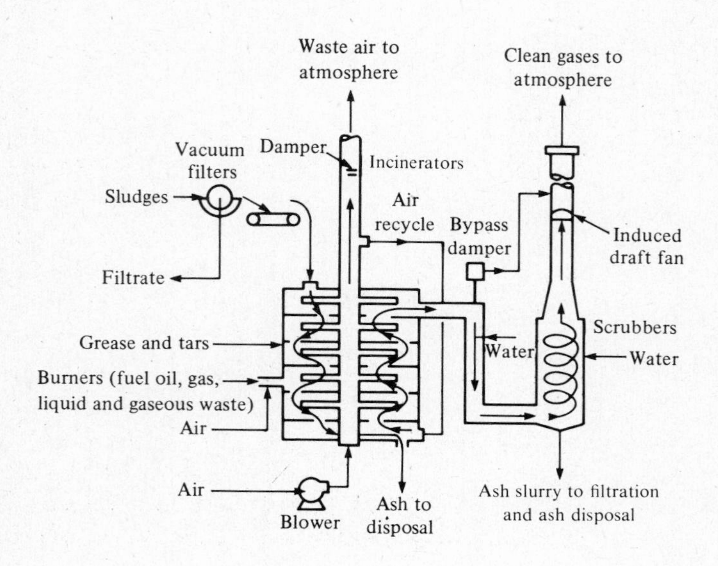

Obviously, no emission abatement is possible and considerable black smoke along with CO, HCl, SO_2 and NO_X can be evolved owing to lack of control of combustion temperature, availability of oxygen and mixing.

Controlled pit burner

A sophisticated form of blowing air into a hole in the ground: but it is surprisingly effective. Illustrated diagrammatically in Fig. 5 it is seen to comprise a box-shaped shell of steel or concrete either sunk into the ground and refractory-lined or purpose built upon a pedestal. A typical pit would be 5 m long, 2.5 m wide and 2.8 m deep.

Air is provided by a large-capacity fan or fans and is directed over the top

Fig. 4. Moving grate incinerator. (Source: Reference [4].)

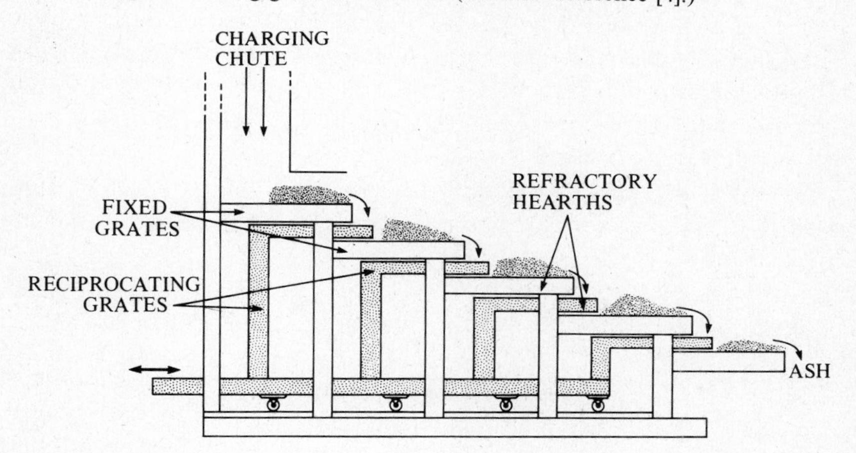

Fig. 5. 'Air-curtain' controlled pit burner. (Source: after Reference [2].)

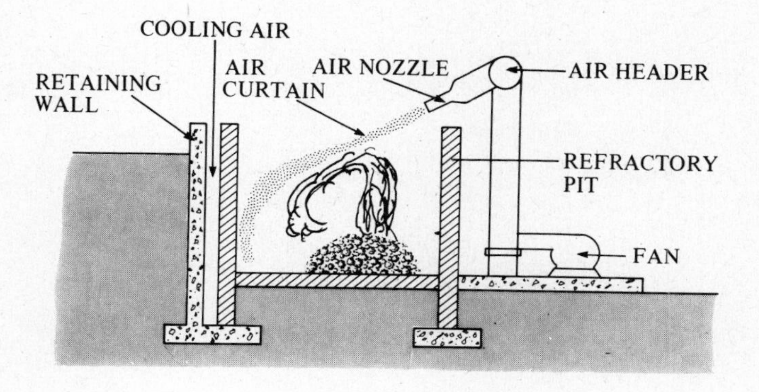

of the pit from a manifold throat that runs the whole length. Thus an 'air-curtain' is formed in the overfire position creating maximum turbulence and re-circulation of the combustion gases. The curtain angle is between 25° and 35° across the pit and this high-velocity air ensures an excess of oxygen. Particulates and combustion gases are thus largely returned to the burning zone and emissions are surprisingly low unless and until the curtain is disrupted, for example during re-feeding the incinerator which is usually and simply accomplished by hand or mechanical delivery into the top of the pit.

Both Union Carbide and DuPont have developed such incinerators in the USA and at least one such burner exists in the UK.

Fluidised bed incinerators

Fluidised bed units are versatile and can be used to dispose of solid, liquid or gaseous wastes: the principal disadvantage is that a constant feedstock is required. Although recent in development, they are relatively common in the petro–chemical and paper industry and have even been used for the processing of nuclear wastes in the USA [2].

The principle is illustrated in Fig. 6. Air is introduced into a plenum beneath the combustion chamber and is forced through a distributor plate

Fig. 6. Fluidised bed incinerator. (Source: Reference [2].)

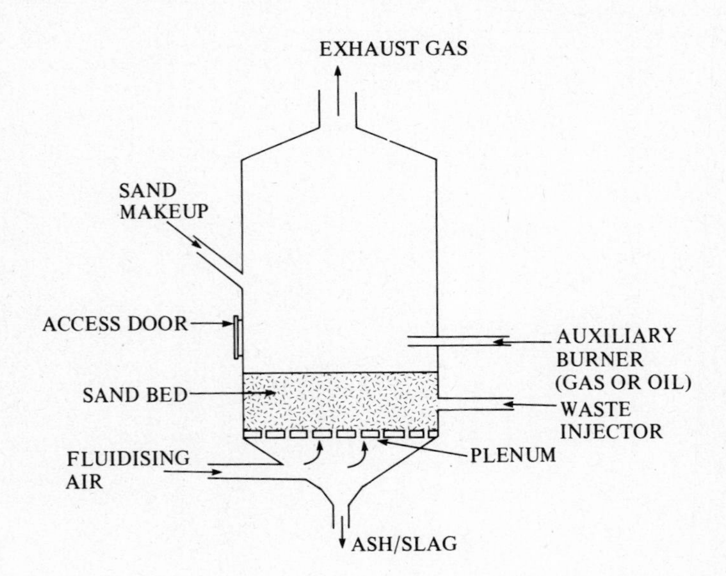

into a bed of sand. Air velocity is maintained such that the sand is kept mobile or 'fluid' and has the appearance of boiling.

Solid wastes need to be first shredded or pulverised and then fed into the combustion chamber, falling to the bed: liquid wastes are injected directly into the bed. The intimate mixing of air with the waste and the enriched oxygen content ensures complete combustion: the agitation of the waste by the bed particles also helps to improve the efficiency of such units. The efficiency is such that combustion is usually achieved at lower temperatures, which restricts NO_X emissions, and this is essential since the sand bed can slag and fuse over 1090°C. Limestone can be added to the bed to reduce acid gas emissions. Operating costs have proved to be high in fluidised bed units but there is considerable scope for reduction owing to the ease with which heat exchanger tubes can be inserted into the bed for heat recovery.

Multiple cell incinerators

There are two basic types, retort and in-line, being distinguished by the path of combustion gas travel. The retort-type incinerator turns the combustion gases into a second combustion chamber by 90° turns in both lateral and vertical directions whereas the in-line system turns the gases in the vertical plane only. Fig. 7 illustrates an in-line multiple cell incinerator. The retort-type, with its two-directional path change, has the advantage of being physically a smaller unit.

The primary chamber is where solid waste combustion occurs and gaseous combustion products are carried over through a flame port into the

Fig. 7. Multiple-cell in-line incinerator.

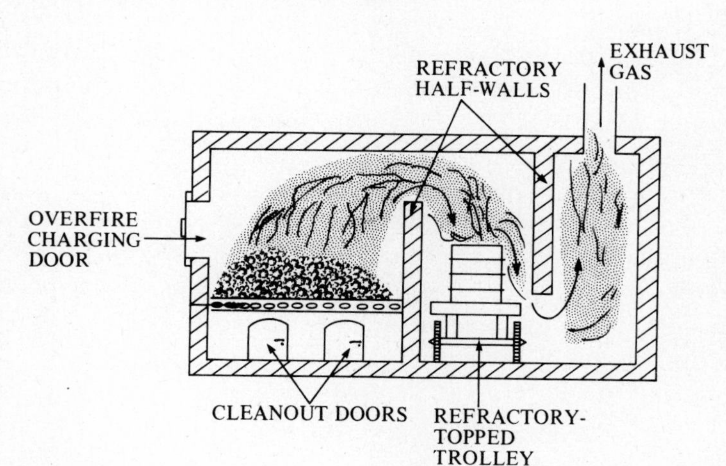

secondary stage which itself comprises two chambers: an initial downdraft or mixing chamber and an up-pass expansion chamber.

The multiple cell concept has been adapted to provide drum burning facilities such that a refractory-topped trolley, upon which sit one or more drums of waste, can be fed into the secondary chamber. The hot gases from the primary combustion chamber are drawn down onto the drums and burning from the surface of the contents takes place.

Liquid combustors

There are two basic configurations for liquid waste incinerators – vertical and horizontal. Basically, they comprise a chamber into which the liquid waste is atomised, and into which air is injected, being equipped with an ignition device.

Before a liquid can be burned it must be converted to the gaseous state and this occurs within the combustion chamber requiring transfer of energy, as heat, from the hot combustion gases. In order to increase this heat transfer it is desirable to effect rapid vapourisation of the liquid feed thereby increasing the surface area available. This is performed by 'atomisation' to 40 μm size or smaller, usually by pressure nozzles but sometimes using a rotary cup device. Atomisation of the feedstock is best performed in the liquid burner at the point of fuel and air mixing.

Liquid burners require considerably more turbulence and time to complete combustion than do simple gas cumbustors. Thus, most liquid incinerators use some form of forced-air ventilation and this is particularly important where viscous liquids are concerned since they are so difficult to vapourise. Pre-heating will have to be employed for many non-mobile wastes or dissolution in less viscous solvents prior to incineration.

Wastes that are self-sustaining in combustion can be injected through the primary burner but aqueous wastes and other partially combustible liquids are best separately injected into the secondary zone of the combustion chamber.

For vertical incinerators it is possible to introduce air tangentially through tuyères disposed around the perimeter of the combustion chamber such that a cyclonic vortex is produced [3]. This increases combustion efficiency significantly but the high heat-release phenomenon of this type of incinerator can cause frequent slagging. Fig. 8 illustrates a cyclonic-type incinerator.

Gas combustors

For gaseous materials, many simple incinerators exist comprising a valved injector, a burner throat into which air is also injected and a combustion chamber. They are direct-flame, thermal incinerator devices. They can either be vertical or horizontal and a typical configuration is shown in Fig. 9.

Stack flares for waste gases are fairly common but are becoming fewer in these energy-conscious times. Typical configurations are shown in Figs. 10, 11 and 12. Steam is usually provided to create turbulence and to inspirate air for the combustion process.

Fig. 8. Cyclone-path liquid incinerator. (Source: after Reference [2].)

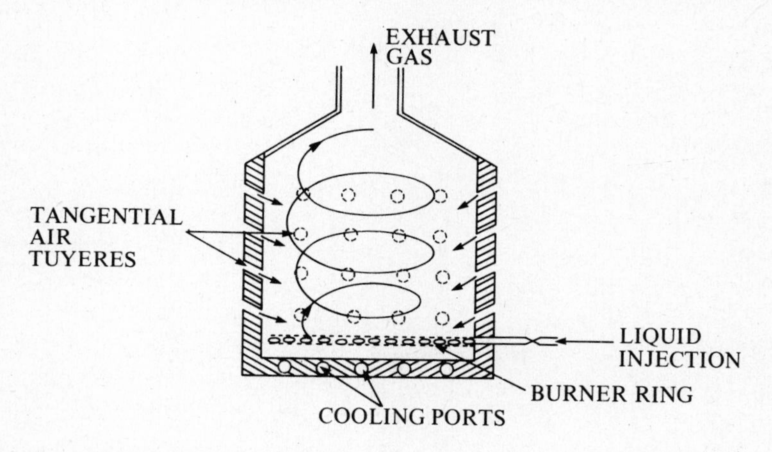

Fig. 9. Direct-flame gas combustor.

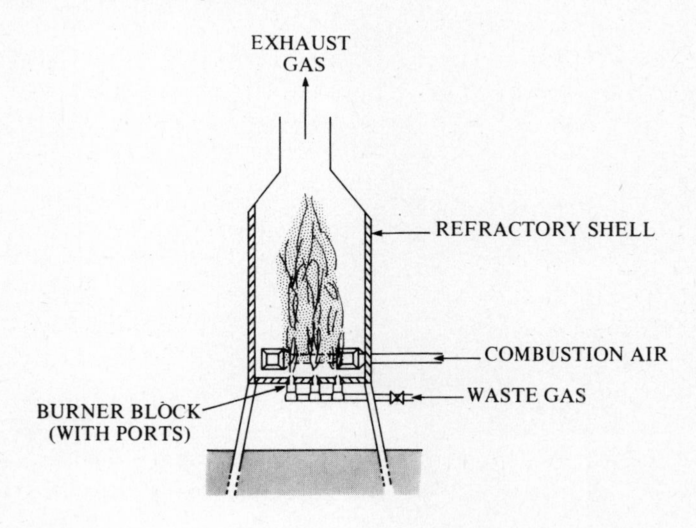

Fig. 10. Stack flare with mixer nozzles. (Source: after Reference [2].)

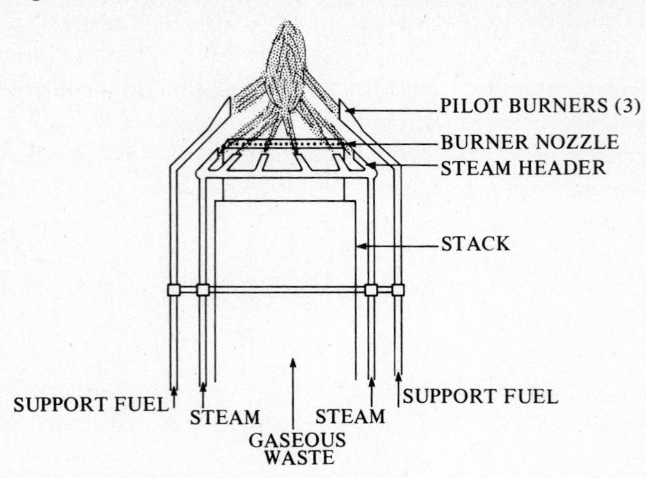

Fig. 11. 'Esso'-type stack flare. (Source: after Reference [2].)

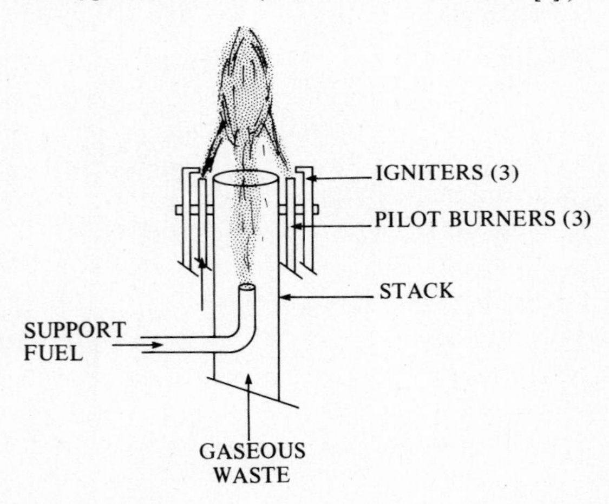

Fig. 12. Venturi-type flare. (Source: after Reference [2].)

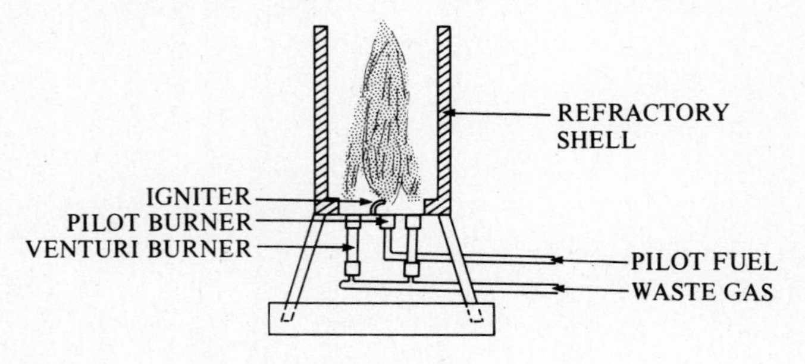

Catalytic combustors

Catalytic oxidation is reserved for combustion of low-concentration materials in a gaseous state. The technique has been successfully used in chemical process industries for odour control and other air pollution control applications.

Efficiency depends upon reaction temperature, waste material concentration, excess oxygen, and the chemical composition and the geometrical configuration of the catalyst.

Catalysts such as the noble metals platinum, palladium and rhodium and the oxides of copper, chromium, manganese, nickel and cobalt have all been used to good effect.

Basically, a catalytic incinerator consists of an afterburner housing containing a pre-heat section and a catalyst chamber. Waste gases are pre-heated before being passed through the suspended catalyst.

A unique maintenance element exists in catalytic incinerators in that the catalyst becomes less efficient with use and must be regenerated and possibly even replaced with time. Catalysts are prone to poisoning, to suppression and to fouling by various contaminants and so correct choice at the outset is of obvious import.

Afterburners

Afterburners are secondary combustion devices designed to incinerate flue gases from the primary incinerator. They comprise a refractory combustion chamber, equipped with burners, through which the primary exhaust gases are passed. Very high temperatures are usually required to effect complete combustion of the exhaust and, even with pre-heating, the support fuel requirements are very high rendering their operation expensive.

A type of afterburner which has recently been developed is the 'molten salt' device. In effect, the combustible exhaust gas is ignited beneath the surface of a pool of molten alkali metal salts, usually a mixture of sodium carbonate and sulphate. The molten salt acts as a catalyst and induces complete oxidation of the materials at temperatures below those of normal combustion. Excess oxygen can be introduced by air injection into the bath, or by chemical oxidation of the bath salts releasing nascent oxygen, or by electrolysis of the nitrate salt. Fig. 13 illustrates the molten salt type afterburner.

Organic materials are oxidised completely to carbon dioxide and water and acid gases are immediately neutralised in the alkali salts. Phosphorus, arsenic and sulphur also react with the molten salts to form sodium phosphate, arsenate and sulphate respectively. These non-volatile salts are then retained in the melt.

Skimming and dredging apparatus are required to remove both floating debris and heavy ash.

Recent advances have identified, as being particularly effective in this process, the eutectic mixtures of sodium/potassium hydroxide and of lithium/sodium/potassium carbonate.

Gas cleaning equipment

As has been noted, it is required that a wastes incinerator is equipped with some form of exhaust gas cleaning equipment. More especially, when halogenated or sulphurous wastes are incinerated this equipment must include some form of wet scrubbing.

There are three basic systems

particulate reduction;
soluble gas removal;
steam plume abatement.

Where no acid gases are produced the problem resolves itself to the single issue of particulate reduction. This can be achieved quite simply by impingement plate arrestors, cyclonic arrestors, bag filters and many other standard dust control techniques. For finer particulates, electrostatic precipitation is the best method.

Wet systems, however, serve the dual rôle of reducing particulates and absorbing acid gases. They can be installed independently but often follow a dry collection system such as a hot cyclone so as to minimise abrasive

Fig. 13. Molten salt afterburner. (Source: after Reference [2].)

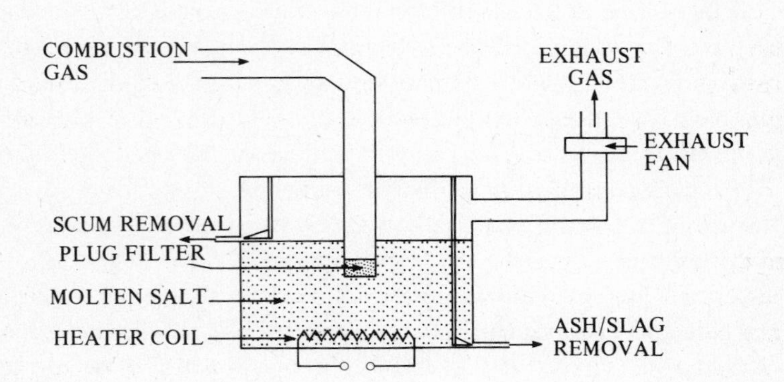

wear within the scrubber system and also to cool down the exhaust gases prior to scrubbing.

There are five basic types of wet scrubber:

(i) Wash towers

The simple wash tower is effective at removing particulates down to 3 or 5 μm and, with packing, down to 1.5–2.0 μm. Random plastic modules are used for packing but there is a danger of clogging unless free movement is allowed between the packing elements: for this reason plastic spheres are often used because they are self-cleaning as they rotate in the gas and water streams.

(ii) Gas-induced scrubbers

These rely upon the kinetic energy of the gas stream itself to atomise the scrubber water into the gas path and are effective down to 0.7–1.0 μm diameter particulates.

(iii) Injection scrubbers

In this system water is atomised and injected into a constricted chamber within the gas stream. The injection of water is independent of the gas, unlike the gas-induced scrubber system, and this type of equipment is very efficient at removing soluble gases and particulates down to 0.6 μm.

(iv) Disintegrator scrubbers

These operate using a system of spinning discs or cages to atomise the water input and do, therefore, require a separate power supply. Efficiency is good with removal of particulates down to 0.4 μm.

(v) Venturi scrubbers

Without doubt Venturi scrubbers are the most efficient of wet de-dusters, being capable of particulate removal down to less than 0.01 μm. Principal disadvantages are the high pressure losses and the high power input. Fig. 14 illustrates these five systems diagrammatically [4, 5].

Materials of construction

The refractory lining of the incinerator must be able to withstand the temperatures of combustion and also be resistant to thermal shocks. This latter point is particularly important with waste-fired incinerators since constancy of feedstock will not prevail.

At one time, refractory bricks were the first choice for incinerators but the jointing compounds very often allowed easy penetration by acid gases, particularly HCl. Bricks are still preferred for rotary kilns where high-abrasion resistance is required but newer, chemically bonded, mouldable refractories are giving superior performance in arduous conditions where abrasion is not so severe.

The presence of alkali metals in their molten oxide state causes spalling and flaking of refractories as they penetrate, with the so-formed flux 'freezing' at the lowest melting eutectic.

Wet scrubbers that remove acid gases do so, of course, by dissolving the gases in the provided water to form the acid. Thus, the scrubbers must be of acid–resistant materials. Stainless steel suffers from stress–corrosion cracking and pitting and has proved unsatisfactory.

Mild steel venturis lined with glass-reinforced polyester have proved satisfactory as have titanium metal and high chrome–nickel alloys such as 'Inconel', 'Incoloy' and 'Hastelloy'.

It is usual to re-circulate the scrubber liquor and so neutralisation of the increasing acidity with use is essential. However, particulate loading of the liquors eventually necessitates a 'clean-out' and replacement otherwise erosive corrosion problems set in.

Fig. 14. Wet collectors. (Source: after Reference [5]. Courtesy Incinerator Co. Ltd.)

WASH TOWER	SCRUBBER	INJECTOR	DISINTEGRATOR	VENTURI
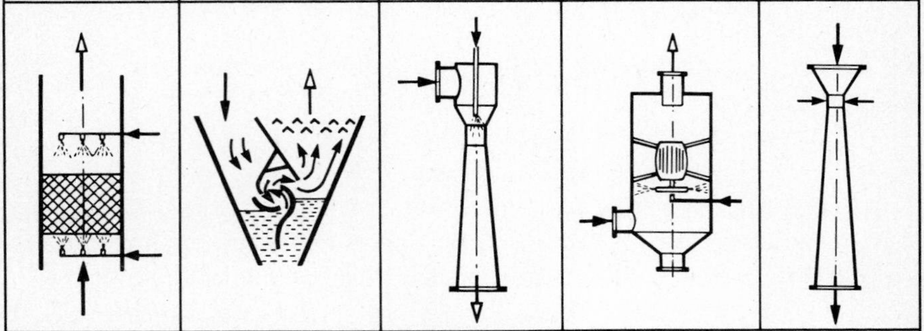				
CUT SIZE FOR sp.gr. 2.6				
> 1.1 μm	0.7–1.0 μm	0.6–0.9 μm	0.4–0.6 μm	< 0.01–0.04 μm
PRESSURE DROP mm H_2O				
20–200	180–280	–	40–100	300–2000

As far as the chimney stack itself is concerned, the residual acid content of the exhaust gas should be low and so potential corrosion risks slight. Nevertheless, the stack can act as a reflux condenser if not insulated so as to minimise the cooling of waste gases and condensation on internal surfaces. In some cases, collected stack condensate has been found to have a pH approaching zero: multi-coated epoxy resins on carefully prepared metal surfaces are the only practical solution where acid exhausts will be experienced.

The practical aspects of incineration of chemical wastes and its role in the disposal industry are discussed elsewhere by Dunn [6, 7].

References

[1] Baker, L.E., 'Non-Landfill Methods of Hazardous Waste Disposal', *in Developments in Environmental Control and Public Health – 2*, A. Porteous, (Ed.). Applied Science Publishers, London, 1980.

[2] Powers, P. W. *How to Dispose of Toxic Substances and Industrial Wastes.* Noyes Data Corporation, London and New Jersey 1976.

[3] Stribling, J.B., 'Steam Output Trebled by Scrap-tyre Incinerator Development', *Environmental Pollution Management*, Vol. 7, No. 2, March–April 1977.

[4] Dunn, K.S., 'Incineration of Toxic Industrial Wastes – Environmental Success or Failure', *Environmental Pollution Management*, Vol. 11, No. 1. January–February 1981.

[5] Dunn, K.S., 'Incineration of Toxic and Dangerous Wastes – Practical Aspects of Disposals by Local Authorities and Industry', *48th Annual Conference: National Society for Clean Air*, 5–8 October 1981. Brighton.

[6] Dunn, K.S., 'Problems and Practicalities of Incinerating Chemical Wastes' *The Chemical Engineer*, November 1979.

[7] Dunn, K.S., 'Incineration's Role in Ultimate Disposal of Process Wastes', *Chemical Engineering/Deskbook Issue*, 6 October 1975.

12

Risk assessment, cost–benefit analysis and future needs

Risk assessment

Industrial managers have become accustomed to the assessment of risk to employees in terms of the fatal accident frequency rate FAFR. FAFR is defined as the number of fatalities occuring in 10^8 working hours and this concept is justified by most workers in this field by the belief that it enables risks to be compared. Kletz [1] argues that, within an industry, the proper allocation of resources is best achieved to ensure that no employee should be exposed to an FAFR greater than the best achieved over a ten-year period. Turning to the assessment of risks to the population at large, other factors need to be taken into account. It becomes necessary to differentiate between imposed and voluntary risks. A man is able to exercise a personal option in choosing to work for a particular employer in a particular industry whereas members of the public have risks imposed on them without their consent or knowledge.

Lord Ashby [2, 3] has argued that risks of one in a million are of no great concern to the average person; though when risks rise to one in 10 000 the public are willing to incur expenditure to reduce the risks, unless the risk is voluntary; and at one in 1000 a risk becomes unacceptable to the general public and there is strong pressure to reduce the risks. Such generalisations are subject to considerable criticisms and Lord Rothschild [4] among others finds them 'hard to stomach'. Absent from such considerations are key elements such as the duration of the risk and the facility available to the public to express acceptance or rejection of the risk under consideration.

Even when our best endeavours are made to quantify risks of this type to enable us to allocate resources appropriately we have to recognise that rational statistical analysis is always incomplete. In the end we delegate the responsibility of risk assessment to elected members of central and local government in a democratic society and we must recognise that under these circumstances, in Lord Rothschild's words, 'hunch is respectable' [4].

But it is necessary to analyse how our elected representatives form the views they hold. They are naturally subjected to representations and lobbying from all quarters and consequently their task is not an enviable one.

The average layman perceives risk in quite a different fashion from our industrialist. He is unconvinced by arguments related to FAFR and need hardly be blamed for this reaction. It can be argued that the layman's assessment of risk is more closely related to harmful effect unrelated to the question of lethality. We noted in Chapter 9 that lethality is only one of the many possible effects of exposure to hazardous chemicals. It is somewhat absolute and therefore easy to quantify and this explains its wide use. But our layman has every right to be concerned about effects such as skin rashes, eye complaints, headaches, sore throats which he perceives have been related to waste disposal operations. These are not unqualified fears since not only are they reported frequently in local newspapers and the national media but they have been substantiated in more authoritative publications [5, 6, 7]. In some cases the reported health effects are more serious. No one should pretend that the assessment of risks in relation to a particular waste disposal operation is an easy task.

The General Accounting Office, Washington DC [8] reported recently on the progress of the US Environmental Protection Agency in tackling the hazardous waste problem. It said

> the dangers posed by uncontrolled waste disposal sites can be divided into short-term (acute) health effects, long-term (chronic) health effects and enviromental damage. The impact of environmental and health damage can be devastating, including serious economic loss, high health care costs, compensation to affected individuals, and property loss, as well as the indirect effects of human suffering and the long-term loss of valuable natural resources. . . .USEPA's risk assessment activities have provided some limited information on the environmental damage caused by hazardous waste sites but the true extent and nature of the threat to human health are virtually unknown. Current scientific knowledge is critically deficient in several vital areas including determination of how hazardous wastes move through the environment, how much of the waste human populations are exposed to and the degree of health hazard these amounts represent . . . though some knowledge is available of the effects of pure chemicals, virtually nothing is known about the effects of chemical mixtures on health and the environment.

These limitations to our present-day knowledge are now recognized, especially in the USA where research expenditure in hazardous waste studies is being increased. Table 1 shows the research expenditure controlled by USEPA but, in addition to this, the US Department of Health and Human Services (HHS) conducts its own research. HHS is concerned with the toxic effects of chemicals, mostly pure chemicals but it has some interest in hazardous waste disposal sites. By 1979 HHS had spent $1 billion directly or indirectly on toxic substances research and during FY 1980 its budget included:

$125 million for toxicity testing;
$114 million for basic toxicity research;
$28 million for studying human epidemiology of chemically related diseases;
$23 million for methods development.

Despite this expenditure it has concluded [8]:

> Whilst at this time it is impossible to determine the magnitude of the toxic chemical risk, it is clear that it is a major and growing health problem. Efforts to define the magnitude of the problem more precisely are hampered by two factors. First the long latent period that frequently exists between chemical exposure and chemically induced diseases and second, the newness of the science

Table 1. *Estimated expenditure on all types of hazardous waste research in USA*

	Research expenditure ($ millions)			
	FY78	FY79	FY80	FY81
(1) Sampling and analytical procedures	0.1	0.7	0.7	4.5
(2) Toxicity assessment and health effects	0.2	0.2	0	3.0
(3) Fate and transport	0.5	0.5	0.5	(a)
(4) Technologies for clean up of existing sites	0	0.5	0.8	3.2
(5) Technologies for managing current and future wastes	2.4	3.4	3.4	14.4
(6) Other expenditure	0	0	0	1.3
Total	3.2	5.3	5.4	26.4

(a) included in (4) and (5).
Source: Reference [8].

of environmental toxicology. Thus, as the problems of toxic chemical waste dumps and acquifer contamination have shown us, we are currently in the very early stages of a health problem which may take years to assess fully.

Under these circumstances, who can blame the average layman for his scepticism and his concern about hazardous waste management. To him arguments about FAFR are cold comfort for his perception of the risks incurred are far more real and personal than fatality statistics. In the end he is simply not prepared to endure any health risks by having a hazardous waste dump at the bottom of his garden. He perceives that even if he and his family escape without detriment to health then he can not overcome the fall in property value which would ensue from such a development.

Cost–benefit analysis

The principle of cost-benefit analysis is applied to many forms of human endeavour and there is no fundamental reason why it should not be applied to waste management. There are, however, a number of inherent problems.

Balancing the cost on one side of the equation with the benefits on the other requires that we employ consistent units throughout the sum.

The people whose benefits we measure must be the same people whose cost we assess. This leads us to consider what we mean by cost. An industrial manufacturer considers his waste disposal cost as the amount he has to pay to be rid of each waste consignment. A local inhabitant residing near to a disposal facility might measure cost in terms of nuisance. A government officer might measure cost to include remedial work incurred some considerable time after the disposal operation which was found to be necessary to restore the area to its former state.

Not only are these cost considerations viewed from a different objective but they produce cost figures which can vary by several orders of magnitude.

The US Environmental Protection Agency includes a department called Land Damage Assessment with responsibility for identifying damage costs and quantifying them. Its findings are published regularly (for examples see [9] and [10]) and specific case histories are published separately (for examples see [11], [12] and [13]). As a result a substantial data base has been established from which the magnitude of the problem has been assessed and the policy of the controlling authority developed.

Dietrich [14] of the USEPA reported in 1979:

> Moreover preliminary assessment of the documented instances of pollution damage from improper disposal of hazardous waste reveals that 'AN OUNCE OF PREVENTION IS WORTH A POUND OF CURE'. In other words, the cost of remedying predictive errors under an assimilation/degradation approach could easily exceed the total cost of prevention for all facilities under the full containment approach. . . . Not too many years ago, only the microbiological properties of drinking water governed its health-related acceptability. Now, previously unrecognised organic pollutants are beginning to govern the adequacy of drinking water quality. . . . Finally, a less-than-full containment regulatory strategy would not be publicly acceptable. Currently, the US public is very apprehensive and even fearful about the location of disposal facilities, particularly hazardous waste disposal facilities in their neighbourhoods. Whether or not their fears and objections are emotional and ill-founded is not relevant. What is relevant is that they perceive that disposal sites leak and emit health-debilitating pollutants. If any progress is to be made in calming these fears, gaining the public confidence and reducing opposition to the siting of needed, adequate hazardous waste disposal facilities, it will only be on the basis of convincing the public that such facilities are capable of fully containing the hazardous wastes, and, therefore, will not leak or emit pollutants. . . . There are professional estimates that over 30 000 closed disposal sites containing hazardous materials that potentially could cause public health or environmental damage. . . . Typically, the cost of remedying an abandoned facility problem is extremely high and exceeds the financial capacity of the responsible party. . . . The cost of permanently correcting an individual abandoned site ranges from several hundred thousand dollars to several tens of millions of dollars.

It would appear that recognition of the magnitude of the cost of remedial work influenced the USEPA to adopt the policy that only full containment sites for hazardous wastes are publicly acceptable.

In the United Kingdom very little evidence has been published concerning the cost of remedial work found necessary as a result of improper waste disposal. Long [15] reported that remedial work at a chemical landfill in Cheshire cost the taxpayers some £300 000. At Ravenfield, near Rotherham £215 000 was spent in making a waste disposal site safe and a further £350 000 on additional landscape work [16].

In Derbyshire the County Surveyor has reported that remedial work on a waste tip will cost more than £1.5 million [17]. There are many other instances where remedial work has been undertaken but little information has been published. Though much of this work is financed by central government there are no published reviews of the grants and subsidies made to overcome problems of this kind. Nevertheless, it is clear that remedial work is costly and this factor needs to be included in any cost-benefit analysis of hazardous waste landfill options. When this is done the landfill of hazardous waste may no longer be the lowest cost option available.

Hansen & Rishell [18, 19] examined unit costs for 16 treatment and five disposal techniques applicable to hazardous wastes from the organic chemical, inorganic chemical and metal finishing industries. Each technology was evaluated by computer-linked models developed for calculating capital and operating costs (including indirect and maintenance costs). The comparisons are presented over the life cycle of the process and the relationship with plant throughput is identified. In each case the treatment technology was required to comply with the latest US standards for the protection of human health and the environment. Risk assessments ranging from catastrophic external events to downtime risks resulting from component malfunction were included. Disposal technologies studied included incineration, land disposal, chemical fixation, encapsulation and evaporation and the range of waste materials which can be handled by these technologies was identified. Since no single technology can be applied to all wastes comparisons have to be made on a case-by-case basis and there is no simple 'best buy' solution.

Arthur D. Little Inc. [20], back in 1973, evaluated the technical, economic risk and legal grounds for various alternatives for managing hazardous wastes. The alternatives were on-site processing, off-site processing and on-site pretreatment with off-site treatment and disposal. The waste categories studied included pesticides, chlorinated organic solvents, cyanides, heavy metal wastes and other miscellaneous chemical species (radioactive and explosive wastes were excluded). The economic analyses were based on graphical decision maps and then the state and federal laws were examined to see if legal or institutional factors mitigated against the conclusions made on economic grounds. Finally, each alternative was evaluated to determine whether risk factors affecting the general public mitigated against conclusions made on economic grounds. The study concluded that off-site treatment and disposal at a regional facility was the preferred approach.

Such a facility would be in the economic and environmental interests of the waste producers as well as the ordinary members of society. A facility of this kind benefited from the economies of scale which outweighed the transportation cost incurred in conveying waste to the site. Secondary benefits such as the ability to treat one waste with another and the use of waste oils as fuel were also recognised. It was shown that waste treatment inherently reduced societal risks even when risks associated with the transport of wastes were factored into the considerations. The benefits of containing all the unit processes on one site serving an industrial region outweighed any attractions that mobile equipment might offer.

In summary, we have seen that cost–benefit analysis is a difficult assessment to make on a strictly objective basis. Cost figures for remedial work need to be factored into the equation. The sparseness of data on remedial costs in Britain is due to factors which exist in our society which deter the publication of historical data and should not be taken to suggest that in Britain we are not faced with significant hidden costs. Rational analysis by overseas authorities has indicated that the treatment of hazardous wastes at a regional complex is the preferred option for both waste producers and ordinary members of the community.

Our future needs

In years gone by our forefathers developed and financed the construction of sewage purification works which are now an accepted component of our western society. No doubt there were arguments about the high capital and running costs and local issues about the siting of such facilities on particular pieces of land. Nevertheless, these installations were built and we now recognise that they perform a vital function protecting our water supplies and thereby improving public health. In an unglamorous and easily forgotten fashion they contribute to the quality of our life.

Now we live in a sophisticated industrialised society which has recognised new risks to public health and the environment. Though we have not yet developed scientific methods to quantify these risks in any widely accepted manner, the public at large recognises their reality and seeks to minimise them. To meet such demands our elected governments face the task of providing the legislative framework and the appropriate vehicles to ensure progress is made.

Our industrial producers, who create our wealth within our society, find hazardous waste production unavoidable and demand reasonable recognition of their plight. Equally, the public at large perceive that landfill of hazardous waste has the undeniable potential to cause persistent damage.

Research studies reviewed in this book have identified some of the limits of our scientific knowledge. We have recognised that there is a grave danger that the knowledge we have gained could be misinterpreted and applied unrealistically. We do not currently possess sufficient confidence to rely upon geochemical factors to prevent the release of harmful pollutants into our environment.

Surely then, like our forefathers years ago, we must develop and finance the necessary facilities to treat hazardous waste. Considerations of economic aspects, and environmental risks show that Regional Treatment Plants provide the best option.

Many of the unit processes which such facilities would operate are already in operation and have been described earlier in this book. These facilities would allow us to concentrate and contain hazardous waste and monitor its effect on the environment. The alternative techniques of dilution and dispersal of hazardous waste magnify the cost of scientific monitoring of the short-term and long-term effects and, consequently, monitoring must always be inadequate and the risks unfathomable.

The question arises as to who should finance such facilities. In Britain hazardous waste disposal has long been the prerogative of private enterprise and domestic refuse disposal the prerogative of the public sector. Such demarcations have been jealously guarded and the private sector has been accused of excessive profiteering resulting from lack of care about environmental standards. Conversely, the municipal authorities can be accused of inefficiencies in domestic refuse collection and disposal and the enforcement of dual standards in their supervisory role in hazardous waste disposal. Claims and counterclaims are exploited by politicians with an eye on the ballot box. But surely the time has come to end such petty squabbles for the magnitude of the hazardous waste management task is so great that only cooperative efforts can hope to provide the solution that our society now demands. Such cooperation has been brought about in some countries. In Denmark and West Germany, Regional Hazardous Waste Treatment Centres have been financed using combinations of central government, regional government and industrial funds. Similar arrangements could be made in Britain if the political will was apparent.

The regulatory authorities could enforce uniformly high environmental standards on such operations and insist on the added protection of performance bonds and environmental impairment insurance. The overall cost of such a strategy would be passed back to the industrial producer and thence to the consumer in accordance with the 'polluter pays principle'. On a wider perspective it seems unlikely that this strategy would add to the community cost of our industrial production. Instead the costs would be redistributed with the biggest potential polluters paying the most for the disposal of their hazardous waste, which is preferable to imposing these costs on tax payers and ratepayers.

Such a strategy demands the establishment of national standards for the disposal of hazardous waste and rigorous enforcement of these standards. Without such a safeguard, hazardous wastes would simply be transported across territory boundaries to seek out low cost disposal options associated with loopholes in the law. Once such a system was established our regulatory officers could resume their role of policemen and enforcement officers in the certain knowledge they had the support of the general public and the industrial manufacturers reinforced by a positive legislative framework. Such is the vision of our future progress.

References

[1] Kletz, T.A., 'The Application of Hazard Analysis to Risks to the Public at Large', Proc. World Congress of Chemical Engineering, Environment and Human Activities, Amsterdam, July 1976.

[2] Ashby, Lord, 'Protection of the Environment: the human dimension', Proc. Roy. Soc. Med. pp. 721–30, 1969.

[3] Ashby, Lord, 'The Subjective Side of Assessing Risk', *New Scientist*, 398–400, 1974.

[4] Rothschild, Lord, 'Risk', *Atom* **268**, 30–5, February 1979.

[5] Hansard, House of Lords Official Report for 17 November 1981.

[6] Read, W. & W. George, 'A Case History of a Toxic Waste Tip in Buckinghamshire', Report to House of Lords Select Committee on Science and Technology. Available in House of Lords Records Office.

[7] Brown, Michael, *Laying Wastes*, Pantheon Books, New York and Oyez IBC, Norwich St, London.

[8] General Accounting Office, *Hazardous Waste Sites Pose Investigation, Evaluation, Scientific and Legal Problems*, PB81–197725 (GAO/CED–81–57), Washington DC, 75 pp., April 1981.

[9] *Hazardous Waste Disposal Damage Reports*, Environmental Protection Publication, SW–151, USEPA, Washington DC, 8 pp., June 1975.

[10] *Hazardous Waste Disposal Damage Reports*, Environmental Protection Publication, SW 151–3, USEPA, Washington DC, 11 pp., December 1975.

[11] Shuster, K.A., *Leachate Damage Assessment: Case Study of Sayville Solid Waste Disposal Site in Islip, Long Island*, Environmental Protection Publication, SW 509, USEPA, Washington DC, 18 pp., June 1976.

[12] Lazar, E.C., 'Summary of Damage Incidents from Improper Land Disposal', Proc. National Conference on Management and Disposal of Residues from the Treatment of Industrial Waste Waters, Washington DC, pp. 253–7, February 1975.

[13] Shuster, K.A., *Leachate Damage Assessment: Case Study of Peoples Avenue Solid Waste Disposal Site in Rochford, Illinois*, Environmental Protection Publication, SW 517, USEPA, Washington DC, 25 pp, June 1976.

[14] Dietrich, C.N., 'Ultimate Disposal of Hazardous Waste', ACS/CSJ Chemical Congress, Honolulu, April 1979.
[15] Long, J. 'The Reclamation of Malkin Bank Toxic Wastes Tip', Symposium on Toxic Waste Disposal, Harwell, May 1973.
[16] South Yorkshire County Council Environment Committee as reported in the local press.
[17] Underwood, C.V., Letter to *The Surveyor*, 9 October 1980.
[18] Hansen, W.G. & H.L. Rishell, *Cost Comparisons of Treatment and Disposal Alternatives for Hazardous Wastes*, Vol. I, PB 81–128, December 1980.
[19] Hansen, W.G. & H.L. Rishell, *Cost Comparisons of Treatment and Disposal Alternatives for Hazardous Wastes*, Vol. II, PB 81–128252, December 1980.
[20] Little, Arthur D. Inc., *The Alternatives to The Management of Hazardous Wastes at National Disposal Sites*, PB 225–164, Cambridge, Mass. 1973.

APPENDIX

Suggested protocol for the collection and analysis of leachate samples

Background

Leachate is the liquid material which collects at the base of a landfill site generated by the natural degradation of the waste and the infiltration of water and liquids from external sources. It is highly polluting and contains significant concentrations of organic and inorganic constituents. Leachate generally exists in an anaerobic (anoxic) condition and, therefore, in order to study its composition and properties it is essential to maintain this condition during the sampling process. This suggested protocol is designed for this purpose.

Leachate should not be confused with surface waters which may also be found on a landfill site. Surface waters may also be contaminated and may even contain some leachate but they are generally less polluting and have been exposed to partial or complete oxidation. Consequently, the composition and properties of contaminated surface waters differs from that of landfill leachate.

Boreholes

Although detailed methods of construction of boreholes are beyond the scope of this work it is important to recognise that borehole construction may itself change the chemical constitution of the waste materials and leachate which the drill encounters. Rotary air-flush drilling or percussion drilling can expose materials to oxidising conditions which are atypical of the natural conditions occurring beneath the landfill. For this reason it is necessary to allow sufficient time after the installation of the sample collector for the environment around the sample collector to return to its natural condition. The use of drilling mud should be avoided if possible.

Perhaps it should be emphasised that 'bucket-on-a-string' techniques are totally unsuitable for this work.

The sample collector

Following the construction of a borehole it is necessary to introduce into it a sampling device. A suitable sampling device has been described by Joseph [1, 2] and is now available commercially in Britain [3].

The construction of the unit is shown in Fig. 1 and it is fabricated either from stainless steel or UPVC. The stainless steel unit has a 1-litre capacity and is suitable for most applications. The sample collector is supported by a nylon rope and employs two Teflon sample tubes. The sample collector is lowered into the borehole as the borehole casing is withdrawn.

In cases where it is necessary to ensure that vertical mixing of the leachate

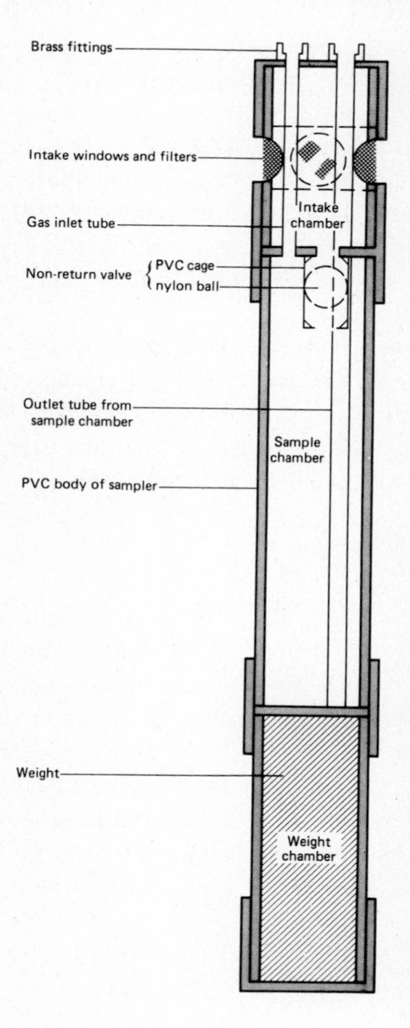

Fig. 1. The *in situ* groundwater sampler. (Courtesy of Wuidart Engineering Ltd.)

is avoided it is necessary to place sealants above and below the sample collector. The sealant needs to be chosen with care in order to avoid sample contamination. Mineral kaolinite is a low cation-exchange clay available from most potteries in fine powder form. It needs to be heated to 300°C to sterilise it. It is then prepared into a thick slurry form and poured into the borehole where it settles to form an impermeable base seal. The sealing process is messy and time consuming since an allowance has to be made for fine clay particles to settle out. Alternatively, bentonite clay can be used. Unlike kaolinite, bentonite swells in contact with water to form a gelatinous plug, sealing the borehole and moving into pores and fissures. Sodium bentonite is available in convenient pellet form and needs to be heated to 200°C for sterilisation before use. After the addition of bentonite pellets it is desirable to wait at least 4 hours before the next stage. It should be noted that bentonite has a high cation-exchange capability which necessitates thorough purging of the entire sampling facility before representative samples are obtained. The sample collector should be covered with about 0.3 metres of fine gravel and a further sealant layer placed above this.

Sample collection

The sample collector is connected by Teflon tubing to the sampling stream (Fig. 2) which is developed from the studies of Weiss & Colombo [4]. It is desirable to use compressed gas from a cylinder to drive the liquid sample from the sample collector to the sampling stream. Argon is the preferred gas. Nitrogen, although cheaper, may effect the nitrate

Fig. 2. Schematic diagram depicting the apparatus for anoxic filtration of water. (Source: Reference [4].)

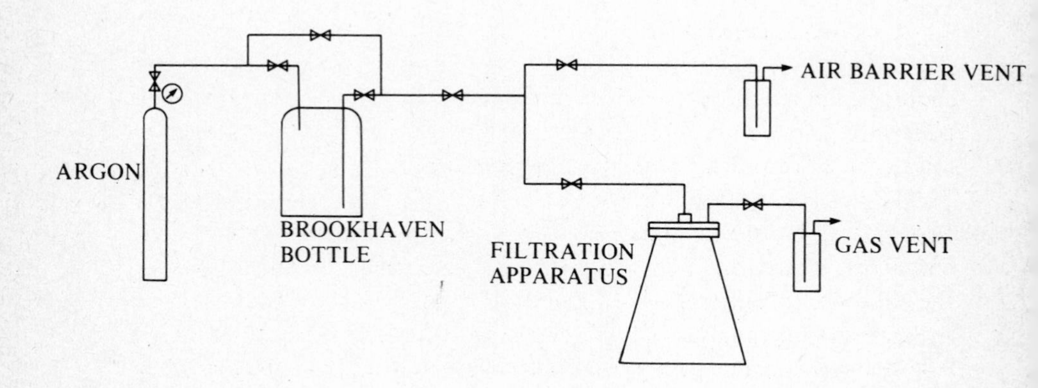

content of the leachate. Compressed air or oxygen should not be used.

The sampling train should be sterilised and purged with argon prior to use. Air barrier coils are included to prevent air entering the system. Measurements of specific conductance, pH and temperature should be made with in-line instruments during the collection of the samples. Clearly it is desirable to purge the sampling stream thoroughly (minimum of 3 volumes) before collecting samples for analysis. Leachate samples are expected to be free from brown ferric hydroxide precipitates if anoxic conditions are maintained satisfactorily.

Sample containers

Weiss & Colombo [4] have demonstrated that the material used for the construction of the leachate container can have an impact on the analytical results obtained. They recommend the use of borosilicate glass containers where leachates are to be examined for organic components and plastic (polypropylene or Teflon) bottles for inorganic analyses. In all cases the bottles are sterilised and argon filled before use and after filling the bottles are sealed in plastic bags and transported in an ice-chest at 4°C to an off-site laboratory. Analysts should be aware that partial degradation of organic components occurs even when samples are stored at these temperatures. Chian & DeWalle [5, 6] showed that leachate could lose 55% of its COD within 1–3 weeks. Consequently it is desirable to report the time interval which occurs between sampling and analysis.

It is desirable to employ a sample container which subsequently permits the transfer of aliquots of the leachate sample to the analytical apparatus whilst maintaining anoxic conditions. The Brookhaven Bottle (see Fig. 6, Ch. 6) is one such device. It is a modified 4-litre borosilicate glass vessel fitted with two high vacuum-type stopcock valves made from Teflon.

Sample preparation

On arrival at the off-site analytical laboratory the leachate samples should be filtered through 0.45 μm membrane filter to remove sediments. It is essential to maintain anoxic conditions during this filtration process. For this purpose Weiss & Colombo [4] employed the equipment depicted in Fig. 7, Chapter 6. Argon gas is used to transfer leachate samples from the Brookhaven Bottle to a stainless steel sanitary sterilising filter holder containing a tared 147 mm 0.45 μm Millipore filter. Bottles to be used for the collection of filtrates which are not to be acidified are kept on ice.

Samples to be analysed for organic components are filtered through a 0.45 μm silver metal membrane to achieve initial filter sterilisation. The

Fig. 3. Protocol for the preparation of samples of leachate for analysis.

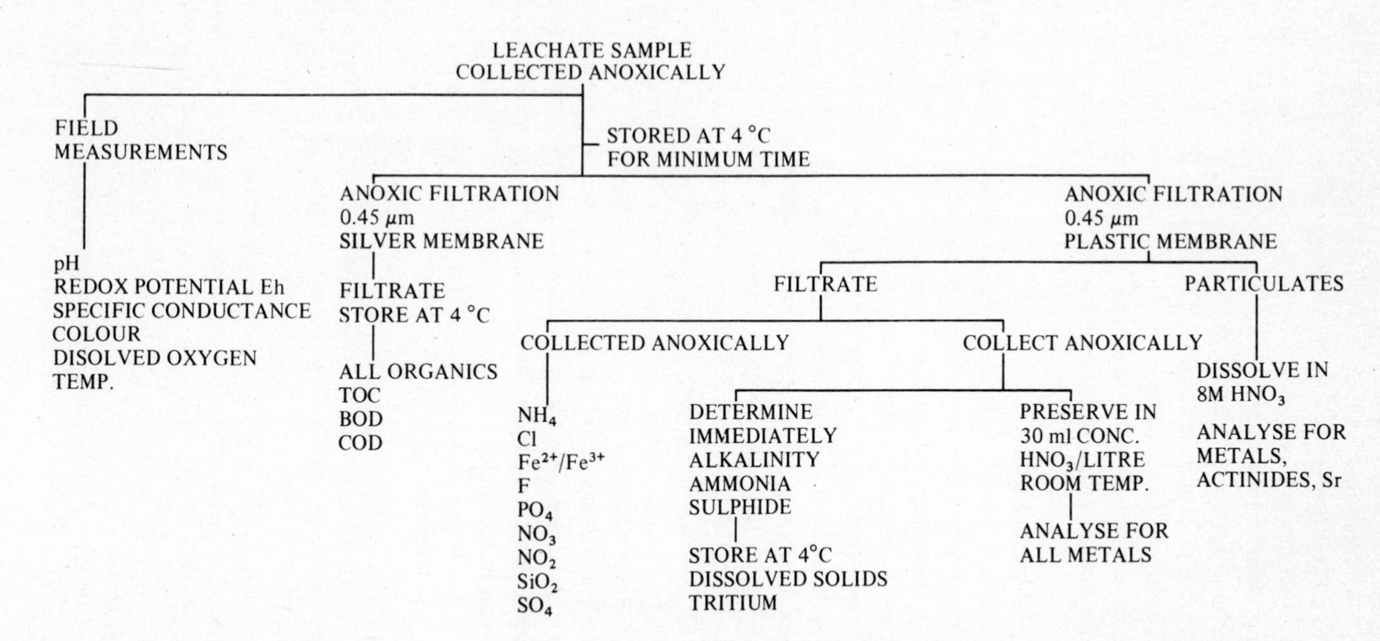

bacteriostatic action of silver inhibits bacterial regrowth in the filtrate. In this way post filtration changes in the organic composition of the filtrate due to bacterial metabolism are precluded.

It is preferable that sample preparation procedures are carried out in a glove box with an argon atmosphere.

The samples obtained in this way are then prepared and analysed as shown in Fig. 3. The acidification of samples with 30 ml conc. $HNO_3\ l^{-1}$ (where indicated) stabilises the leachates and permits the determination of inorganic metals by conventional analytical techniques.

Comment

It will be apparent from the above that the precautions necessary during the collection and preparation of samples are very elaborate; but they are necessary if meaningful data relating to the conditions and properties of leachate which exist at the base of landfills are to be obtained. In the absence of these precautions samples will loose volatile organic acids, increase in pH, absorb oxygen, and form a gelatinous precipitate of iron and manganese hydroxides which causes the co-precipitation and adsorption of many metals and organic components. Furthermore, the absence of special precautions during the storage of leachate samples leads to rapid biodegradation of organic components.

Acknowledgements

The author wishes to pay tribute and acknowledgement to the pioneering studies of Allen Weiss and Peter Colombo (Brookhaven National Laboratory) and Edward Chian and Foppe DeWalle (University of Illinois) in this field.

References

[1] Joseph, J.B., *Installation and Operating Manual for the WRC In-Situ Groundwater Sampler*, Water Research Centre, Report 27 M., Medmenham Laboratory, Marlow, December 1980.

[2] British Patent No. 1527751, Groundwater Sampler.

[3] Available from Wuidart Engineering Ltd, Clifton Road, Shefford, Bedfordshire, SG17 5AB.

[4] Weiss, A.J. & P. Colombo, 'Evaluation of Isotope Migration Land Burial', NUREG/CR -- 1289, Brookhaven National Lab., New York, July 1980.

[5] Chian, E.S.K. & F.B. DeWalle, *Evaluation of Leachate Treatment*, Vol. I., *Characterisation of Leachate*, USEPA, EPA–600/2–77–186*a*, September 1977.

[6] Chian, E.S.K. & F.B. DeWalle, 'Characterisation and Treatment of Leachates Generated from Landfills', *AIChE, Symposium Series*, Vol. 71, No. 145, pp. 319–27, 1975.

Index